From Scientific Philosophy to Myth: An Unquiet Philosophical Journey

Volume 1

Poznań Studies in the Philosophy of the Sciences and the Humanities

VOLUME 125

The titles published in this series are listed at *brill.com/ps*

From Scientific Philosophy to Myth: An Unquiet Philosophical Journey

Volume 1: Studies Inspired by Francesco Coniglione

Edited by

Giacomo Borbone
Krzysztof Brzechczyn

BRILL

ISSN 0303-8157
ISBN 978-90-04-72771-7
ISBN 978-90-04-75649-6 (PDF)
DOI https://doi.org/10.1163/9789004756496

Library of Congress Control Number: 2026936044

Brill and Koninklijke Brill BV are part of De Gruyter Brill.
www.degruyterbrill.com

Questions about General Product Safety Regulation: productsafety@degruyterbrill.com

Cover illustration: Francesco Coniglione, Mississippi State University, 2005. Photo taken by Michelangelo Caponetto.

Contents

Preface VII
List of Figures XII
Notes on Contributors XIII

An Unquiet Mind: the Philosophical Journey of Francesco Coniglione 1

PART 1
Reason as an Inspiration for Philosophy

1 The Problem of Critical Realism in the School of Milan 47
Fabio Minazzi

2 Science and Philosophy between "Open Society" and "Depersonalized Society": from Claude Bernard to Karl Popper 73
Riccardo Roni

3 Francesco Coniglione on Ernst Cassirer's Theory of Concept-Formation: Some Critical Remarks 89
Giacomo Borbone

4 The Development of the Theory of Revolution in Comparative Historical Sociology
An Attempt at a Methodological Analysis of Skocpol and Goldstone's Approaches to Revolution 107
Krzysztof Brzechczyn

5 The University "System" Has Long Been "Dead": Who Were the "Perpetrators" of Its Demise? 129
Francesco Alfieri

6 The Catania MIRRORS Project and the AI Transparency Problem 142
Thomas Nickles

7 How Do You See? Ludwik Fleck's Theory of Vision as a Form of Visual Semiotics 173
Emanuele Fadda

PART 2
Myth as an Inspiration for Philosophy

8 Ontological Realism and Quantum Mechanics 191
Alberto Giovanni Biuso

9 Elements of Solar Theology in Proclus: between Astronomy and Theurgy 203
R. Loredana Cardullo

10 Interpreting Francis of Assisi 223
Riccardo Pozzo

11 Some Considerations on Silence, Music, and the Essence of Things: an Opportunity to Explore Francesco Coniglione's "Southern Passage" 231
Emanuele Coco

12 How Myths Help to Deal with Complexity 250
Santo Di Nuovo

13 Objectivity, Knowledge, and Rights 265
Fabrizio Sciacca

Index of Names 279
Subject Index 283

Preface

Philosophy has always been suspended between *mythos* and *logos*. Plato used myth to convey truths and theses that lie beyond the reach of human reason. However, in different historical periods and philosophical traditions, the balance between these two components has been configured in varying ways. For example, Hegel regarded myth as a sign of the weakness of human thought, which is unable to fully articulate itself. In contrast, Heidegger saw myth as capable of transcending the limitations of *logos*, particularly in the realm of metaphysics.

This first volume is more directly engaged with the scientific work of Professor Coniglione. The essays it includes address topics discussed in his writings, reflect upon them, and develop them in a creative way. By contrast, the essays collected in the second volume – while devoted to themes that overlap with those explored by Professor Coniglione – do not engage with his work explicitly, opting instead to pursue an autonomous path within their respective fields.

More specifically, the first volume is constructed around these two poles of thought, not in opposition to one another, but rather in a complementary relationship. Accordingly, this collection of essays presents two equally important sources of philosophical inquiry. As illustrated in the opening essay of this volume by Giacomo Borbone and Krzysztof Brzechczyn, according to Francesco Coniglione, the structure of human knowledge resembles an iceberg. Concepts, definitions, models, algorithms, theories, and paradigms constitute the visible tip. Beneath the surface, however, lies *tacit knowledge* – in the sense attributed to it by Michael Polanyi and Ludwik Fleck – comprising beliefs, myths, parables, and metaphors that provide a holistic vision of the world.

In this preface, we will not dwell on the biographical details or the scientific work of Professor Coniglione, as these aspects are thoroughly addressed in the comprehensive essay by Borbone and Brzechczyn previously mentioned, to which we refer the reader. Rather, we will offer a concise overview of the contributions included in this volume – contributions that collectively embrace both reason and myth as key sources of inspiration in Coniglione's philosophical journey, while also addressing other topics in which he has shown particular interest.

The papers in the first part – entitled "Reason as an Inspiration for Philosophy" – explore the rational foundations and inspirations of philosophical thought. In their contributions, Fabio Minazzi and Ricardo Roni explicitly reference Karl Popper as a key source of inspiration. In his paper "The Problem of Critical Realism in the School of Milan," Fabio Minazzi compares the tradition of rationalism cultivated by the Philosophical School of Milan with Popper's critical rationalism. It is worth noting that this school was founded in the early 20th century by two philosophers: Piero Martinetti and Antonio Banfi. In the 1930s, its core figures included Enzo Paci, Remo Cantoni, Giulio Preti, and Dino Formaggio. The School of Milan exerted a significant influence not only on philosophy but also on various fields of science and art in Italy. Nevertheless, there are surprisingly few references to Preti's critique of the Vienna Circle's neo-positivism and to his notion of rationalism, which often closely aligns with Popper's approach. Roni, in turn, in his article "Science and Philosophy between 'Open Society' and 'Depersonalized Society': from Claude Bernard to Karl Popper," compares the views of the Austrian philosopher with the principles of falsificationism, which he identifies in the works of Claude Bernard (1813–1878), considered the father of experimental medicine.

Giacomo Borbone, in his article "Francesco Coniglione on Ernst Cassirer's Theory of Concept-Formation: Some Critical Remarks," presents Coniglione's interpretation of Ernst Cassirer's thought, highlighting how Coniglione emphasized the crucial distinction made by the philosopher from Breslau between Aristotelian-derived abstraction and idealization, as more recently formulated by the so-called Poznań School. According to Borbone, Coniglione's merit lies in having been the first to underscore the importance of idealizational processes in Cassirer's philosophy, revealing their affinity with the epistemological proposals of the Poznań School.

Krzysztof Brzechczyn, in his article "The Development of the Theory of Revolution in Comparative Historical Sociology: An Attempt at a Methodological Analysis of Skocpol and Goldstone's Approaches to Revolution," applies Coniglione's typology of approaches to idealization to the modeling procedures employed in theories of revolution within comparative historical sociology. According to Coniglione, four basic methodological attitudes toward idealization can be distinguished: (1) Unconscious users – scholars who employ idealization in scientific practice without being methodologically aware of it, or who misinterpret their own scientific behavior; (2) Conscious users – those who use idealization and are fully aware of its methodological role in science; (3) Non-users – scientists who neither apply idealization nor recognize its methodological potential; (4) Cultural expanders – those who advocate for the application of the method in broader cultural domains beyond science. In light of this typology, comparative historical sociologists such as Skocpol and Goldstone can be classified as belonging to the first group.

The next three contributions focus on the social context in which science operates, developing themes that Coniglione has explored with particular interest throughout his multifaceted intellectual journey. Francesco Alfieri, in his article "The University 'System' Has Long Been 'Dead': Who Were the 'Perpetrators' of Its Demise?" offers a pessimistic reflection on the Italian academic system and its opaque employment practices. His analysis is inspired by a short volume written by Coniglione (2011), as well as by other writings in which Coniglione critically addressed the same topic.

In "The Catania MIRRORS Project and the AI Transparency Problem," Thomas Nickles reflects on the threats posed by artificial intelligence to a knowledge-based society. His contribution is grounded in research carried out under the coordination of Coniglione, to which Nickles also actively contributed, as part of the European Union's Seventh Framework Programme. The project, titled MIRRORS (*Monitoring Ideas Regarding Research Organizations and Reasons in Science*), resulted in a major research report on knowledge-based societies *Through the Mirrors of Science. New Challenges for Knowledge-based Societies* (Coniglione, ed., 2010), as well as numerous other articles and papers.

Emanuele Fadda, in his article "How Do You See? Ludwik Fleck's Theory of Vision as a Form of Visual Semiotics," draws upon the Italian edition of Fleck's essays, edited by Coniglione, to examine Fleck's concept of thought styles (or thought collectives), and how these influence the interpretation of facts and the construction of scientific representations of reality.

The articles in the part entitled "Myth as an Inspiration for Philosophy" explore the mystical and mythical foundations of philosophical inquiry. Serving as a bridge between the two parts of this volume is the opening paper, "Ontological Realism and Quantum Mechanics," by Alberto Giovanni Biuso. The author, engaging also with some theses advanced by Coniglione, examines the physical world through the lens of quantum physics and, in particular, the so-called "Copenhagen interpretation." In criticizing this interpretation, Biuso draws on the realist epistemology of Lee Smolin, starting from the central claim that time is synonymous with being, time as the foundation of reality. By doing so, the author shows how a fruitful dialogue with metaphysics is possible. The paper also offers several examples illustrating this dialogue and critically discusses Smolin's philosophical perspective.

R. Loredana Cardullo's article, "Elements of Solar Theology in Proclus: between Astronomy and Theurgy," enters into close dialogue with Coniglione's theses on the esoteric dimension of thought, particularly as developed in his extensive essay on Plato, (Coniglione 2018, pp. 5–51). Cardullo deepens the analysis of the esoteric dimension of Plato's philosophy – especially as manifested in his unwritten doctrines – and explores the associated doctrine of the Sun, seen as the sensible and divine representation of the Good: the supreme Idea, the ultimate end and purpose of all beings. The article further shows how, in late antiquity, the Neoplatonists developed a form of astro-theology by grafting Chaldean and Orphic religious beliefs onto the foundations of

Platonic mathematical physics, harmonizing them with traditional Olympian religion, often in opposition to Christianity. The contribution concludes by focusing on Proclus as a representative of a solar theology that emerges from the perfect harmonization of Platonic astro-ontology with the theurgical and oracular wisdom of theology. Riccardo Pozzo, in his contribution "Interpreting Francis of Assisi," offers a reading of Francis's thought through the lens of naturalistic spiritualism, drawing upon Francesco Coniglione's recent work devoted to the Saint of Assisi (Coniglione 2022). In "Some Considerations on Silence, Music, and the Essence of Things: an Opportunity to Explore Francesco Coniglione's 'Southern Passage'", Emanuele Coco identifies a unifying thread that runs through many of Coniglione's writings. This thread may be understood as the relationship between the individual and reality, explored from scientific-epistemological, mythical-mystical, and socially engaged perspectives. Coco refers metaphorically to the "Southern Passage" (in contrast to the famed Northwest Passage), alluding to Coniglione's southern cultural roots. Through an analysis of the thought of Francis of Assisi, Blaise Pascal, and Arthur Schopenhauer, the article reflects on the epistemological and mystical significance of silence.

The final two authors featured in this part reflect on the role of myth in understanding the human condition and its function within democratic societies. In the chapter entitled "How Myths Help to Deal with Complexity," Santo Di Nuovo argues that myths assist in making sense of complex realities, realities that resist explanation through logic, reductionism, or causal reasoning. Di Nuovo draws on a co-authored essay written with Coniglione on this very topic, as well as on several of Coniglione's other writings, to offer additional perspectives from a psychological point of view. He also includes some critical reflections on Coniglione's position, examining how myths attempt to reconstruct realities that are otherwise unspeakable or unexplainable, using cognitive and linguistic tools such as imagination and metaphor. Di Nuovo supports his argument with an analysis of Carl Jung's *Red Book*.

In "Objectivity, Knowledge, and Rights," Sciacca revisits Coniglione's thematic strand of inquiry into the philosophy of knowledge and its practical implications in politics and law. He attempts to draw a comparison between myth and narrative: while myths portray the world as a realm of unresolved mysteries and riddles, narratives organize facts through temporal sequencing.

Giacomo Borbone and Krzysztof Brzechczyn

References

Coniglione, F. (2011). *Maledetta università. Fantasie e realtà sul sistema della ricerca in Italia*, Trapani: Di Girolamo.

Coniglione, F., (ed.). (2010). *Through the Mirrors of Science. New Challenges for Knowledge-based Societies*. Heusenstamm: Ontos Verlag.

Coniglione, F. (2018). Some Remarks on the Meaning of Esotericism and Plato's Unwritten Doctrines., *Mondi. Movimenti simbolici e sociali dell'uomo,* vol. 1, n. 1, pp. 5–51.

Coniglione, F. (2022). *L'uomo venuto da un altro mondo. Francesco d'Assisi*, Acireale–Rome: Bonanno.

Figures

0.1 The iceberg metaphor 19

4.1 Skocpol's model of revolution. *SR* – social revolution, *Ip* – international pressure, *Pc* – political crisis, *Pr* – people's revolts, *Oe* – political elites of opposition, arrows symbolize influence 120

4.2 Goldstone's model of revolution. Explanations: *SR* – social revolution, *Pc* – political crisis, *Pr* – people's revolts, Gp – population growth, *Ce* – rivalry among elites, arrows symbolize influence 121

Notes on Contributors

Francesco Alfieri
is Research Fellow at the Albert-Ludwigs-Universität, Freiburg im Breisgau, Germany.

Alberto Giovanni Biuso
is Full Professor at the Department of Humanities, Catania University, Italy.

Giacomo Borbone
Ph.D. in Human Sciences, Catania University, Italy.

Krzysztof Brzechczyn
is Full Professor at the Faculty of Philosophy, Adam Mickiewicz University, Poznań, Poland.

R. Loredana Cardullo
is Full Professor at the Department of Formative Processes, Catania University, Italy.

Emanuele Coco
is Associate Professor at the Department of Formative Processes, Catania University, Italy.

Santo Di Nuovo
is Emeritus at the Department of Formative Processes, Catania University, Italy.

Emanuele Fadda
is Associate Professor at the Department of Humanities, Università della Calabria, Cosenza, Italy.

Fabio Minazzi
is Full Professor at the Department of Theoretical and Applied Sciences, Università degli Studi dell'Insubria, Varese, Italy.

Thomas Nickles
is Foundation Professor, Emeritus at the Department of Philosophy, University of Nevada, Reno, U.S.A.

Riccardo Pozzo
is Full Professor at the Department of Humanities, Università di Roma Tor Vergata, Roma, Italy.

Riccardo Roni
is Research Fellow at the Department for Humanistic, Scientific, and Social Innovation, Università degli Studi della Basilicata, Potenza, Italy.

Fabrizio Sciacca
is Associate Professor at the Department of Political and Social Sciences, Catania University, Italy.

An Unquiet Mind: the Philosophical Journey of Francesco Coniglione

Giacomo Borbone and Krzysztof Brzechczyn

1 Some Biographical Information

With regard to the "external" coordinates of his life, it may suffice to note that Professor Francesco Coniglione was born in Catania on December 12, 1949, and earned his degree in Philosophy from the University of Catania on February 22, 1972, graduating with highest honors. His early career included a research fellowship at ISVI (Institute of Political Science) in Catania (1974–1977), followed by a tenure as high school professor of History and Philosophy (1977–1986). A pivotal moment came in November 1986 when he secured a permanent university research position through national competitive examination. Beginning in 1992, he expanded his academic work, teaching core philosophical disciplines – including Theoretical Philosophy, Logic, and Philosophy/History of Science – at both the University of Catania and Kore University of Enna. During these years of university life, a figure who was always close to him and helped and advised him both in research and from a human and affective point of view was Anna Escher Di Stefano, who also held the chair of History of Philosophy at the University of Catania and recently passed away (2024). His distinguished academic career concluded with retirement on November 1, 2020, from his position as Full Professor at the University of Catania. He is married to Olimpia Casamassima and has two children: Emanuele and Marialuisa.

Professor Coniglione has served in several significant editorial roles, reflecting his broad scholarly engagement. He is currently a member of the Advisory Committee of *Poznań Studies in the Philosophy of the Sciences and the Humanities* (Rodopi Publisher, now Brill), the *Polish Journal of Philosophy*, and the journals *Paradigmi*, *Il Protagora*, and *Historia Magistra*. Previously, he was also part of the editorial board of *PL.IT – Rassegna italiana di argomenti polacchi*.

It is worth mentioning that Coniglione has actively participated in Polish academic life. He has published in philosophical journals such as *Nauka*, *Polish Journal of Philosophy*, *Poznańskie Studia z Filozofii Nauki*, *Przegląd Filozoficzny. Seria Nowa*, and *Zagadnienia Naukoznawstwa*. In September 2015, he was invited to deliver a plenary lecture at the Tenth Polish Philosophical Congress in Poznań. Coniglione is a member of the Scientific Committee of the quarterly *Science*, and serves on the Advisory Board of the *Polish Journal of Philosophy*.

 | HTTPS://DOI.ORG/10.1163/9789004756496-002

He is also a member of the Italian Society for Logic and Philosophy of Science (SILFS), of the Gioenia Academy (founded in 1823), and of the Italian Philosophical Society (SFI), for which he served as National President from 2013 to 2016 and as a board member for the following three years. He was also a member of the National Scientific Qualification (ASN) commission in the sector of Theoretical Philosophy. Furthermore, he was a member of the Scientific Council (Rada Kuratorów) of the Polish Academy of Sciences from 2016 to 2022 and was awarded a fellowship for the 2018–2019 academic year for a study and research stay at the PIASt (Polish Institute of Advanced Studies) in Warsaw. In addition to these roles, he has actively participated in the governance of his own university as well as in other national and international associations and institutions. Over the years, he has developed numerous lines of research (which will be discussed later) and established scientific collaborations with several foreign universities, including Mississippi State University, the University of Nevada, Reno, and Adam Mickiewicz University in Poznań.

Coniglione also played a key role in popularizing Polish philosophy in Italy and abroad (in the latter case, through the aforementioned *Polish Philosophy Page*). He also edited, translated, and published works by Polish authors, for instance *Oltre Marx* (Nowak 1987) a work of Kazimierz Ajdukiewicz's (1989) and of Ludwik Fleck's works (2019). More, with Roberto Poli, he edited a collection of articles by authors from the Poznań School of Methodology. He translated Kazimierz Ajdukiewicz's, Jerzy Brzeziński's, Nowak's, and Piotr Przybysz's works into Italian (Coniglione, Poli 2004). His collaboration with Professor Leszek Nowak led to the publication of three volumes of *Poznań Studies* dedicated to idealization: *Idealization I: General Problems* (Brzeziński *et al.*, 1990a), *Idealization II: Forms and Applications* (Brzeziński *et al.*, 1990b), and *Idealization XI: Historical Studies on Abstraction* (Coniglione *et al.*, 2004). Coniglione is also a member of the Advisory Committee of the book series *Poznań Studies in the Philosophy of the Sciences and the Humanities*, published by Rodopi (now Brill). A particularly noteworthy contribution was his editorship of the Polish Philosophy Page, developed also in collaboration with Arianna Betti. This pioneering digital resource – recognized with *EpistemeLinks.com*'s "Gems of the Web Award" in January 1999 – provided essential access to a philosophical tradition largely overlooked in Western academia due to language barriers. Though currently inactive (due to technical issues), its revival in some form would be immensely valuable, given its historical role as a key reference for scholars navigating Polish philosophy's rich but linguistically restricted corpus.

2 The Formative Years

To examine Coniglione's intellectual trajectory, we must first recognize that his entire scholarly output embodies the restless inquiry of a perpetually inquisitive yet systematically synthetic mind. Even a cursory review of his works reveals an exceptional breadth – spanning the history of philosophy, philosophy of science, epistemological challenges of the knowledge society, higher education policy, and the study of myth and "rejected knowledge." This remarkable range reflects not mere academic opportunism, but rather a deliberate methodological commitment: the pursuit of truth through dialectical engagement across disciplinary boundaries. What distinguishes Coniglione's approach is his unwavering holistic vision, a sophisticated (yet never merely eclectic) understanding of reality as an interconnected whole. This can be seen by considering his very first scientific production, devoted to a topic on which Coniglione spent most of his intellectual energies. We refer to the idea of "scientific philosophy" (Russell, Neopositivism and so on) and, more specifically, to the critical rationalism of Karl R. Popper (see below § 4. Studies on Scientific Philosophy).

However, as probably few people know, this "turn" toward issues related to scientific philosophy was originally prompted by an academic environment that had shown little interest in the dissertation defended by Coniglione in 1972. The dissertation, which remained unpublished, was dedicated to the Italian philosopher Julius Evola and to Theodor W. Adorno, and was entitled *Julius Evola e Th. W. Adorno: saggi alternativi di autocritica dell'idealismo* [Julius Evola and Th. W. Adorno: Alternative Essays on the Self-Critique of Idealism]. It was assigned to him by Mariano Cristaldi, then professor of theoretical philosophy at the University of Catania, on the basis of needs expressed to him still quite generic and motivated essentially by the desire – in the turbulent years following 1968 – to understand the origins of such diametrically opposed worldviews.

The dissertation's two subjects embodied diametrical ideological poles: Evola, the intellectual cornerstone of radical traditionalism, stood opposed to Adorno, whose Frankfurt School – particularly through Marcuse – had become the left's philosophical lodestar. Yet as Coniglione acknowledges in an autobiographical note, his initial encounter with Evola was almost accidental. During a 1968 hitchhiking trip (that quintessential post-graduation ritual), his driver – a right-wing student – cited Evola as his intellectual reference when prompted. This revelation struck young Coniglione as paradoxical: in an era when leftist thought dominated university discourse, and the right was stereotyped as all brass knuckles and Gentilean echoes, here was evidence of a sophisticated reactionary tradition. During the militant university years that followed, when anti-fascist clashes prioritized physical confrontation over dialectical debate, Evola remained a peripheral figure in Coniglione's intellectual landscape. Like most of his generation, he immersed himself instead in the era's thriving marketplace of radical thought, where theoretical nuances spawned political factions with a sectarian intensity rivaling medieval quarrels over the *filioque* clause.

This radical dichotomy haunted the young Coniglione, presenting an intellectual puzzle that demanded resolution. The prevailing Marxist explanations – which reductively framed right-wing activism as capitalism's enforcers – struck him as inadequate. While he recognized that economic elites had historically weaponized reactionary politics, he discerned a crucial distinction: between the pragmatic right of bourgeois interests (with its instrumental, often superficial engagement with ideology) and the radical right's metaphysical revolt, a total worldview drawing sustenance from symbolic universes and philosophical systems diametrically opposed to leftist thought. Evola epitomized this latter tendency. His works constructed an anti-modern cosmology so comprehensive that it mirrored – in inverted form – the systemic critiques of the Frankfurt School. For Coniglione's leftist peers, authors like Marcuse provided the conceptual armor for their protests; Evola's traditionalism offered an equally robust, if antithetical, framework for his interlocutors on the right. This realization – that political commitments could spring from deep philosophical wells rather than mere class positioning – marked a pivotal moment in Coniglione's intellectual maturation.

This intellectual trajectory explains why, when formulating his thesis, Coniglione proposed applying a Proppian morphological analysis (Propp [1928] 2009) to compare right-wing and radical left cultural frameworks – with Evola emerging as the natural focal point for examining the former. The choice gained further relevance from the period's intense engagement with Hegelian thought, particularly through Frankfurt School interpretations and Marxist reworkings of dialectical materialism. In this climate, investigating how divergent readings of Hegel could yield such diametrically opposed worldviews – one revolutionary, the other reactionary – seemed particularly interesting. Professor Cristaldi, while primarily a theorist with limited investment in right-wing philosophy, recognized the value in contrasting Evola's interpretation of Gentile's actualist idealism with Adorno's negative dialectics ([1963] 1993). Under this guidance, the young Coniglione embarked on reconstructing both systems: the synthetic, unifying dialectic of Italian idealism versus the relentlessly critical German approach. His ambitious goal was to demonstrate how these methodological differences could "necessarily" generate antithetical conceptions of society, history, and human nature. While we cannot here evaluate the scholarly merits of a 22-year-old's first major academic work, the project's fate proved telling: Cristaldi, after proposing the comparative framework, showed diminishing interest in the results, providing scant supervision and ultimately leaving his student to work primarily with an assistant.

Following his graduation, after some half-hearted attempts to continue exploring themes related to his dissertation, Coniglione ultimately abandoned this line of inquiry entirely. His decision to distance himself from these subjects – despite having approached them from a rigorously critical perspective, far removed from either contemporary "Evola devotees" or the ideological circles that still champion such ideas – stemmed from two key factors: first, the professional risks associated with studying so-called "forbidden" topics (those dismissed as intellectually illegitimate, historically dubious pseudo-scholarship), and second, the complete absence of academic mentors willing to guide a young researcher through these intellectually perilous waters. The climate of intolerance was so pronounced that a professor of Roman history – an acquaintance of Coniglione's – once remarked that, had he been on the examination committee, he would have blocked the thesis defense outright, solely due to the perceived ideological and scholarly irredeemability of engaging with a figure like Evola.

Thus, through a confluence of personal circumstances and growing recognition of the practical impossibilities surrounding such research, Coniglione's intellectual focus shifted decisively toward another pole of longstanding fascination. Since his *liceo* years as an avid science fiction reader, he had been drawn to scientific thought and its philosophical underpinnings. What ultimately catalyzed this transition was a fundamental epistemological question: What grants scientific knowledge its singular character, that distinctive quality setting it apart from both everyday cognition and traditional humanistic inquiry?

3 The Polish Connection

This explains the "diurnal" side of his later cultural and academic journey, the one that came to define his official academic profile and eventually led – after teaching high school history and philosophy for nine years – to his appointment as professor of the history of philosophy at the University of Catania

(Sicily). The most significant milestone in this intellectual itinerary – following an initial period in which he engaged with Popperian and post-Popperian epistemology – was his long intellectual apprenticeship in Poland.

At the close of the 1970s, Coniglione encountered the writings of Leszek Nowak and other exponents of the Poznań School of Methodology, following a reading – undertaken with great interest – of one of Nowak's works in its Italian translation (Nowak 1977). What particularly struck him was the apparent success with which it reconciled a steadfast adherence to the foundational principles of Marxism with the demands of scientific rigor. Coniglione was in fact deeply convinced that Marxism ought to follow this path – an intuition he had already formed during his youth and university years through his engagement with the work of the Italian Marxist Galvano della Volpe, whose thought continued to exert a lasting influence on him. The publication of two substantial articles in *Rinascita* – the official weekly magazine of the Italian Communist Party – presenting the thought of the Polish school, became the occasion for establishing contact with Nowak's wife, Izabella Nowakowa. She informed Coniglione that her husband was at the time imprisoned in an internment camp for his opposition to the regime of General Jaruzelski, who had imposed martial law in Poland in 1981. Following Nowak's release and return to his university teaching, he was able to extend an invitation to Coniglione to visit Poznań in 1984. The following year, and again during 1985–1986, Nowak facilitated a nine-month fellowship for him at Adam Mickiewicz University in Poznań.

For the young Coniglione, this was an opportunity to break free from teaching in secondary school and to fully dedicate himself to scientific research, with the aim of exploring the relationship between Marxism and science, drawing inspiration from the teachings of the Polish school. During his time in Poznań, he began learning Polish and publishing studies on idealization and its historical roots. He also established contact with many other Polish philosophers, maintaining those connections to this day. There, he deepened his understanding of the methodological issues addressed by the Poznań School and of the so-called Idealizational Conception of Science (ICS), which not only provided a scientifically acceptable version of Marxism, but also – according to Coniglione – offered a way out of the aporias and deadlocks into which post-Popperian reflection had fallen after Kuhn and Feyerabend. He applied this approach in the conclusions of his first major work (Coniglione 1978), which was among the first in Italy to explore the contributions of Kuhn, Lakatos, and Feyerabend. The leading figure of Polish methodological school – who played a pivotal role in shaping Coniglione's conception of science – was Leszek Nowak (1940–2009), to whom he has always expressed his gratitude for countless insights and enlightening discussions.

The research undertaken in Poland was subsequently deepened and broadened by Coniglione upon his return to Italy – most notably after 1986, when he assumed the research position at the University of Catania – and it culminated in a comprehensive presentation of the Poznań School in the substantial volume *Realtà e astrazione* (Coniglione [1990] 2010), which includes an extensive bibliography in Polish as well as all the essential references. This book is the first – and so far, the only – monograph on the Poznań School of Methodology available on the Italian market, as well as on the Polish and international markets. According to Coniglione, the Poznań School of Methodology has its roots in the achievements of the Lvov-Warsaw School at the turn of the 19th and 20th centuries, the philosophical movement that shaped the Polish tradition of scientific philosophy and which also sparked Coniglione's interest, as evidenced by numerous articles. This influence is blended with the traditions of Polish non-orthodox Marxism, so that, in practice, Coniglione's book offers a broad overview of Polish philosophy in the first half of the 20th century: although it focuses primarily on the methodological school of Poznań, it also provides extensive insight into the other major currents of Polish philosophical thought.

The perspective developed by Nowak and by those scholars of the School most strongly influenced by his teachings (such as Andrzej Klawiter, Krzysztof Łastowski, Jerzy Brzeziński, Jolanta Burbelka, Marek Gaul, Piotr Buczkowski, Anna Jasińska, Andrzej Kupracz, Barbara Tuchańska, Renata Zielińska, Krzysztof Brzechczyn, Piotr Przybysz, and many others) was entirely rationalistic and, in some respects, "scientistic." However, in the final phase of Nowak's life, it opened up toward a form of unitarian metaphysics – an evolution that deserves greater attention than it has so far received (Nowak 1998; 2004; 2007; see Kiedrowski 2013; Przybysz 2013). What intrigued Coniglione, however, was not only this epistemologically compelling reconstruction of Marxism, embedded as it was in the domain of scientific thought and perceived by him as a more methodologically rigorous and sophisticated realization of the "moral Galileanism" proposed by the Italian philosopher Galvano della Volpe (Coniglione 1987, pp. 87–113; 2020). More importantly, what fascinated him was the concept of science based on the construction of ideal models, not derived through abstraction from empirical data, but rather through the distinctive process of idealization, which produces theories that maintain only a mediated and indirect relationship with reality.

It was precisely in light of the concept of idealization that Coniglione was the first in Italy to point out the limits of Popper's falsificationism. As an example, Popper claims that all concepts are dispositional; however, as Coniglione rightly observes, when one considers an ideal entity such as a perfectly round circle, it becomes clear that it is not a dispositional concept at all – and this holds true for all ideal concepts. Still guided by the idealizational approach to science, Coniglione also highlighted the limitations of the Neopositivist perspective – championed by Carnap, Neurath, Schlick, and Hempel – which sought to reduce all theoretical terms to purely observational ones. Such a program was doomed to failure, given that ideal entities like a "perfect gas" are, by their very nature, not directly observable.

The study of the idealizational conception of science developed by the Poznań School also provided Coniglione with a framework for re-examining key ideas in other thinkers who had profoundly influenced him. Chief among these was Ernst Cassirer ([1910] 1923; [1937] 1956), whose works had been instrumental in shaping Coniglione's early post-graduate understanding of science and reality (see Borbone's contribution in this volume). Later, Husserl's phenomenology ([1901–02] 2001) would further enrich this intellectual foundation. What distinguished the idealizational approach, in his view, was its methodological rigor and systematic perspective, an innovative framework within the landscape of Western philosophy. Through this lens, he sought to trace an intellectual lineage connecting the Poznań School's insights with other, often overlooked thinkers and scientists who had explored similar epistemological terrain. These figures, whose contributions had previously lacked proper theoretical contextualization, now found their place within Poznań School's comprehensive epistemological approach.

Coniglione exemplifies this perspective through his systematic engagement with key figures across philosophical traditions – from classical thinkers like Plato and Aristotle to Enlightenment pioneers such as Galileo, Locke, Hume, and Berkeley, through to modern systematizers including Hegel, Cassirer, and Husserl. His analysis extends equally to 20th century epistemologists and scientists – Russell, Carnap, Hempel, Nagel, Popper, Thom, Prigogine, and Mandelbrot among them. This comprehensive intellectual archaeology demonstrates how the "Polish School" of scientific philosophy – represented successively by the Lvov-Warsaw School and later the Poznań School – rediscovered and refined a model of scientific rationality that had long been overlooked. As Coniglione's work (2004a) highlights, it is largely through the Poznań School's contributions in recent decades that this sophisticated approach has regained scholarly attention and recognition.

Through this examination, Coniglione elucidates several pivotal insights that reshape our understanding of philosophy's engagement with scientific knowledge: (a) he demonstrates how the Aristotelian model of abstraction dominated epistemological thought for centuries, establishing a conceptual paradigm that would long endure. (b) he reveals how the Scientific Revolution quietly birthed an alternative approach – what we now term "idealization" – which many practicing scientists instinctively adopted, though often without explicit methodological justification. This new paradigm emerged despite (c) the constraining influence of empiricist and inductive philosophies, which left many scientists operating with inadequate epistemological self-awareness. Paradoxically (d), the most sophisticated philosophical treatments of this alternative approach frequently appeared outside mainstream epistemology, sometimes even within frameworks harboring anti-scientific tendencies. Finally, (e) Coniglione documents the contemporary resurgence of interest in idealization, highlighting the seminal contributions of the Poznań School in systematically developing this neglected tradition into a rigorous methodological framework.

Coniglione's engagement with the idealizational conception of science led him to revisit – through a historiographical lens – also certain foundational debates in Marxist philosophy that, while less prominent today, once sparked intense scholarly controversy. A central focus of this reexamination is the complex relationship between Marx and Hegel, particularly regarding Marx's purported "overthrow" of Hegelian dialectics, his "turning Hegel right side up", a topic surprisingly overlooked by the Poznań School despite its methodological significance. By applying the notion of idealization as an interpretive key, Coniglione sheds new light on Marx's often ambiguous references to Hegel's "method" in his mature scientific works (Coniglione 1990; 2012). This approach challenges superficial readings that reduce Hegel's influence to vague notions like "historical process" or "systemic interconnection," or that reinterpret dialectics through a crude naturalistic lens as mere reflections of material laws. At the heart of Coniglione's interpretation lies a crucial insight: Marx identified in Hegel a radical alternative to empiricist abstraction. By discarding what Marx termed the "mystical shell" of Hegel's system while preserving its "rational kernel," we discover that Hegel's methodological approach – far from being mere speculative idealism – mirrored the very procedures Marx observed in the natural sciences and subsequently adapted for his own social scientific research. As Coniglione demonstrates, this epistemological continuity reveals deeper affinities between Hegel's dialectic and scientific methodology than traditionally acknowledged:

> Marx found in Hegel and especially in his *Logic* the guidelines for a scientific method using a completely different concept of abstraction from that traditionally held by empiricist philosophy and common among contemporary positivists. Hence his great appreciation of Hegel's work

> and the inspiration he clearly found for his *Capital* (as is shown in his letters to Engels on the subject). Moreover, when Hegel speaks of the true speculative method in dealing with the doctrine of essence, Marx saw, once it was materialistically interpreted, the method followed by theoretical science to build its theories, again different from that proposed by contemporary positivist culture. The method concerned is the essentialist-idealizational one which the idealizational conception of science was the first to point out clearly and with methodological accuracy
> CONIGLIONE 1990, p. 75

Marx's "overthrow" carries profound dual significance. Ontologically, it substitutes the dialectic of ideas with that of material-historical reality. More crucially, methodologically, it rejects speculative philosophy in favor of what Coniglione identifies as the authentic scientific approach – the idealizational and essentialist method – transplanted from Hegel's realm of Spirit to empirical science, whether in physics or political economy. Coniglione terms this the "methodological inversion," a two-stage epistemic revolution: First, it identifies in Hegel's Logic the idealizational method in embryonic form. Second, it reclaims this method for science – the very discipline Hegel had deemed inferior to speculative philosophy – while exposing the limitations of empiricist abstraction. In this way, Coniglione sought to elucidate a question that had received scant attention within the Poznań School, as it was generally regarded as the domain of humanistic Marxist currents – currents toward which the School maintained a critical stance, privileging instead a rigorously scientific orientation.

This interpretative framework was likewise applied to the analysis and clarification of major figures in the history of philosophical and Marxist thought, with particular attention to the Italian tradition. Among the most prominent thinkers examined are Galileo Galilei (Coniglione 2016a), Antonio Labriola (Coniglione 1987, pp. 17–38), Antonio Gramsci (Coniglione 2008c), Benedetto Croce and Giovanni Gentile (Coniglione 1997; 2009; 2016b), as well as the aforementioned Galvano della Volpe (Coniglione 2020).

4 Studies on Scientific Philosophy

Familiarity with Polish epistemology as a whole – and especially with the Lvov-Warsaw School – led Coniglione to clarify the meaning and provide a more precise historiographical placement of an old topic: that of "scientific philosophy." Alongside his direct engagement with the Poznań School, this area has occupied his attention in the years following the 1990 volume and has remained one of the central fields of his scholarly activity, both in addressing philosophers and intellectual currents on the international stage (Coniglione 1999; 2002a; 2005; 2007; 2008a) and in relation to the Italian context (Coniglione 2008b; 2011a; 2014a; 2016; 2018a).

He conceives of "scientific philosophy" as a distinct way of understanding philosophy, one that is not synonymous with either epistemology or the philosophy of science. According to Coniglione, the concept originated in the work of Franz Brentano and was further developed in Poland by Kazimierz Twardowski, even before becoming the emblem of Bertrand Russell's philosophical vision in the West. Scientific philosophy, as Coniglione sees it, is a *sui generis* project. It is not to be conflated with Neopositivism or the Vienna Circle tout court, nor with "linguistic philosophy," "analytic philosophy," or the philosophy of science. As Coniglione himself states:

> Scientific philosophy is the idea that imbues the Cartesian project and acquires distinct awareness with Hume's intention of building a science of human nature inspired by the model of the Newtonian science of the physical world. It is such an idea that lies behind the demand for an exact, rigorous language, similar to that of Mathematics as much as possible. We can see this demand becoming urgent in Leibniz, Condillac and all the thinkers who looked at mathematics or natural science as a source of inspiration to bring forward a new, eventually mature philosophy, not tangled anymore in the never-ending traditional disputes.
>
> CONIGLIONE 2007, p. 9

More specifically, Coniglione examines the particular meaning of scientific philosophy – understood, in his view, as its most distinctive form – as defined by the Polish historian Władysław Tatarkiewicz: a philosophy that takes science as its model and that "must formulate and solve its problems using the same methods and criteria, the same standards of precision, as the particular sciences" (Tatarkiewicz [1958] 1988, p. 263). This is scientific philosophy in the strict sense of the term, one that has its own subject matter, distinct from that of science, yet is capable of producing knowledge that, while different from scientific knowledge, is obtained by emulating the methods and procedures (understood broadly) of science.

Scientific philosophy was practiced during the early stage of the Vienna Circle before the influence of Wittgenstein's thought became decisive. Afterwards, there was a quick transition to philosophy intended as subsidiary to science, as a mere clarification of meaning, coming, in the end, to its liquidation with Carnap's logical syntax: "the logic of science takes the place of the inextricable tangle of problems which is known as philosophy" (Carnap [1936] 1971, p. 279). However – and in this aspect harbors its originality – The Lvov-Warsaw School took a different path, adhering to Brentano's initial program outlined in the incisive statement: "Vera philosophiae methodus nulla alia nisi scientiae naturalis est" (Brentano [1929] 1968, p. 136), and never relinquished the notion of achieving a scientific philosophy. Decisive, here, was the absence of Wittgenstein's influence and the utter irrelevance of that of Mach (Coniglione 2007).

However, the history of scientific philosophy does not come to an end with the Lvov-Warsaw School in the interwar period, nor with the dissolution of the Vienna Circle and its diaspora. According to Coniglione, scientific philosophy has continued to evolve in the postwar era while preserving its distinct identity. In Poland, it remains deeply rooted, influencing many polish philosophers (indeed, the Poznań School owes much to this tradition). But this ideal has not disappeared from contemporary Anglophone philosophy either. Thinkers such as Mario Bunge are revitalizing the project of constructing a "good philosophy," one "capable of tackling interesting philosophical questions in the light of the best available factual knowledge, and with the help of precision tools forged by formal sciences" (Bunge 2001, p. 10). The ideal of "philosophy as science" is also advanced by other scholars, such as William Lycan (1988), who – following in the footsteps of Spencer – assigns to philosophy the task of constructing highly abstract theories, from which the most promising can be selected using criteria that are highly plausible, even if not entirely identical to those used in the empirical sciences. In Italy, Paolo Parrini has also embraced this view, proposing his own idiosyncratic version of positivism understood as scientific philosophy (Parrini 1995; 1999). Thus, scientific philosophy endures; although it now leads a rather precarious existence, caught between the decline of a once self-assured philosophy of science, the assertiveness of hermeneutic, deconstructionist, and postmodern approaches, and the risk of losing its identity within the expansive and inclusive domain of analytic philosophy.

Moreover, the overall assessment expressed by Coniglione in his numerous contributions on the subject has, in more recent years, shifted toward a more disenchanted view of the project represented by scientific philosophy. This change is partly the result of his in-depth studies on the evolution of thought about science in the post-Popperian era (Coniglione 2005), and partly due to the difficulty of identifying the Method to which philosophy was expected to entrust itself ("the search for the Holy Grail," as he calls it – 2025, pp. 28–30). That solid foundation on which philosophy was once thought to build its fortunes – science and its method – no longer enjoys the undisputed status of *faktum* once attributed to it by Kant and upheld throughout much of the 20th century. It no longer seems to possess the prestige or play the paradigmatic role of genuine knowledge, the very qualities that once made it an object of both envy and emulation. As Coniglione asserts:

> Under the blows of the new philosophy of science, wounded and humiliated in its imperialistic pretensions by the likes of Hanson, Kuhn, Lakatos, Feyerabend, Morin, and Latour, and delegitimized in its search for a method by those who now declare such a goal unattainable, a fresh noumenon that eludes the cognitive ambitions of metascience. Science now seems to be searching for a new *raison d'être*, and today appears more in need of philosophy than philosophy ever was of science. The proposed remedy has not only failed to cure philosophy's age-old infirmity but risks becoming a veritable pandemic, killing the doctor along with the patient.
>
> CONIGLIONE 2008a, p. 172

In the end, as Coniglione argues, scientific philosophy was nothing more than a "beautiful illusion": the belief that philosophy could finally be freed from its age-old errors and fruitless meanderings, and set back on a rational, well-defined path marked by scientific rigor. This was to be achieved through the use of the most advanced intellectual tools – mathematical logic and the scientific method – regarded as the long-sought remedy for a condition once thought incurable. Yet, this aspiration ultimately proved to be in vain:

> The language that analytic thinkers sought to tame into a docile servant of knowledge has revealed itself instead to be a labyrinth, from which no Ariadne's thread can lead us out. Logic – once expected to be the guiding light protecting philosophy from error and deception, placing it on a straight path toward a life of clarity and intellectual health – has proven either too impoverished to capture the intricate nuances of philosophical discourse, or too remote and frigid in its formalism, increasingly unintelligible and unusable for a philosophy that still aspires to speak to humanity. Condemned to wander through corridors and convoluted passages – illuminated only at times by flickers of poetry and intuition – the philosopher finds himself exhausted, exposed, his gaze clouded by the nostalgia for a project that now can only look backward for meaning. In the future lies only the vast, horizonless sea, one that no Ulysses seems able to cross.
>
> CONIGLIONE 2008a, p. 174

This new disenchanted view of scientific philosophy and its project – carried forward by many 20th-century philosophers, beginning with the Brentano school in its Polish branch and later by scholars who, in various ways, aligned themselves with neo-positivism and analytic philosophy – is articulated with great care and precision in Coniglione's most recent and extensive work, aptly titled *Lontano da Popper* (*Far from Popper* – Coniglione 2025). In it, he partly revises the assessment he had given in the first edition, *Popper addio* (*Farewell to Popper* – Coniglione 2008d), in which the trajectory of English-language post-Popperian epistemology is concisely outlined. These two works (the first and second editions) also offer a history of rationality, tracing its roots back to the ancient Greek world, a subject to which Coniglione has consistently devoted attention, producing several studies on the topic. Referring to *Popper addio*, Wolters (2015, p. 84) writes that it offers "a splendid overview of the development of scientific rationality," so that "the Italian reader for once needs not take recourse to the holy language of modern philosophy, i.e. English."

5 Toward a Pluralist View of Knowledge

The exploration and analysis of scientific philosophy's development necessarily requires engaging with the history of philosophical reflections on science – namely, epistemology and philosophy of science in their historical dimensions. Characteristically, Coniglione approaches this subject not through formal or logical reconstructions of scientific theories, but through rigorous historiography. As he explicitly states in the preface to his most recent work:

> This work cannot be classified as that of a "philosopher of science" for two fundamental reasons: first, because this designation has acquired such specialized meaning that it demands competencies I do not claim to fully master; second, because I have no pretension of vying with the established authorities in this field, who remain quite protective of their disciplinary boundaries. Rather, these pages represent the labor of a historian of philosophy who has dedicated much of his academic life to examining philosophy's engagement with science and its broader intellectual influence, particularly in recent years, focusing on the complex interplay between scientific rationality and what traditionally falls beyond its demarcated frontiers. The reflections offered here on various thinkers and traditions constitute, in this spirit, a philosophical intervention: an invitation to reconsider contemporary epistemology's trajectory, and indeed, to ponder whether practicing philosophy remains

> meaningful – to whatever extent such an endeavor remains possible within one's limitations
>
> CONIGLIONE 2025, p. 9

Thus, his scholarly focus gravitates toward thinkers who placed the epistemic nature of science and its cognitive revolution at the heart of philosophical inquiry, along with the attendant questions radiating into wider domains of human understanding. This orientation naturally privileges certain intellectual traditions: the rigorous formalists of the Vienna Circle and Lvov-Warsaw School; the architects of the Received View and their Popperian challengers; and later iconoclasts like Kuhn, Lakatos, and Feyerabend, whose heterodox critiques exposed the methodological fault lines beneath orthodox accounts of scientific rationality. These figures constitute the vital constellation through which his investigations unfold.

Indeed, Coniglione's first major scholarly contribution was devoted to popperism and its dissolution (Coniglione 1978). This groundbreaking work represented the first systematic Italian engagement with Feyerabend's radical critique, framing it as both a consequence and an indictment of Popperian epistemology's inherent tensions. By situating Feyerabend's arguments alongside Kuhn's paradigm shifts and Lakatos's research programs, Coniglione constructed a cohesive historiographical narrative of post-Popperian thought. Significantly, the volume's concluding section proposed the methodological framework of the Poznań School – particularly as articulated in Nowak's works (1977), then available in Italian – as a potential resolution to the epistemological crises exposed by Feyerabend's critique. Building on this work, Coniglione's scholarship bifurcated into two complementary trajectories: rigorous studies of the Poznań School's conceptual apparatus, and critical analyses of major figures in 20th-century epistemology (see the bibliography placed at the end of second volume). These investigations unfolded both through comprehensive, didactically oriented surveys (Coniglione 2004b) and specialized essays on pivotal epistemologists. Ultimately, these parallel strands of research converged into a synthesized framework in his two definitive works (Coniglione 2008d; 2025), marking both a personal and methodological culmination.

This investigation into scientific rationality gained new dimensions through the rediscovery of marginalized contributions, most notably the work of Polish epistemologist Ludwik Fleck. While the Lvov-Warsaw School and its successor had long dominated discussions of scientific philosophy in Poland, Fleck's heterodox perspectives emerged as a vital corrective as Coniglione's thought matured. Fleck's insights – which challenged both the scientistic tendencies of Polish analytical philosophy and Viennese Neopositivism – anticipated key themes later developed by Kuhn, who openly acknowledged Fleck's influence (Fleck 1935). Coniglione's engagement with Michael Polanyi's concept of the "tacit dimension" ([1958] 1962; 1966; 2015a) further deepened this reappraisal. By highlighting the non-rational and linguistically ineffable aspects of scientific practice, Polanyi's framework allowed Coniglione to reinterpret Kuhn's epistemology and reassess what he had once termed Feyerabend's "epistemological mysticism" (Coniglione 1992). This line of inquiry would later crystallize in a major study (Coniglione 2024a; 2024c), where he identified these themes as a unifying thread in post-Popperian thought. Coniglione's scholarly commitment to Fleck culminated during his fellowship at PIASt (Polish Institute of Advanced Studies – Warsaw), where he edited a comprehensive volume of Fleck's epistemological writings (Fleck 2019). His extensive introduction to the volume (Coniglione 2019a) offers a thorough analysis of the distinct place occupied by the Polish microbiologist in the contemporary epistemological debate.

As previously noted, these evolving perspectives prompted Coniglione to substantially revise arguments first advanced in *Popper addio* (2008d). The new edition (Coniglione 2025) not only expands upon earlier analyses but incorporates fresh chapters addressing previously neglected dimensions, including an exhaustive reconstruction of the Idealizational Conception of Science (Coniglione 2025, pp. 300–320). Significantly, the work now foregrounds those aspects of scientific practice that destabilize the Received View's logicist framework, particularly the irreducible non-rational elements permeating scientific activity. However – as Coniglione emphasizes – this evolution in his thinking does not negate science's unique epistemic status as humanity's most reliable guide to what Wittgenstein termed the "world of facts." Rather, it crystallizes into a pluralistic epistemology (2022b) that creatively synthesizes two fundamental insights: the modeling-centric approach of the Idealizational Conception with Feyerabend's late-career recognition of nature's irreducible "abundance" (1999). This theoretical fusion consciously resists what Coniglione has critiqued as the "epistemic imperialism" (2022a) of monolithic rationality. Notably, this framework has led to a reappraisal of complexity science – a field he initially approached with reservation (1996) – which he now recognizes as conceptually coherent within his model-based epistemology (Coniglione 2015b; 2022b; 2024b).

Coniglione now identifies in complexity theory a crucial epistemological lesson: the necessity of cognitive pluralism as an antidote to what he terms "theoretical monotheism", a stance mirroring religious absolutism in its claim that "there is but one true sacred theory, one religion reflecting the very nature of God and the ontological order He established" (Coniglione 2022b, p. 68). Complexity theory fundamentally challenges this monolithic vision, demonstrating the impossibility of compressing reality's totality into what Coniglione sardonically calls a "zipped" universal formula. Building on physicist Ignazio Licata's framework (Coniglione 2015, pp. 118–119) – whose collected essays he edited – Coniglione dismantles the conventional narrative of science as progressive, cumulative enlightenment. Whereas traditional epistemology envisions a pyramidal ascent toward an increasingly transparent grasp of reality – an ideal whereby the *logos* of human discourse progressively encompasses ever-wider domains of the real, absorbing earlier theories along the way (Coniglione 2022b, pp. 69–70) – Coniglione proposes, by contrast, a topography of distinct epistemic summits: some converging, others remaining solitary, each offering its own model of reality from a unique vantage point. This radical reconceptualization, as Coniglione argues, reflects science's authentic pluralistic nature: "Within any given discipline – physics being paradigmatic – multiple theoretical perspectives legitimately coexist, each addressing research questions through its unique explanatory prism" (Coniglione 2022b, p. 46):

> This perspective moves beyond merely acknowledging that each science carves its distinct object from a shared reality – a notion already well-established in philosophical discourse, often (and somewhat reductively) associated with relativistic approaches. It makes the far more radical claim that even within a single discipline, irreducibly distinct theoretical models can legitimately coexist. Each offers its own way of sectioning reality through different epistemic incisions, embracing unique phenomenal domains, and employing varying degrees of precision. Consider physics: both classical and relativistic mechanics describe the same physical world, yet they operate through fundamentally different conceptual "prisms" – their divergence rooted in core parameters like velocity. Neither theory's explanatory power negates the other's validity within its proper sphere of application.
>
> CONIGLIONE 2008a, p. 174

In Coniglione's view, this represents the crucial insight reached by the most recent developments in epistemology – an intuition already latent in the teachings of the Poznań School through its idealizational conception of science. Yet this perspective had been anticipated by earlier philosophers and epistemologists, and it is now garnering growing attention across diverse fields. These include sociology, exemplified by the work of David Bloor; radical constructivism, as elaborated by von Glasersfeld, von Foerster, Maturana, and Varela; and the domain of non-equilibrium thermodynamics, notably in the contributions of Ilya Prigogine.

However, for Coniglione, this does not imply an endorsement of nihilism, but rather a distinctive formulation of relativism. The slices and simplifications that science imposes upon the boundless richness of reality give rise to explanatory models – each theory valid within its own domain and capable of accommodating genuine cognitive progress. Yet these theories are not universally true, contrary to the imperialist conception of science that envisions a linear ascent toward a final, all-encompassing "theory of everything." From this vantage point, Coniglione aligns himself with the critical and deconstructive thrust of Richard Rorty's thought, although he distances himself from Rorty's overtly nihilistic and "conversational" conclusions (Coniglione 2025, pp. 484–505). Ultimately, Protagoras's famous assertion that "man is the measure of all things" may be upheld – so long as "man" is taken to mean humanity as a species, rather than the individual. This outlook underscores the inevitable conditionality of knowledge: it does not emanate from the mind of God, but from that of a particular species of animal, filtered through its cognitive faculties, shaped by idiosyncratic cultural traditions, and articulated in a language that is mutable, often ambiguous, and in any case, always partial. In this light, Coniglione finds particularly compelling certain strands of Indian philosophy – most notably Jainism and aspects of Buddhism – which advocate for a "modest" form of relativism (Coniglione 2024c). Far from being timid, this perspective is in fact courageous, especially when contrasted with the general condemnation such views receive from a Western philosophical tradition enamored with its own grand cathedrals of thought. To deny this situatedness of knowledge would be to pursue an illusion: that of a divine or "God's-eye" perspective – a notion decisively critiqued by Hilary Putnam (1981, pp. 49–50) and later by Richard Rorty (1992, p. 15). As von Glasersfeld aptly observed, "a human being's view of the world is necessarily a human view. Unless you claim some form of direct mystical revelation, whatever you call knowledge – your ideas or concepts, the relations that connect them, your images of yourself and the world – will be human, because the way you have produced them was yours, and you, whether you like it or not, are bound by the human ways" (von Glasersfeld 1995, p. 27).

This epistemological stance compels Coniglione to seriously engage with alternative modes of human engagement with nature (Coniglione 2022a). He maintains that:

> The analytical exercise of reason [...] cannot be regarded as either exclusive or exhaustive of the entirety of the human personality and its cognitive activities. It necessitates, as its counterpart and indispensable complement, the dimension of the anti-*logos* – most powerfully

> expressed in mythical thought. Within this lies the "unitive" perspective: a comprehensive and integrative gaze which, even after the decline of the mythic age, has remained foundational to manifold expressions of religiosity and mysticism across civilizations often vastly removed from one another in time and space.
>
> CONIGLIONE 2022a, p. 30

In articulating his epistemological vision, Coniglione invoked the image of an iceberg – a compelling metaphor through which to depict the layered structure of human knowledge (Coniglione 2022b, p. 65).

In this figure, the iceberg serves as a metaphor for the totality of scientific endeavor, or more broadly, the expanse of human cognitive activity. Its emergent tip – visible above the waterline – represents that portion of knowledge susceptible to explicit rational articulation: theories, algorithms, and mathematical models, the domain conventionally identified as "knowledge." By contrast, the submerged portion – concealed beneath the surface – corresponds

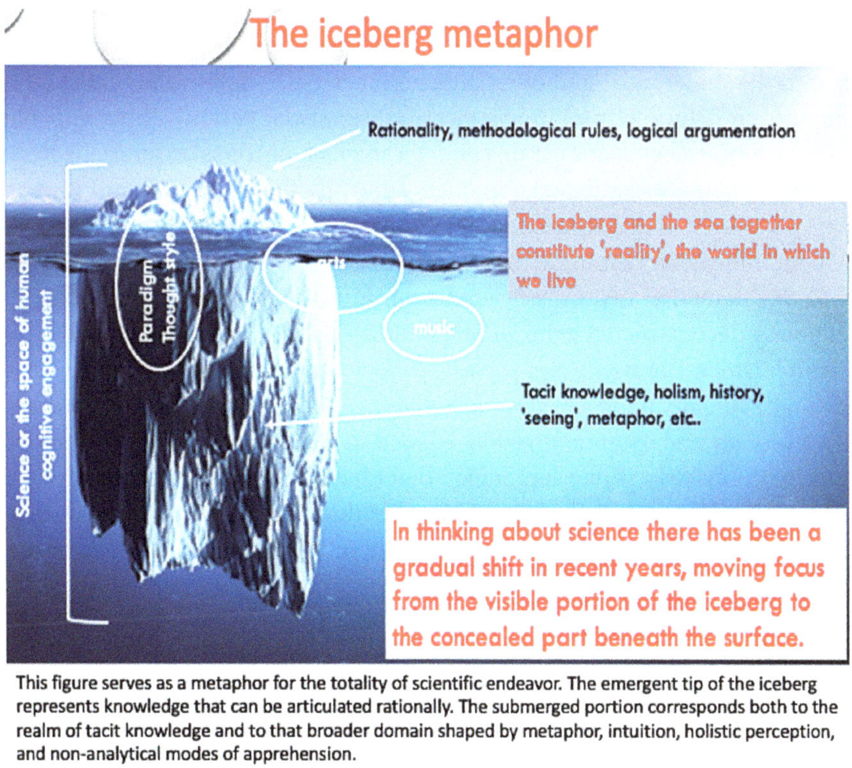

FIGURE 0.1 The iceberg metaphor

both to the realm of *tacit knowledge* and to that broader domain shaped by metaphor, intuition, holistic perception, and non-analytical modes of apprehension – what might more aptly be called *wisdom*, in the richer sense conveyed by the Italian word *sapienza*. The Received View and the dominant currents of 20th-century epistemology directed their focus almost exclusively to this visible segment, presuming it sufficient to account for science in its entirety, as well as for all human cognitive endeavors, which were expected to conform to its explicit, formalized models. Yet, as Coniglione (2025) contends, post-Popperian epistemology has increasingly recognized that knowledge is more often transmitted tacitly than explicitly, more through lived exemplars than through methodical treatises or rigid prescriptions. It is Kuhn's "paradigmatic" examples that shape the competent researcher, and, as Fleck and Feyerabend suggest, it is only through sustained engagement with the practices of exemplary scientists that one truly grasps what science is and how it ought to be pursued.

In essence, this renewed emphasis on a frequently neglected dimension of human cognition resonates – albeit implicitly – with the stance adopted by certain classical thinkers who contested the primacy of the *definitional logos*, later canonized by Socratic and Platonic philosophy. Among these figures stands the "heterodox" and often underappreciated Gorgias, who resisted the impulse to define "virtue" in abstract and exhaustive terms – those capable of subsuming all virtuous actions under a singular, universal essence. Instead, he championed a contextual approach, adjusting the concept to suit diverse actions, stages of life, times, and circumstances. For Gorgias, any claim to a comprehensive definition presupposed the existence of an ontological and metatemporal *logos* dissociated from human subjectivity – an idea he famously deconstructed in his principal theoretical work, *On Non-Being or On Nature* (*Περὶ τοῦ μὴ ὄντος ἢ περὶ φύσεως*). In so doing, he sought to confer epistemological dignity upon a traditional understanding of virtue rooted in archaic society – an understanding vividly echoed in Theognis of the sixth century BCE, who advised a young companion that excellence and upright conduct are not acquired through abstract reasoning, but through the imitation of virtuous and noble men, just as one learns to engage in "good science." As Coniglione observes,

> This dimension of knowledge finds its roots above all in the later Wittgenstein, who discouraged the quest for conceptual definitions – such as what constitutes a game – and instead invited his imagined interlocutor to refrain from thinking and simply look: "denk nicht, sondern schau!" This injunction points to a form of cognition that transcends conceptual

> abstraction and precise definition. It is grounded, rather, in practice: a kind of tacit learning shaped through repetition and the refinement of one's gaze. Such knowledge is not attained through abstract reasoning, but through active engagement in – or attentive observation of – multiple language games. It does not consist in conceptual mastery, but in skill: a trained ability to discern affinities and resemblances, rather than to uncover the "hidden identities" pursued by metaphysical speculation.
>
> CONIGLIONE 2019b, p. 135

It is the same line of thought that Coniglione has sought to illuminate in the aforementioned authors – Fleck, Polanyi, Kuhn, and finally Feyerabend, to name the most prominent. In doing so, he has ventured into uncharted terrain, traversing the realm of myth (Coniglione 2019c; 2021a), the domain of knowledge marginalized by the triumphant march of science (Coniglione 2004c; 2014b; 2017), and returning to scrutinize the origins of Western rationality by revisiting pivotal moments in classical thought (Coniglione 2018b; 2018c; 2019b). Ultimately, this constitutes an effort to reckon with an aspect of humanity's confrontation with nature that eludes the grasp of scientific cognition – a theme that had already beckoned Coniglione in his earliest inquiries.

In sum, Coniglione's conception of cognition unfolds along several key propositions. First, he contends that humanity's epistemic engagement with the natural world cannot be confined to the scientific paradigm alone. While fully acknowledging science's unmatched capacity to model physical reality, he underscores the need to account for the phenomenological dimensions of lived experience, the symbolic systems through which cultures mediate their understanding of nature, and the tacit forms of knowledge embedded in artisanal and traditional practices. Second, he illustrates how these "non-scientific" approaches often operate within their own internally coherent and rigorously structured epistemic frameworks. Whereas scientific idealization abstracts from the particular to construct general models, these alternative modes employ what he terms "contextual rationality," a mode of reasoning that preserves – rather than erases – situational complexity. Finally, Coniglione embraces this epistemological pluralism not as a challenge to scientific authority, but as its indispensable counterpart. As he writes: "The map is not the territory, and no single projection can capture every contour of reality's landscape" (2022a, p. 112). It is this recognition – that diverse forms of knowledge illuminate distinct facets of the world – that constitutes the philosophical foundation of his mature thought.

In this light, we are brought to yet another horizon of his intellectual odyssey.

6 The Other Side of Reason

Running parallel to the lines of research previously examined, which belong to the "diurnal history" of Coniglione's philosophical pursuits, there existed a "nocturnal" dimension that remained largely concealed. This aspect – often implicit and frequently veiled over long periods – can be traced back to his youthful interests, as noted in § 2. Throughout much of his scholarly work, this dimension expressed itself primarily through a particular sensitivity to certain themes and the occasional reading of books which Coniglione, however, found immediately off-putting due to their imprecision, lack of rigor, and the amateurish tendency to blend together divergent and irreconcilable levels of discourse. This refers to the vast domain of esotericism and its related disciplines, a realm of expansive libraries, bookstore shelves more densely populated than those of any other "scientific" subject, and a multitude of individuals engaged in its practice. These individuals continue to hold their beliefs while simultaneously applying reason in daily life, even pursuing scientific research in fields where Western *logos* has achieved its greatest successes, such as physics, logic, and mathematics.

This boundless universe, which never ceased to unsettle Coniglione, stirred in him two contrasting responses. On one hand, there was a rejection of what he regarded as a tangled mass of notions, superstitions, reveries, and astonishing flights of imagination, lacking in control, logical coherence, and rational discipline. Such a mélange was entirely incompatible with a mindset steeped in Enlightenment principles, in science, logic, and rational argumentation, and shaped by the rigorous, precise formalism of the Poznań School. On the other hand, he felt a compelling need to grasp the "logic" underpinning this cultural cosmos, to understand the deep-rooted necessity from which it springs. This desire retraced the steps of that youthful curiosity which had once led him to the very themes explored in his dissertation.

Inevitably, Coniglione found himself compelled to ask a series of questions on the matter. Should these disciplines and subjects be seen as a form of intellectual ailment, a malady of reason, a pathology of the *logos*, or perhaps an intellectual frailty? Are they expressions of a lyrical sensibility that, lacking more suitable avenues of expression, manifests itself in this way, echoing Carnap's famous thesis on metaphysics (Carnap [1939] 1959, p. 78)? Should we relegate to the realm of sentiment and irrationality those ponderous treatises which – at least in the most serious and culturally sophisticated cases, leaving aside the sensationalism of occult supermarket literature – offer structured arguments and detailed analyses? Should we thus deem them irrelevant to the human condition, or even harmful to an enlightened reason and to humanity's quest to free itself from superstition and vanity? In other words, is humanity grappling with a kind of systemic schizophrenia? Are those who act rationally or devote themselves to disciplines grounded in the principles of *logos* schizoid beings, torn between conflicting dimensions of the self, uncertain of which to embrace and surrender to?

Not only that: alongside the universe of esotericism and its related disciplines, there exists also the vast domain of Eastern thought and spirituality, which had captivated Coniglione in the years immediately following his graduation, particularly through Hegel's interpretation of it and the perceived affinities between his ideas and those of the Indian philosopher and mystic Sri Aurobindo. This interest, however, proved fleeting: tentative efforts to seek support or encouragement from a philosophy professor at the University of Catania – whom Coniglione held in some esteem – were met with polite but firm disinterest, both toward him and the subject he proposed. Nevertheless, these issues concern billions of people. Are they merely the result of missteps in human progress, cultural cul-de-sacs or regressive modes of engaging with the world? Should we maintain the familiar Western cultural and philosophical posture toward them, epitomized by Dante's famous dictum, "let us not speak of them, but look and pass on" (*non ragioniam di loro, ma guarda e passa, Inferno*, III, 51 – trans. R.T. Durling)? Should we continue to follow Hegel's explicit Eurocentrism, which placed Eastern culture in the infancy of humanity, judging its philosophy as still entangled with religion and mythology, and therefore at the initial stage of the long journey toward fully developed rational self-consciousness?

By dismissing these experiences and spiritual-philosophical worldviews, do we not merely persist in erecting our baroque cathedrals of thought, intricate yet aloof, or to weave our fatigued meditations within specialized, overly compartmentalized conceptual microcosms? Do we continue to act as though the fundamental questions that have haunted Western thought for millennia have found no echo elsewhere, as though they have never been probed, never contemplated, never wrestled with in other civilizations' examinations of human existence? Is this not, perhaps, the lingering echo of that *hybris* which, in less self-aware times, found justification in the myth of the "white man's burden"?

Yet, as Coniglione argues, this is precisely the prevailing attitude: we proceed as though these modes of spiritual experience and philosophical inquiry had never arisen, as though they held no claim on reason's domain. We cling to the illusion that the totality of human questions can be resolved, or at least incrementally advanced, through hyper-specialized fragmentation. We advance, myopically constrained within a single cultural lineage – however illustrious – while consigning the study of alternative traditions to the "orientalists": a discipline now institutionalized and sanitized, rendered philosophically inert against the dominant currents of Western thought.

For Coniglione, this approach has borne significant consequences, its influence echoing through the study of cultural and philosophical history, a history largely framed as exclusively "Western" and governed by an unbroken, self-evident continuity. In perpetuating what Richard King called the "myth of cultural homogeneity" (King 1999, p. 8), it has reinforced the notion of a distilled "essence" of Western philosophy, counterposed against an equally reductive "essence" of Indian or Eastern thought. Such framing suggests the possibility of a seamless, linear narrative devoid of substantive ruptures, even as it arbitrarily includes or excludes thinkers according to shifting intellectual trends, personal biases, or cultural orthodoxies. Philosophers once marginalized are periodically "rediscovered," while others, once influential, fade into deliberate obscurity. While Rorty rightly critiques the nebulous boundaries of "Western philosophy" as a classificatory concept (Rorty 1984), Coniglione maintains that philosophy has nevertheless been instrumental in forging the identity of European and Western civilization, precisely by defining itself in opposition to other cultural traditions. In this process, it has enshrined a particular model of rationality as the universal standard against which all other forms of reflection or world-conceptualization are measured. As Coniglione argues,

> This systematic concealment – which has very deep motivations and above all reflects "the Grand Polemical Narrative" carried out by organized religion, especially during the era of the witch-hunt – has relegated this field of interest to a mere and inevitable slag produced by the advance of knowledge, the indispensable friction and energy inevitably generated and dissipated by the advance of progress. The rationalism following the scientific revolution has also helped to precipitate an entire sector of cognitive and practical wisdom in a kind of dark dungeon of European conscience, in karst current that sometimes manifests itself on the surface by challenging the official culture, sometimes revealing and expressing itself within the thoughts of "respectable" thinkers and scientists as a kind of tic of reason, immediately isolated and quarantined by official historiography, which has always judged it as the result of human weakness and the consequence of psychological failures.
>
> CONIGLIONE 2017, p. 207

The consequence has been the relegation of these disciplines to impassioned enthusiasts – many lacking the technical rigor or hermeneutic subtlety such traditions demand – who plunder, reinterpret, and uncritically recycle texts and testimonies from these diverse intellectual worlds. Their efforts often seek to unearth in these materials an alternative to institutionalized "high culture," that is, to the sanctioned knowledge production of universities and academies. Frequently alienated by – or dissatisfied with – the perceived inadequacy of official intellectual culture in addressing existential and spiritual yearnings, these practitioners turn to fragmented reinterpretations, often distorting the very traditions they aim to elevate.

Over time, Coniglione came to recognize the possibility of navigating these conceptual territories without succumbing to their latent pitfalls. This realization crystallized upon encountering a seminal text that marked a turning point in his intellectual journey – a work he discovered after engaging with another essay by the same author for an academic examination. The book, *The Greeks and the Irrational* by Eric R. Dodds, struck him with the force of a cultural earthquake. For a mind educated in the Classical Lyceum tradition, steeped in reverence for Greek rationality and its celebrated role in the civilizing and enlightening mission of the West – Dodds' work delivered nothing short of a revelation. It laid bare the shamanic undercurrents in Pythagoras and Empedocles, while illuminating the shadowed, irrational dimensions of the Greek world: a cosmos irreducible to the sanitized pantheon of philosophical and literary heroes.

He came to perceive in Greek culture a dual pathway for accessing and engaging with reality, for "taming" it, and in some sense rendering it harmless or even favorable. One path was rooted in myth, mystery, and the sapiential or shamanic practices belonging to the "hereditary conglomerate" evoked by Dodds; the other emerged from the detachment and autonomous unfolding of the *logos*, which would lead to the birth of philosophy and the rational spirit as we still understand, practice, and celebrate them today. Coniglione was struck by the situation that could be summed up in what he later found written by Walter F. Otto (2000, p. 30): "The Greeks, this strange people who developed the λόγος to astonishing heights and anticipated our scientific thinking by providing an exact meaning of it, are simultaneously the people of myth, and (despite all criticism) have always remained so!" What astonished and fascinated him was the Greeks' distinctive ability to hold together – albeit not without tensions and inner conflicts – these two divergent yet deeply human modes of relating to the world, the ability to balance between various dimensions of human spirituality, which seemed to him to represent, as Nietzsche had so brilliantly intuited, the very hallmark of that originality which rendered Greek civilization a *chef-d'œuvre* still admired by people today.

The trajectory illuminated by Dodds led Coniglione to a deepened historiographical engagement with the realm of "rejected knowledge." Frances A. Yates's studies on Giordano Bruno and Renaissance Hermeticism opened up to scholarly attention a territory that had previously been scarcely explored or left to the incursions of amateurs from various backgrounds. Over time, this field has been substantively advanced by a growing cadre of scholars exploring these themes across diverse historical epochs. Foremost among these figures is Antoine Faivre, who in 1979 became the first scholar to occupy a European academic chair dedicated to the history of Western esotericism at the École Pratique des Hautes Études (Sorbonne). Equally significant is Nicholas Goodrick-Clarke (1953–2012), who not only directed the Exeter Centre for the Study of Esotericism (EXESESO) from 2002 but also held the chair in Western esoteric traditions at the University of Exeter from 2005. Arthur Versluis, professor of Religious Studies at Michigan State University, is another distinguished figure in this field. Together, these thinkers have transformed esotericism from a peripheral curiosity into a rigorous interdisciplinary field, bridging history, religious studies, and philosophy.

Yet for Coniglione, the most consequential contribution emerged from the Dutch school centered in Amsterdam under Wouter J. Hanegraaff's leadership. Appointed to the chair of History of Hermetic Philosophy and Related Currents in 1999 and serving as president of the European Society for the Study of Western Esotericism (2002–2006), Hanegraaff spearheaded a methodological revolution. Through prolific monographic output and the landmark *Dictionary of Gnosis and Western Esotericism* (2005, 2 vols.) – co-edited with Faivre, van den Broek, and Brach – his school established rigorous scholarly standards that normalized esotericism studies within mainstream academia. This institutional legitimacy is now evidenced by dedicated peer-reviewed journals, monograph series, international conferences, and degree programs (Hanegraaff, Pijnenburg 2009).

Equally noteworthy is Giorgio Galli's Italian intervention (Coniglione 2021d), though with a distinct focus on political theory and institutional history. Galli's work highlights how Western civilization developed through the dialectical tension between its dominant paradigms – "from the Greek miracle to the computer age" – and the subterranean "alternative values" that persistently accompanied them. His framework interrogates "the relationship between victorious cultures that shape our lived reality and defeated cultures that nevertheless left indelible marks on social organization" (Galli 2014, pp. 53–54).

This is neither the place to recount in detail the historical developments that established this field of research (for which see Coniglione 2017, pp. 208–11 and Hanegraaff's concluding chapter [2012]), nor to examine the nuances of

Galli's theoretical positions. What has been said thus far is sufficient to clarify the motivations that led Coniglione to direct his attention to this area of inquiry. Indeed, these developments liberated the study of esotericism from the grip of amateurism, elevating it to the status of a serious academic pursuit, to be conducted with the same rigor demanded by other disciplines in the humanities. It was precisely these authoritative frameworks, combined with the more substantive works emerging from the mainstream of Traditional thought, that enabled Coniglione to move beyond his initial rationalist skepticism and return to the fundamental questions that had first captivated him as a young scholar:

> why does mankind have two very different cognitive approaches, the "rationalist" and the "esoteric"? Why does the human species, and often also single scholars, have two such different attitudes to the world? What are the needs to which they answer? Perhaps to the usual ones of "reason" and "heart"? Or do they have a deeper meaning in the evolutionary history of man which is key to explaining certain aspects of the culture and human experience in different times and places?
>
> CONIGLIONE 2017, p. 211

These very questions have formed the cornerstone of Coniglione's sustained investigation into the dynamic interplay between myth and scientific thought. His scholarly engagement with this nexus has manifested not only through numerous publications but also through collaborative research initiatives. Most notably, together with Professor R. Loredana Cardullo, he co-directed the two-year research project *Rational and A-rational in the History of Thought: The Ongoing Conflict between Apollo and Dionysus, from Ancient Philosophy to Neurosciences* (2015–2017). This ambitious undertaking brought together an international team of distinguished scholars, culminating in the publication of its proceedings (Cardullo, Coniglione, 2017). Building on this foundation, Coniglione and Cardullo subsequently spearheaded another significant research endeavor titled *Future Archetypes: Forms, Functions, Meanings and Styles in the Relationship between Mythos and Logos*. This project yielded two substantial published volumes (Coco, Vasta, 2019; Cardullo, Coniglione, 2021), further expanding the interdisciplinary dialogue between mythological studies and contemporary scientific paradigms

Coniglione's investigations in this domain remain ongoing, though significant contributions have already emerged in his recent essays addressing these persistent questions. In his study of "rejected knowledge" (Coniglione 2017), while engaging with the aforementioned issues, he shifts focus away from the substantive content of esoteric doctrines – which he regards as fantastical constructs lacking rational scaffolding – toward their *functional role* in human cognition. As he articulates, these systems demand examination in terms of "function they fulfill in human thought and the personality of man understood holistically without neglecting rationality nor emotionality and sentimentality" (Coniglione 2017, p. 211). Developing this functional analysis, Coniglione advances a hypothesis: esoteric traditions, mystical thought – previously examined in relation to mathematics in the thought of Wittgenstein and Russell (Coniglione 2002b) – and mythic consciousness, reassessed in the light of Cassirer's legacy (Coniglione 2019c, pp. 115–16), collectively perform what he terms an *anagogical function*. Rather than offering cognitive claims about reality, they initiate pathways toward alternative, synthetic worldviews, complementary rather than competitive with scientific understanding. This reflects a fundamental orientation of the psyche that manifests diversely while operating in parallel to analytical modes of thought. Indeed, as numerous studies in anthropology and the history of religions have shown – sources to which Coniglione explicitly refers (citing Tylor, Frazer, Lévy-Bruhl, Lévi-Strauss, Tambiah, Eliade, Vernant, and Cassirer) – we may discern two distinct mental architectures: one historically (if problematically) termed "primitive" or "archaic," and another embodied in Western logical-operational paradigms. These represent "radically divergent mechanisms for perceiving and conceptualizing reality" (Coniglione 2019c, p. 214), not as evolutionary stages, but as coexisting modalities of human sense-making.

Furthermore, drawing on the latest findings in neurophysiology, Coniglione seeks to trace the roots of these two distinct mentalities to the evolutionary history of the human species and to the development of functionally differentiated brain operations, often, though not always accurately, attributed to the activity of the two cerebral hemispheres. The psychologist Jerome Bruner had already posited that "there are two modes of cognitive functioning, two modes of thought, each providing distinctive ways of ordering experience, of constructing reality. The two (though complementary) are irreducible to one another. Efforts to reduce one mode to the other or to ignore one at the expense of the other inevitably fail to capture the rich diversity of thought" (Bruner 1986, p. 11). This line of thought has been further developed through extensive experimental research by contemporary neuroscientists such as Giorgio Vallortigara and collaborators (2013; Rogers, Vallortigara 2017), Elkhonon Goldberg (2009), and Daniel Kahneman (2011), among others, all of whom are discussed in Coniglione's aforementioned essay. These studies provide compelling evidence for the existence of two complementary, yet distinct, cognitive modes, thereby reinforcing the hypothesis of a dual epistemological structure inherent to human rationality. This, in turn, leads Coniglione to the conclusion that

> when diagnosed at an anthropological level and contrasted with the mythical-esoteric thought, the birth of the scientific method can be convincingly explained by the genesis of the brain functions, of which one part has been specializing over time to process the information and the knowledge of the world using quite rigid and consolidated categories and schematic models, which provide an abstract and formal vision of reality but which are extremely potent in their ability to make us classify and recognize situations which fall within the normality of daily life, for ordinary survival purposes, as in the case of scientific laws (conceptualizing according to the idealizational approach to science). Whereas the other part of the brain conveys a more plastic holistic approach, that allows humans to deal with those novelties that are only subsequently consolidated into new models, into new thought patterns that lie – to the extent that they are effective – within the cognitive stock available to Man.
>
> CONIGLIONE 2017, p. 222

The dynamic interplay between these two modes of brain function aptly illustrates the deep-seated tension between esoteric and scientific thought, and more broadly, between what has traditionally been termed "primitive" versus "civilized" mentalities. In essence, this continuous shift "depicts the way in which, according to Thomas S. Kuhn, normal science proceeds through the construction of schematic models that are concretized and approximated to reality; so, the explanation of a phenomenon is in the first instance adaptively incorporated into one of the models already available" (Coniglione 2017, p. 223). At its core, this oscillating dynamic represents more than an epistemological mechanism. It constitutes an essential adaptive strategy for human cognition and cultural evolution, the normal way in which humans live, and a prerequisite for both psychological equilibrium and constructive engagement with reality.

The healthy functioning of both individual psyches and civilizations ultimately depends on this symbiotic coexistence between mythic and rational modes. Far from being antagonistic, the mythic imagination provides the necessary substrate that makes rigorous rationality possible, without which the mind risks descending into pathological rigidity or sterile abstraction. This insight sheds new light on what Coniglione has consistently identified in his studies of scientific philosophy and the history of epistemology as the mystical or irrational components within science itself. These are not contaminants threatening science's rational enterprise, but rather vital complements that fuel its creative potential.

For genuine innovation to occur, the mind must first immerse itself in formal structures, only to subsequently transcend them, a cyclical process where the right hemisphere's holistic capacities periodically disrupt established paradigms, giving rise to those revolutionary breakthroughs that propel knowledge forward. To prevent dissolution in the labyrinth of unbounded imagination, creative thought must first immerse itself in the discipline of formal structures and rational frameworks, only to emerge again, empowered to dismantle and reconfigure those very constraints. This dialectical rhythm between containment and transcendence mirrors what Coniglione identifies as the essential pulse of intellectual advancement: the systole and diastole of cognitive evolution, where temporary submission to form becomes the precondition for its revolutionary overcoming. In this perpetual dance between structure and inspiration – between what Coniglione might call the Apollonian and Dionysian poles of cognition – lies the engine of intellectual and cultural vitality. This reveals how the healthy functioning of both individual cognition and cultural evolution fundamentally depends on the symbiotic coexistence of mythic and rational modes of thought. Far from being antagonistic, the mythic imagination provides the essential substrate that enables and sustains rational cognition, without which the mind risks descending into pathological rigidity or sterile abstraction. In short, as Coniglione well says:

> Reason exercised exclusively on an analytical-rational basis would lead to madness if not compensated for by uninhibited holistic moments. Conversely, the mere holistic disinhibition would lead to the loss of reason and thus to the diminuendo of humanity if there were no rational procedures, able somehow to concretize and make holistic insights operational. Analytical reason is driven by holistic creativity to new horizons, which are, however, accessible only to the extent that they become rationally manageable. And vice versa, explosive creativity fertilizes analytical reason, which would otherwise wither by eternally repeating the identical. In the same way as wakefulness needs sleep, sleep must be translated into an effective and operational wakefulness. Consequently, a dual relationship is required: reason implies mystic intuition, the holistic disinhibition, a dimension of the sacred. The mystical and sacred provide holistic feedback on the reason, allowing creativity and preventing repetitive and tautological sterility. It is this dual relationship, in our view, which is the most authentic legacy of the Western logos (distinguishing it from that of any other people) and the road to the future. Only this is

> given to Western Man, walking on the razor thin blade separating two abysses into he can plunge: the hubris of reason celebrating itself and, in this exaltation, goes blind, versus the ecstatic sentimentality of a slimy mysticism that soils and tangles everything.
>
> CONIGLIONE 2017, pp. 224–5

From what has been said, it becomes clear that in the more recent phases of his research, Coniglione seeks to reconcile the various "journeys" his thought has taken, without renouncing any of them. Scientific philosophy and its concern with the rationality inherent in it are acknowledged for their specificity and cognitive value with regard to the real world, that is, to our concrete experience of reality. At the same time, the opposite pole of human thought – often dismissed as irrational – is not relegated to the margins as mere slag to be purged in the radiant ascent of logic and science. Rather, it is recognized as an essential dimension of the human psyche, the mortification of which not only hinders the full unfolding of rational inquiry, but risks engendering madness, fragmentation, and alienation of the self. Moreover, the realm of the anti-*logos* – as Coniglione frequently calls it – holds an intrinsic significance for the spiritual development of humanity or, in more secular terms, for the shaping of one's inner world and moral character. So, Coniglione recognizes these non-rational elements as constitutive features of human cognition – vital forces whose suppression not only impoverishes scientific reason itself, but risks precipitating psychological fragmentation and existential alienation. What he frequently terms the "anti-logos" represents not just a counterpoint to rational thought, but a fundamental modality of human understanding with profound implications for spiritual development, or, in secular terms, for the cultivation of a fully integrated mind and character. This dialectical vision resists both the reductionism of scientism and the obscurantism of pure irrationalism, charting instead what might be called an "ecology of mind".

This philosophical orientation illuminates Coniglione's sustained engagement with St. Francis of Assisi during his later years – a focal point revealing profound continuities across humanity's spiritual traditions. In the Franciscan experience, he identifies what might be termed the perennial dimension of religious consciousness: that transformative impulse manifest in mystical and esoteric currents alike, which cultivates inner refinement through disciplined praxis and the harmonious integration of human nature's instinctual foundations. Though emerging from the distant world of the late Middle Ages – from "another world" – Francis's example speaks with startling urgency to our contemporary condition. Our era, while technologically sophisticated in developing external competencies, remains strikingly deficient in what the ancients called *ars animae*, the art of soul-formation. Where modernity excels at instrumental mastery, the Franciscan tradition offers an antidote to our spiritual atrophy, reminding us that true human flourishing requires the careful tending of interiority no less than the cultivation of technical skill. In effect, according to Coniglione (2017, p. 19), the enduring relevance of Francis's message transcends both historical context and doctrinal particulars. What renders his teachings profoundly contemporary is their uncompromising insistence on the *iter difficilior*, that arduous path which alone leads to authentic human growth. This disciplined journey forms the bedrock of all genuine intellectual refinement, moral development, and spiritual pedagogy, transforming individuals into integrated beings rather than fractured composites. Francis's wisdom offers a piercing indictment of the modern condition, in which sophisticated scientific-technical mastery often coexists with stunted ethical sensibilities, producing what might be called a "centaur culture": intellectually advanced yet morally rudimentary, theoretically complex yet dogmatically rigid in its unexamined convictions, often steeped in religious fideistic fundamentalism. In this light, the Franciscan path offers not antiquated piety, but an antidote to the spiritual imbalance of our age.

This cultivation of interiority is not solely fostered by scientific disciplines, – despite Heidegger's reductive dismissal of them as mere "calculating thought" (*Denken als Rechnen*) –, but equally by the humanities, particularly those domains Coniglione has sought to illuminate: poetry, literature, the arts, and esoteric, mystical and religious traditions. These are not ancillary curiosities but foundational expressions of human consciousness, capable of shaping character in a harmoniously integral way and finding vibrant interplay within the minds of those rare individuals who marry scientific rigor with mystical insight. Eminent figures such as Newton, Gödel, Brouwer, Wittgenstein, Russell, and others – frequently referenced in Coniglione's work – epitomize this phenomenon. Yet, while he often cites them in support of his arguments, Coniglione has engaged in thorough analysis only of a select few.

This remains an open intellectual journey. In a recent essay, Coniglione (2024b) – drawing upon Russell and Schlick's notion of "acquaintance," along with the previously examined concepts of tacit knowledge and learning through practice – proposes that mystical, esoteric, and religious traditions might be understood a daring yet perilous attempt to bridge the fundamental epistemic gap: the chasm between reality's infinite manifestations and the necessarily bounded, contextual models human language can construct. Mystics have traditionally articulated this through metaphors of divinity's "infinite names" or manifold "faces." From this perspective, what conventional theology terms "God" might be reconceived as *natura naturans*, or what Spinoza called *deus sive natura*. Yet it transcends *natura naturata* (the realm of empirical facts), as it encompasses infinite attributes, while human understanding perceives only two (thought and extension). The conflation of this God of infinite attributes (faces, names, etc.) with the limited nature displayed to us through a few discernible attributes was one of the reasons Spinoza was accused of pantheism.

The accusation of pantheism arises from conflating two distinct concepts that Spinoza carefully distinguishes: while God includes nature as an expression of His essence, He transcends its finite manifestations and is not reducible to *natura naturata*. In this view, Spinoza effectively revisits Giordano Bruno's doctrine of a God who is simultaneously *mens insita omnibus* (immanent in all things) and *mens super omnia* (transcending all things). The divine permeates the world yet infinitely surpasses it, a paradox articulated across traditions, whether through theistic transcendence or through secular frameworks emphasizing reality's inexhaustibility (as in Feyerabend's thought). This incommensurable excess defies linguistic capture: no sequence of names or enumeration of attributes can contain divine essence. As Bruno observes, "God, then, must be signified either by no name or by all names. He is all in all, for it is to all things that He grants being; yet He is none of them, surpassing each in essence, nobility, and universal virtue" (quoted in Coniglione 2024b, p. 134). To grasp this requires abandoning the illusion of language's absolute transparency – the vain hope that reality might be exhausted by any single doctrine or "theory of everything." Instead, we must embrace epistemological humility: accepting contingency, limitation, and imperfection while remaining open to infinite, renewing articulations. Only through the totality of partial expressions can we approach the whole, always acknowledging an insurmountable gap between human representation and the ineffable plenitude it seeks to signify.

This is not a plea to forsake the scientific paradigm – whose rigorous methodology has demonstrably shaped modern civilization through its technological and cognitive achievements – in favor of some purportedly superior interpretation of reality. Such a replacement would naively assume that mystical revelation could offer more refined or reliable knowledge than secular science, an illusion quickly dispelled by the contradictory plurality of these "revelations." As Coniglione's study of St. Francis demonstrates (2022c, Ch. III), the insights of mystics are invariably shaped by their cultural milieu, symbolic universe, and the inherent limitations of their linguistic tradition. The value of mysticism lies not in any claim to cognitive superiority, but precisely in its complementary role to rationality, what we might call the *anagogical function* discussed earlier. Rather than competing with scientific understanding, these traditions serve as necessary counterweights, reminding us that human knowledge systems thrive not through absolutism, but through the dynamic tension between empirical rigor and intuitive grasp, between analytical precision and synthetic vision. Therefore, the objective is to harmoniously integrate mystical insight and scientific construction, illustrating how

> each has its own field of validity, beyond which one falls into the perilous hubris of reason or sentiment that leads humanity astray. It is crucial to recognize that scientific theories are only models of reality, which cannot be exhaustively grasped by any of them, in line with the logic of complementarity proposed by Bohr. As Bateson [1972, § 3.5.1] warned in referring to the "two great opposite poles of mental process, both lethal when taken alone [...], rigor alone is death by paralysis, but imagination alone is madness". The key is to attain a state of "grace" where "the reasons of the heart must be integrated with the reasons of reason"
>
> CONIGLIONE 2024b, p. 149

In essence, Coniglione – following Wittgenstein and Schlick's epistemological legacy – maintains that all genuine knowledge is linguistically constituted, forged through intersubjective discourse, argumentative persuasion, and the dialectical confrontation of competing frameworks. As he asserts, "There is no knowledge other than that constructed within language's scaffolding, through communicative exchange, the imperative to persuade, and the capacity to counterpose discourses, a paradigm exemplified most rigorously (though not exclusively) by science, which remains the foremost heir of the Hellenic λόγος" (Coniglione 2019d, pp. 79–80). Yet at the same time, he recognizes

> that this rationality – when pursued to its furthest extent and lived with that intensity and passion proper to the intellect, which the heart cannot feel (inverting Pascal's maxim) – can only emerge through sustained engagement with the ineffable. This other dimension has shadowed the λόγος throughout its historical development, not as its antagonist but as its necessary counterpart. What emerges is neither a synthesis nor a hierarchy, but an irreducible duality woven into humanity's cognitive fabric: the rational and the numinous perpetually co-constituting one another. This interlacing – transcending historical epochs and defying temporal categorization – forms the very warp of human intellectual history, where the measurable and the mysterious remain inextricably entwined.
>
> CONIGLIONE 2019d, p. 81

Reinterpreting ancient yet frequently misconstrued concepts – particularly the distinction between knowledge (ἐπιστήμη), wisdom (σοφία), and the archaic English notion of *wiseness* (φρόνησις) – Coniglione proposes an Aristotelian revitalization of *wiseness* as *integral reason*. This synthesis harmoniously unites technique (knowledge and science as currently conceived) with metaphysics (wisdom as a vision of the whole, conveyed not by the *logos* but by esoteric and mystical insight). This gives rise to a threefold articulation of the human relationship with the world: *knowledge* (technical-cognitive), *wisdom* (esoteric or contemplative), and *wiseness* (practical-synthetic). The latter "would entail the capacity to live in the right balanced state between knowledge and wisdom, assigning each its rightful role and scope without categorical confusion, while recognizing the indispensability of both and their mutual complementarity" (Coniglione 2022d, p. 14).

7 The Civic Engagement

Any comprehensive account of Francesco Coniglione's intellectual trajectory must recognize yet another dimension of his scholarly engagement: his profound civic commitment. This facet of his work has manifested through vigorous participation in the political and cultural debates of his residing nation, with substantive contributions appearing as essays (including several monograph-length treatments), opinion pieces, and critical interventions published across national and regional newspapers, digital platforms, and edited volumes. While an exhaustive catalogue falls beyond this scope, this corpus stands as a testament to his civic engagement and multifaceted intellectual pursuits.

A particularly significant episode in Coniglione's scholarly career was his participation in the European Commission's Seventh Framework Programme research initiative (Grant Agreement No. 217796). The MIRRORS project (Monitoring Ideas Regarding Research Organizations and Reasons in Science) sought "to build an efficient and democratic society based on knowledge so as to stimulate a harmonious integration of scientific and technological subjects and relative research policies in Euro-pean society" (Coniglione 2010, p. 9). This collaborative endeavor yielded multiple scholarly outputs under Coniglione's direction, featuring contributions from both his research team – including Giacomo Borbone, Emanuele Coco, Enrico Viola, Rosaria Di Tommasi, Anna B. Francese, Cinzia Rizza, and Giuseppe Sapienza – and distinguished international scholars such as Sébastien Charles, Paul Hoyningen-Huene, Eleonora Montuschi, Thomas Nickles, Matthias Kaufmann, Andrzej Klawiter, Ignazio Licata, and Gereon Wolters (Brzechczyn 2011, 2012).

This research initiative provided Coniglione with a crucial platform to systematically examine the structural challenges of the knowledge society and Italy's persistent difficulties in meeting them. His analysis culminated in a sobering assessment of the nation's prospects, which questioned Italy's institutional and cultural capacity to adapt to the demands of this new paradigm (Coniglione 2011b). The diagnosis was unequivocal: systemic constraints were severely limiting Italy's ability to cultivate innovation and respond dynamically to the accelerating transformations of global knowledge economies.

> This analysis yields a grave prognosis: Italy's systematic dismantling of its research and innovation infrastructure is actively cultivating conditions for future geopolitical dependence, marginalization within knowledge economies, and eventual decline from the ranks of advanced industrialized nations. What may endure is merely the reduction of Italy to its touristic simulacrum (the "bel paese"), a nation defined by historical patrimony and environmental assets, its global image increasingly confined to reductive stereotypes: from the musical tropes of "O Sole Mio" and mandolin iconography to culinary clichés of margherita pizza ("which, much like in Naples, nobody knows how to prepare"), and luxury brand fetishism of "made in Italy".
>
> CONIGLIONE 2011b, p. 326

This trajectory mirrors what development economists term the "tourist trap, a fate already endured by many resource-poor developing nations, wherein cultural capital and natural beauty become the sole remaining commodities. As Coniglione's analysis suggests, without urgent intervention, Italy risks completing its transition from industrial power to what Lord Byron [...] once romanticized as a land of melodrama and macaroni. The culmination of this decline would see even Italy's last remaining strongholds – like high fashion – succumbing to global competition, while the nation's physical and institutional landscapes drown under waves of unchecked development, corruption, and ethical decay.

Coniglione's pessimistic prognosis has, in retrospect, demonstrated remarkable prescience – a quality equally evident in his analyses of Italy's political apparatus and societal power structures (Coniglione 2019e, 2021b). His scholarly critique extends to the politicization of science and higher education, as evidenced by his extensive contributions to ROARS (Return On Academic Research and School), a platform dedicated to reforming Italy's academic and research institutions (see https://www.roars.it). This intellectual engagement reflects Coniglione's enduring civic-humanist ethos (Coniglione 2021c, 2022f), which in recent years has been profoundly shaped by his sustained engagement with Franciscan thought (2022d, 2022e, 2023). His body of work thus accomplishes a dual purpose: it not only diagnoses systemic failures in Italian society but also reaffirms the vital role of ethical and humanistic frameworks in shaping collective futures. By synthesizing historical analysis, sociopolitical critique, and Franciscan spirituality, Coniglione models an integrative approach to societal transformation – one that bridges intellectual rigor with moral vision.

•••

A constellation of themes, an intellectual odyssey both vibrant and untamed, at times appearing diffuse, yet unified by what we might characterize as the restlessness of "an unquiet mind." This is a mind that refuses the comfort of settled conclusions, resisting the tranquil contemplation of past achievements even after his 2020 retirement from academic duties. Rather than waning, this intellectual energy has surged with renewed vitality, proving that Francesco Coniglione's philosophical journey remains very much in progress. His career stands as a testament to the power of undisciplined curiosity, a deliberate rejection of narrow specialization in favor of boundary-crossing explorations. With undiminished vigor, he has consistently ventured beyond established frontiers, charting new territories where others might hesitate to tread. As we bring this introductory portrait to a close, I – Giacomo Borbone – find myself compelled to acknowledge the profound debt I owe to his mentorship, a guiding light that has indelibly shaped my own philosophical and scientific formation.

Yet true intellectual discipleship demands more than uncritical admiration – it requires fidelity to the Aristotelian principle: *Amicus Plato, sed magis amica veritas*. While a comprehensive critique of Coniglione's oeuvre lies beyond this essay's scope, certain strategic choices have demonstrably limited the broader reception of his work. Chief among these is his persistent preference for local and specialized publishers, a decision that – whether born of pragmatic constraints or deliberate intellectual independence – has inevitably restricted his visibility in mainstream academic circles. The consequences are measurable: without the imprimatur of major publishing houses (a near-prerequisite in Italy for serious scholarly engagement), his writings have seldom been reviewed in influential journals or media outlets. This marginalization is further exacerbated by three interrelated factors: (1) the lack of systematic compilations organizing his works thematically, resulting in a fragmented corpus that belies its underlying conceptual unity; (2) his enduring affiliation with a regional university excluded from Italy's elite rankings of academic "excellence"; and (3) his principled commitment to solitary scholarship and research over participation in academic politics or the pursuit of academic power and institutional prestige. Together, these elements have created what might be termed – in our view – the "Coniglione paradox": a body of work remarkable for its depth and originality, yet institutionally constrained in its dissemination and influence.

To be sure, Coniglione has achieved meaningful scholarly dialogue nationally and internationally, engaging prominent thinkers across disciplines. Yet the very existence of this dedicated volume, with its multifaceted engagement of his ideas, stands as testament to the untapped potential of his thought when liberated from these circumstantial constraints.

References

Adorno, Th. W. ([1963] 1971). *Tre studi su Hegel*. Bologna: Il Mulino.

Ajdukiewicz, K. (1989). *Problemi e teorie della filosofia*, edited by F. Coniglione. Trento: Reverdito.

Bateson, G. (1972). *Steps to an Ecology of Mind*. Northvale – London: Jason Aronson Inc.

Bateson, G. (1978). *Minds and Nature. A Necessary Unity*. New York: E.P. Dutton.

Brentano, F. ([1929] 1968). *Über die Zukunft der Philosophie*. Felix Meiner: Hamburg.

Bruner, J. (1986). *Actual minds, possible worlds*. Cambridge-London Harvard University Press.

Brzechczyn, K. (2011). Review of Mirrors: Science and Knowledge Society ed. by F. Coniglione. "Epistemologia. An Italian Journal for the Philosophy of Science" vol. 34, no. 1, pp. 145–148.

Brzechczyn, K. (2012), Nauka w społeczeństwie opartym na wiedzy [Science in the Knowledge-Based Society]. "Ruch Prawniczy, Ekonomiczny i Socjologiczny," no. 2, pp. 307–314.

Brzeziński, J., F. Coniglione, L. Nowak, T.A.F. Kuipers (1990a) (eds.), *Idealization I: General Problems*. Amsterdam-Atlanta: Rodopi.

Brzeziński, J., F. Coniglione, L. Nowak, T.A.F. Kuipers (1990b) (eds.), *Idealization II: Forms and Applications*. Amsterdam-Atlanta: Rodopi.

Bunge, M. (2001). *Philosophy in Crisis. The Need for Reconstruction*, Amherst: Prometheus Book.

Cardullo, R.L., F. Coniglione (Eds.) (2017). *Reason and no-reason from ancient philosophy to neurosciences*. Sankt Augustin: Academia Verlag.

Cardullo, R.L., F. Coniglione, (Eds.) (2021). *Mythos e Logos. Tra archetipi antichi e sguardi sul futuro*. Pisa: Edizioni ETS.

Carnap, R. ([1932] 1959). The Elimination of Metaphysics Through Logical Analysis. In: A.J. Ayer (ed.), *Logical Positivism*, pp. 60–81. New York: The Free Press.

Carnap, R. ([1936] 1971). *The Logical Syntax of Language*, trans. by A. Smeaton. London: Routledge and Kegan Paul.

Cassirer, E. ([1910] 1923). *Substance and Function and Einstein's Theory of Relativity.* Chicago: Open Court.

Cassirer, E. ([1937] 1956). *Determinism and Indeterminism in Modern Physics.* New Haven: Yale University Press.

Coco, E., S. Vasta (Eds.) (2019). *Il mito tra filosofia e scienza. Temi e prospettive dall'antichità a oggi.* Catania: Malcor D'.

Coniglione, F. (1978). *La scienza impossibile. Dal popperismo alla critica del razionalismo.* Bologna: Il Mulino.

Coniglione, F. (1987). *Il sentiero interrotto. Il fantasma della scienza nel marxismo teorico italiano.* Catania: Edizioni del Prisma.

Coniglione, F. (1990). Abstraction and Idealization in Hegel and Marx. In: J. Brzezinski, F. Coniglione, L. Nowak, T.A.F. Kuipers (Eds.), *Idealization 1: General Problems*, pp. 61–88. Amsterdam-Atlanta: Rodopi.

Coniglione, F. ([1990] 2010). *Realtà e astrazione. Scuola polacca ed epistemologia postpositivista.* Acireale-Rome: Bonanno.

Coniglione, F. (1992). Irracjonalizm epistemologiczny Paula Feyerabenda [Paul Feyerabend's Epistemological Irrationalism]. *Poznańskie Studia z Filozofii Nauki*, z. 12, pp. 111–132.

Coniglione, F. (1996). Ecologia, caos, complessità. Per una critica delle confusioni epistemologiche. *Alternative*, 3, pp. 107–122.

Coniglione, F. (1997). Astrazione e modelli di scientificità in Croce e Gentile, *Synaxis*, vol. 15, no. 2, pp. 677–727.

Coniglione, F. (1999). *L'idea di filosofia scientifica. Per una storia dell'epistemologia del Novecento.* Catania: CUECM.

Coniglione, F. (2002a). Russell e la nascita dell'idea di filosofia scientifica. In: *Filosofia, scienze, cultura*, edited by G. Bentivegna, S. Burgio and G. Magnano San Lio, pp. 181–218. Soveria Mannelli (CZ): Rubbettino.

Coniglione, F. (2002b). La parola liberatrice. Matematica e misticismo in Russell e Wittgenstein. In: F. Coniglione, *La parola liberatrice. Momenti storici del rapporto tra filosofia e scienza*, pp. 63–139. Catania: CUECM.

Coniglione, F. (2004a). Between Abstraction and Idealization: Scientific Practice and Philosophical Awareness. In: F. Coniglione, R. Poli and R. Rollinger (eds.), *Idealization XI: Historical Studies on Abstraction*, pp. 59–110. Atlanta-Amsterdam: Rodopi.

Coniglione, F. (2004b). *Introduzione alla filosofia della scienza. Un approccio storico.* Acireale-Rome: Bonanno.

Coniglione, F. (2004c). Apoteosi del soggetto e annichilimento della corporeità in Julius Evola. *Magazzino di filosofia*, n. 13, anno V, pp. 67–95.

Coniglione F., R. Poli, (2004) (eds.). *La realtà modellata. L'approccio idealizzazionale e le sue applicazioni nelle scienze umane.* Milan: Franco Angeli.

Coniglione, F., R. Poli and R. Rollinger (2004) (eds.). *Idealization XI: Historical Studies on Abstraction*. Atlanta-Amsterdam: Rodopi.

Coniglione, F. (2005). Per la storia della filosofia scientifica. Il Circolo di Vienna e la Scuola di Leopoli-Varsavia. In: G. Gembillo (Ed.), *Filosofia e scienze. Studi in onore di Girolamo Cotroneo*, pp. 109–141. Soveria Mannelli (CZ): Rubbettino.

Coniglione, F. (2007). The Place of Polish Scientific Philosophy in the European Context. *Polish Journal of Philosophy*, no. 1, pp. 7–27.

Coniglione, F. (2008a). Il pensiero infermo. Origine e destino della filosofia scientifica. In: B. Bonghi and F. Minazzi (Eds.) *Sulla filosofia italiana del Novecento*, pp. 151–174. Milan: Franco Angeli.

Coniglione, F. (2008b). Filosofia scientifica europea e positivismo italiano. In: G. Bentivegna, F. Coniglione and G. Magnano San Lio (Eds.), *Il positivismo italiano: una questione chiusa?*, Atti del Congresso tenutosi a Catania, 11–14 settembre 2007, pp. 39–67. Acireale-Rome: Bonanno Editore.

Coniglione, F. (2008c). Gramsci e il pensiero marxista tra storicismo e scienza. In: S. Salmeri, R.S. Pignato (Eds.), *Gramsci e la formazione dell'uomo. Itinerari educativi per una cultura progressista*, pp. 81–113. Acireale-Rome: Bonanno Editore.

Coniglione, F. (2008d). *Popper addio. Dalla crisi dell'epistemologia alla fine del logos occidentale*. Acireale-Rome: Bonanno Editore.

Coniglione, F. (2009). Gentile, Croce e il Marxismo. *Logoi* (Annali dell'ISIS Castelvetrano), vol. 5, Istituto d'Istruzione Superiore "Giovanni Pantaleo", pp. 21–35.

Coniglione, F. (2010) (Ed.). *Through The Mirrors of Science. New Challenges for Knowledge-based Societies*. Heusenstamm: Ontos Verlag.

Coniglione, F. (2011a). Giulio Preti, Ludovico Geymonat e la filosofia scientifica. In: F. Cambi and G. Mari (Eds.), *Giulio Preti. Intellettuale e critico attuale*, pp. 51–83. Florence: Firenze University Press.

Coniglione, F. (2011b). Una sfida già persa? L'Italia di fronte alla società della conoscenza. *Il Protagora*, anno XXXVIII, luglio–dicembre 2011, n. 16, pp. 299–333.

Coniglione, F. (2012). Il rapporto tra Marx ed Hegel alla luce del concetto di idealizzazione. In: G. Borbone (Ed.), *Scienza, marxismo e metafisica. Leszek Nowak e la scuola metodologica di Poznań*, pp. 23–74. Villasanta: Limina Mentis.

Coniglione, F. (2013). Filosofia, esoterismo e limiti del dicibile in Guénon ed Evola. In: F. Caroti, C. Pace (Eds.), *Io e l'antropologo. Scritti in onore di Paolo Chiozzi*, pp. 47–60. Acireale-Rome: Bonanno Editore.

Coniglione, F. (2014a). Positivismo, filosofia scientifica e scienze umane nell'Italia d'inizio secolo. In: G. Cimino, G.P. Lombardo (Eds.), *La nascita delle "scienze umane" nell'Italia post-unitaria*, pp. 29–45. Milan: Franco Angeli.

Coniglione, F. (2014b). Julius Evola: tradizione e orientalismo. *Civiltà del Mediterraneo*, 25, dicembre, pp. 183–226.

Coniglione, F. (2015a). Quale conoscenza per la "Società della conoscenza"?. *Bollettino della Società Filosofica Italiana*, 216, Settembre–Dicembre, pp. 3–24.

Coniglione, F. (2015b). Introduzione. Complessità del reale, semplicità del pensiero. In: I. Licata, *I Gatti di Wiener. Riflessioni sistemiche sulla complessità*, edited by F. Coniglione, pp. 7–39. Acireale-Rome: Bonanno Editore.

Coniglione, F. (2016a). Galileo and Contemporary Epistemology. Do we still have something to learn from Galileo's 'methodological revolution'?, *Przegląd Filozoficzny. Nowa Seria*, vol. 25, issue 1 (97), pp. 123–140.

Coniglione, F. (2016b). Croce tra scienza e filosofia scientifica. In: C. Tuozzolo (Ed.), *Benedetto Croce. Riflessioni a 150 anni dalla nascita*, pp. 291–310 Canterano (RM): Aracne Editrice.

Coniglione, F. (2017). 'Rejected Knowledge' and the Tradition of Reason. A few Hints on their Necessary and Inextricable Link. In: L.R. Cardullo, F. Coniglione (Eds.), *Reason and no-reason from ancient philosophy to neurosciences*, pp. 211–231. Sankt Augustin: Academia Verlag.

Coniglione, F. (2018a). Una "ragione minuscola". Mario Dal Pra e la filosofia scientifica del Novecento. In: F. Minazzi (Ed.), *Mario Dal Pra nella "Scuola di Milano"*. Milan-Udine: Mimesis, pp. 307–323.

Coniglione, F. (2018b). Hybris illuminista e accesso alternativo al vero nel mondo greco. In: M. Albana, C. Soraci (Eds.), *Amicitia res plurima continet. Omaggio a Febronia Elia*, pp. 139–162. Acireale-Rome: Bonanno Editore.

Coniglione, F. (2018c). Some Remarks on the Meaning of Esotericism and Plato's Unwritten Doctrines., *Mondi. Movimenti simbolici e sociali dell'uomo*, vol. I, n. 1, pp. 5–51.

Coniglione, F. (2019a). Lontano da Vienna, lontano da Leopoli. Ludwik Fleck: l'uomo giusto nel posto e nel tempo sbagliati, Introduzione a Fleck 2019, pp. 7–93.

Coniglione, F. (2019b). La parola che risana. Gorgia, logica e tolleranza. In: di L.R. Cardullo, F. Coniglione (Eds.), *La forza del Logos. Gorgia a 2500 dalla nascita*, pp. 109–154. Acireale-Rome: Bonanno Editore.

Coniglione, F. (2019c). L'intreccio antropologico tra mito e logo: una indispensabile (ma non pacifica) coesistenza. In: E. Coco and S. Vasta (Eds.), *Il mito tra filosofia e scienza. Temi e prospettive dall'antichità a oggi*, pp. 109–140. Catania: Malcor D'.

Coniglione, F. (2019d). I due volti dell'umano. In: E. Frasca (Ed.), *Il valore e la virtù. Studi in onore di Silvana Raffaele*, pp. 69–82. Acireale-Rome: Bonanno Editore.

Coniglione, F. (2019e). La stagione di "Mani pulite" e il nuovo patto sociale, *Magistra*, anno xi, n. 31, pp. 35–55.

Coniglione, F. (2020). Verso un marxismo post-empirista? Il tentativo di Galvano Della Volpe. In: C. Tuozzolo (Ed.), *Marx in Italia. Ricerche nel bicentenario della nascita di Karl Marx*, pp. 673–726. Canterano (RM): Aracne Editrice, pp. 673–726.

Coniglione, F. (2021a). La penultima verità: il mito tra *pistis* e *logos* al tramonto del mondo antico. In: R.L. Cardullo, F. Coniglione (Eds.), *Mythos e Logos. Tra archetipi antichi e sguardi sul futuro*, pp. 178–205. Pisa: Edizioni ETS.

Coniglione, F. (2021b). La fine delle ideologie e la metamorfosi del partito politico. Il caso italiano. *Historia Magistra*, anno xiv, n. 31, pp. 16–45.

Coniglione, F. (2021c). Etica laica ed etica come fraternità. In: P. Amato, A.G. Biuso, V. Bochicchio, M.T. Catena, F. Masi, V. Pinto, N. Russo, S. Venezia (Eds.), *Metafisica dell'immanenza. Scritti per Eugenio Mazzarella*, vol. II, *Etica e religione*, pp. 185–199. Milan-Udine: Mimesis.

Coniglione, F. (2021d). L'illuminismo "aggiornato" di Giorgio Galli. *Mondi. Movimenti simbolici e sociali dell'uomo*, 1, pp. 110–122.

Coniglione, F. (2022a). Dal mito alla scienza e ritorno. Per una visione non imperialistica della conoscenza. In: E. Coco (Ed.), *L'invenzione della realtà. Scienza, mito e immaginario nel dialogo tra psiche e mondo oggettivo. Una prospettiva filosofica*, pp. 11–30. Pisa: Edizioni ETS.

Coniglione, F. (2022b). Complessità e pluralismo cognitivo. *Complessità*, n. 2, luglio–dicembre, pp. 32–72.

Coniglione, F. (2022c). *L'uomo venuto da un altro mondo. Francesco d'Assisi*. Acireale-Rome: Bonanno.

Coniglione, F. (2022d). Introduzione a S.G. Brown, *L'esoterismo in Paolo di Tarso. La sapienza nascosta in 1Cor 2:6–3:4*. Acireale (CT): Tipheret, pp. 7–17.

Coniglione, F. (2022e). Da un mondo all'altro. Ciò che Francesco d'Assisi può ancora dire all'uomo d'oggi. *Mondi*, 1, pp. 18–43.

Coniglione, F. (2022f). Pacifismo radicale. Non c'è giustizia senza pace. Volerelaluna.it, 28–dicembre 2022.

Coniglione, F. (2023). Il lupo di Gubbio e l'orso di Mosca. *Avvenire*, 10 febbraio, p. 3.

Coniglione, F. (2024a). Pluralism and Mysticism in the Thought of Paul K. Feyerabend. *Filozoficzne Aspekty Genezy*, vol. 20, no. 2, pp. 15–54.

Coniglione, F. (2024b). Reincantare la natura. La "strana" convergenza tra complessità e sacro. convergenza tra complessità e sacro. In: Loredana R. Cardullo (ed.), *Uomo e natura, un rapporto da riscrivere. Prospettive interdisciplinari i "cura" di sé e del mondo*. Pisa: Edizioni ETS, pp. 125–149.

Coniglione, F. (2024c). "La signora che mai conosceremo". "Abbondanza" del mondo e misticismo in Paul K. Feyerebend. Paper given at the Conference on the 100th Anniversary of the Birth of P.K. Feyerabend, Palermo 10–11 maggio 2024, in print.

Coniglione, F. (2025). *Lontano da Popper. L'epistemologia post-positivista e le metamorfosi della razionalità*. Pisa: Edizioni ETS.

Feyerabend, P.K. (1999). *Conquest of Abundance. A Tale of Abstraction versus the Richness of Being*. Chicago and London: The University of Chicago Press.

Fleck, L. (1935). *Entstehung und Entwicklung einer wissenschaftlichen Tatsache. Einführung in die Lehre vom Denkstil und Denkkollektiv*. Basel: B. Swhwabe und Co. Verlabuchhandlung.

Fleck, L. (2019). *Stili di pensiero. La conoscenza scientifica come creazione sociale*, edited by F. Coniglione. Milan-Udine: Mimesis.

Galli, G. (2014). *Le ribelli della storia. Baccanti, gnostici e streghe: i vinti della storia e la loro eredità*. Milan: Shake edizioni.

Goldberg, E. (2009). *The new executive brain. Frontal lobes in a complex world*. Oxford *et al.*: Oxford University Press.

Hanegraaff, W.J. (2012). *Esotericism and the Academy. Rejected Knowledge in Western Culture*. Cambridge: Cambridge University Press.

Hanegraaff, W.J. - Pijnenburg, J. (2009) (Eds.). *Hermes in the Academy: Ten Years' Study of Western Esotericism at the University of Amsterdam*. Amsterdam: Amsterdam University Press.

Husserl, E. ([1901–02] 2001). *Logical Investigations*, vol. I. London and New York: Routledge.

Kahneman, D. (2011). *Thinking, Fast and Slow*. New York: Ferrar, Straus and Giroux.

Kiedrowski, K. (2013). *Metodologiczne podstawy negatywistycznej metafizyki unitarnej Leszka Nowaka* [Methodological Foundations of Negativistic Unitarian Metaphysics of Leszek Nowak]. Poznań: AMU Press.

King, R. (1999). *Indian Philosophy. An Introduction to Hindu and Buddhist Thought*. Washington: Georgetown University Press.

Licata, I. (2015). *I Gatti di Wiener. Riflessioni sistemiche sulla complessità*. Acireale-Rome: Bonanno Editore.

Lycan, W. (1988). *Judgement and Justification*. Atlantic Highlands: Humaruties.

Nowak, L. (1977). *La scienza come idealizzazione. I fondamenti della metodologia marxiana*. Bologna: Il Mulino.

Nowak, L. (1987). *Oltre Marx. Per un materialismo storico non-marxiano*. Rome: Armando.

Nowak, L. (1998). *Byt i myśl. U podstaw negatywistycznej metafizyki unitarnej, tom I, Nicość i istnienie* [Being and Thought: Foundations of Negativistic Unitarian Metaphysics, vol. I: Nothingness and Existence]. Poznań: Zysk i S-ka Wydawnictwo.

Nowak, L. (2004). *Byt i myśl. U podstaw negatywistycznej metafizyki unitarnej, tom II, Wieczność i zmiana* [Being and Thought: Foundations of Negativistic Unitarian Metaphysics, vol. II: Eternity and Change]. Poznań: Zysk i S-ka Wydawnictwo.

Nowak, L. (2007). *Byt i myśl. U podstaw negatywistycznej metafizyki unitarnej, tom III, Enigma i rzeczywistości* [Being and Thought: Foundations of Negativistic Unitarian Metaphysics, vol. III: Enigma and Realities]. Poznań: Zysk i S-ka Wydawnictwo.

Otto, W.F. (2000). *Il mito*. Genova: Il Melangolo.

Parrini, P. (1995). *Conoscenza e realtà. Saggio di filosofia positiva.* Rome-Bari: Laterza.

Parrini, P. (1999). *Immanenzgedanken* and Knowledge as Unification. Scientific philosophy and philosophy of science. In: M.C. Galavotti, A. Pagnini (Eds.), *Experience, Reality, and Scientific Explanation: Workshop in Honour of Merrilee and Wesley Salmon.* Dordrecht: Springer Science/Business Media B.V., pp. 15–37.

Polanyi, M. ([1958] 1962). *Personal Knowledge. Towards a Post-Critical Philosophy.* London: Routledge.

Polanyi, M. (1966). *The Tacit Dimension.* Garden City (N.Y.): Doubleday & Company, Inc.

Propp, V. ([1928] 2009). *Morphology of the Folktale.* Austin, TX: University of Texas Press.

Przybysz, P. (2013). Cosa significa *essere* nella metafisica unitaria di Leszek Nowak. In: G. Borbone, ed., *Scienza, marxismo e metafisica. Leszek Nowak e la Scuola Metodologica di Poznań*, pp. 305–315. Villasanta: Limina Mentis.

Putnam, H. (1981). *Reason, Truth and History.* Cambridge: Cambridge University Press.

Rogers, L.J., Vallortigara, G., Andrew, R.J. (2013). *Divided Brains. The Biology and Behaviour of Brain Asymmetries.* Cambridge: Cambridge University Press.

Rogers, L.J., Vallortigara, G. (2017) (Eds.). *Lateralized Brain Functions. Methods in Human and Non-Human Species.* New York: Humana Press.

Rorty, R. (1984). The Historiography of Philosophy: Four Genres. In: R. Rorty, J.B. Schneewind, Q. Skinner (eds), *Philosophy in History. Essays on the historiography of philosophy.* Cambridge: Cambridge University Press.

Rorty, R. (1992). *Philosophy and Social Hope.* London *et al.*: Penguin Books.

Tatarkiewicz, W. ([1958] 1988). *Historia filozofii* (*History of Philosophy*), vol. 3. Warsaw: PWN.

Von Glasersfeld, E. (1995). *Radical Constructivism: A Way of Knowing and Learning.* London - New York: Routledge Falmer.

Wolters, G. (2015). On Having the Last Word: Epistemological and Normative Considerations. *Bollettino della Società Filosofica Italiana*, 216, pp. 69–87.

PART 1

Reason as an Inspiration for Philosophy

CHAPTER 1

The Problem of Critical Realism in the School of Milan

Fabio Minazzi

Abstract

Through a loosely structured narrative, the author reconstructs the history of the *International Insubric Centre*, which preserves the archival and library heritage of the *Milan School*. The *Centre* currently holds approximately thirty philosophical archives, along with seven author libraries. From the work and thought of Piero Martinetti and Antonio Banfi to Banfi's distinguished pupils of the 1930s – Enzo Paci, Remo Cantoni, Giulio Preti, and Dino Formaggio – the Milan School expanded to achieve a breadth unparalleled in the cultural landscape of the twentieth century. The School was not limited to philosophy but also encompassed poets such as Antonia Pozzi, Vittorio Sereni, and Daria Menicanti, as well as musicologists, pedagogues, literary figures, and filmmakers. Today, the legacy of this tradition endures as a research program rooted in a rationalist-critical and phenomenological framework, one that continues to demonstrate significant critical

Keywords

critical rationalism – epistemology – philosophical archives – the International Insubric Centre

This paper investigates the emergence, characteristics, and internal tensions of critical realism as developed within the Milan School tradition. In particular, it focuses on the interplay between the transcendental structures of rationality and the challenges posed by experiential and historical realities. A central problem explored here concerns the limitations of a purely formal rationalism, a difficulty recognized by several key figures in the School. Additionally, the differences among major exponents of the School regarding the conception of critical rationalism are addressed, offering a more articulated picture of its intellectual legacy.

 | HTTPS://DOI.ORG/10.1163/9789004756496-003

In doing so, I place myself within the thematic trajectory that has been central to the contributions of Francesco Coniglione on scientific philosophy, not only within the framework of Italian intellectual culture at the turn of the century, but also in broader European and international contexts; in this way, I seek to offer an additional element for a better understanding of critical rationalism, which cannot be entirely absorbed or represented – as is often the case, especially in international debates – by Popper and his school. In fact, Coniglione's original research program on "scientific philosophy" (Coniglione 1999; 2002; 2005; 2007; 2008a) has, in many respects, become intrinsically linked with my own investigations into the problem of critical rationalism and, more recently, critical realism as developed within the Milan School. Numerous works by Coniglione not only address Popperian rationalism and its fate, but also turn their attention to the theme of Italian scientific philosophy (Coniglione 2008b; 2014a) and to the intellectual legacy of philosophers such as Giulio Preti (1911–1972) and Ludovico Geymonat (1908–1991), situating them within a broader context than that of Italian philosophy alone. In particular, the distinctive merit of his reflection lies above all in his analysis of the role of scepticism in Giulio Preti's epistemology – a central issue to which he has devoted two major contributions (Coniglione 1993; 2015).

Moreover, in pursuing this original line of inquiry – both historically grounded and conceptually refined – Coniglione succeeded in articulating these themes on multiple critical levels, situating the work of Geymonat and Preti firmly within the intellectual and cultural milieu of the early twentieth century. This was a period marked by the decline of nineteenth-century positivism and the ascendance of Italian neo-idealism, with which both philosophers were compelled to engage (Coniglione 2014a). By closely examining these movements, Coniglione was able to situate Italian philosophical debates within the broader and more complex landscape of contemporary European thought, thereby underscoring the European significance of contributions by figures such as Preti and Geymonat. It was, therefore, almost inevitable that my own research program on critical rationalism and critical realism would converge with Coniglione's investigations.

In light of this shared intellectual focus, I would now like to examine more closely the so-called Milan School, building upon the discussion initiated by Coniglione regarding its defining features and positioning it within its broader historical and theoretical framework. However, in doing so, I have proceeded from the conviction that any cultural tradition is destined to wither and stagnate if it is not nourished and enriched by what may be called its essential "material" foundations – namely, the documents, writings, and correspondence that must be made available to scholars wishing to investigate and fully comprehend its history. This is all the more evident when one considers the marginal status of Italian culture in comparison to the dominant Anglophone literature, which – particularly in the eyes of international scholars – has come to represent the hegemonic discourse. As a result, specific national traditions, however intrinsically valuable, are often unjustly dismissed as "provincial" expressions of a main intellectual trend whose centers of gravity lie overseas.

It is in this context that I present, in the opening paragraph of this contribution, the work I have undertaken in establishing and organizing the archives of the Milan School, which are currently curated by the University of Insubria through the International Insubric Centre, founded in the spring of 2009. To date, the Centre has brought together approximately fifteen such archives and, whenever possible, has also acquired the corresponding author libraries alongside the archival holdings. Subsequently (in §2), I will examine the overall features of the Milan School, before turning (in §3) to the teaching of Antonio Banfi as reflected in the work of his numerous students. Finally (in §4), I will attempt to assess Banfi's Enlightenment legacy, also in comparison with Popper's conception of critical rationalism. Inevitably, what I shall discuss is inextricably intertwined with the trajectory of my own intellectual development. This is not intended as a form of self-celebration, but rather as an attempt to give flesh and substance to a narrative that might otherwise risk becoming sterile and flavorless.

1 The Genesis of the International Insubric Centre.

We began organizing the International Insubric Centre with the Preti Collection, which comprises the complete archive of Giulio Preti. This collection does not include any letters, as Preti had the peculiar habit of destroying all incoming correspondence at the end of each day. Nonetheless, as often happens with such idiosyncratic figures, many of the letters he sent to others have been preserved, leaving us with a rather one-sided but nonetheless valuable epistolary record. Another important bequest, also intimately linked to the cultural history of Milan, was entrusted to us in the spring of 2009: the archive of manuscripts by Carlo Cattaneo (1801–1869), covering the period of his exile through to his death, that is, from 1848 to 1869.

This connection between Cattaneo and Preti is also linked to my own intellectual biography. While I was teaching in Lecce (2001–2008), Preti's longtime collaborator – who had preserved and safeguarded all of Preti's papers – passed away. That scholar was Ermanno Migliorini (1924–1999) (Peruzzi 2011), who had taught in Florence, where Preti held the chair of Philosophy until his death in Djerba, Tunisia, on July 28, 1972. Following his death, Migliorini's wife decided to entrust me with the entire collection of papers, which I had previously studied as a young researcher at the Università Statale of Milan. At that time, my thesis advisor, Mario Dal Pra (1914–1992), had assigned me – on behalf of the then newly established Department of Philosophy – the task of conducting an "exploratory mission" on Preti's papers in Florence. I catalogued the collection (Minazzi 1984) and immediately recognized its exceptional value. That same archive would later become the one I inherited from Migliorini. I should also note, as an interesting detail, that several offers were made from abroad to acquire the collection – most notably from the University of Italian Switzerland, where I had taught for five years (1999–2003) – but I declined, as I believed it was essential that the archive remain on Italian soil.

By bringing together these two core collections – those of Cattaneo and Preti – we hoped to generate a kind of "magnet effect," and indeed, this aspiration was soon fulfilled. Beginning in 2011, we launched a cycle of conferences dedicated to prominent figures of the Milan School. The first was held in that same year to mark the centenary of Giulio Preti's birth (see Minazzi 2015a; Coniglione 2015). Subsequent conferences followed in annual succession: in 2012 on Carlo Cattaneo; in 2013 on Antonio Banfi; in 2014 on Mario Dal Pra (Minazzi 2018; Coniglione 2018); in 2015 on European critical rationalism; in 2016 on Piero Martinetti, regarded as the "philosophical grandfather" of the School; in 2017 on Husserl's reception in Milan; and in 2018 on Ludovico Geymonat (see also Coniglione 2011).

As our work progressed, additional archives of considerable value came into our possession. Among the most remarkable was the archive and library of the young poet Antonia Pozzi (1912–1938), which was recently acquired. In 2011, we also received the so-called "Secret Archive" of Antonio Banfi (1886–1957), comprising approximately 5,000 letters spanning from his early studies in Berlin to the final years of his life. The designation "secret" does not refer to confidential content per se, but rather to the politically sensitive nature of certain correspondence – particularly from Banfi's period as a senator of the Italian Communist Party. For example, a recurring figure in many letters is a person cryptically referred to as "Canopo." After consulting Banfi's grandson, we discovered that "Canopo" was in fact a pseudonym for Palmiro Togliatti. The value of this archive is further underscored by its documentation of Banfi's personal encounters with major intellectual figures such as Husserl and Simmel, including his impressions upon reading *Philosophie des Geldes* (The Philosophy of Money). As Fulvio Papi appropriately remarked upon reviewing the collection: "Banfi's intellectual biography will have to be rewritten – now it can truly be reconstructed day by day." Banfi was, in fact, an obsessive letter-writer, often sending three letters daily to his fiancée and later wife, Daria Malaguzzi, and so we have some five thousand of them. There are, of course, historical precedents for such prolific correspondence: at the time, telephones were not in widespread use, and the postal service was remarkably efficient. The archive thus provides a vivid window into Banfi's intellectual and emotional life.

In addition, there is another archive whose key features are worth noting. The entire collection comprises approximately 16,000 volumes distributed across 30 archives. Among them is the archive of Evandro Agazzi, who studied at the Università Cattolica under Gustavo Bontadini and later joined the research group led by Ludovico Geymonat at the National Research Council (CNR), following the publication of his 1961 work on the problems of axiomatics (Agazzi 1961). The archive also includes Agazzi's personal library – a specialized collection focused on mathematical logic and the philosophy of science. It is an extensive library containing major works in multiple languages, reflecting the international scope of Agazzi's intellectual engagement.

There are also several smaller holdings that are no less significant. These include a modest archive of Vittorio Sereni (1913–1983), a small archive of Daria Menicanti (1914–1995), and a collection of materials by Guido Morpurgo Tagliabue (1907–1997). We also preserve the complete set of Giovanni Vailati's papers concerning the Royal Commission for the Reform of the Italian School System, along with the archives and libraries of Bruno Widmar, Augusta (Lella) Monti, Guido Bersellini, and Silvio Ceccato. I'll stop here. Those interested can visit the website of our online Centre, where a short film is available showcasing our archives and libraries. The reference, of course, is to the Centro Internazionale Insubrico "Carlo Cattaneo" and "Giulio Preti."

One important point to emphasize is that all these archives and libraries have been acquired through *donations*. I highlight this fact because, even in an Italy so often described as devastated and frayed, there remains a segment of *civil society* that is fundamentally *sound*. A second point is just as significant: anyone who has been involved in acquiring archives and libraries knows that the relationship rarely ends with a single donation. On the contrary, it often opens the door to further exchanges. Many archives are continuously expanded over time, and for a very understandable reason. The original owners of these collections often maintain a deep emotional attachment to them, and when they make an initial donation, there is typically a "residue" that remains, which later gives rise to subsequent contributions. Behind every archive, there are always *human relationships*. In fact, around these archives and this Centre, a web of personal and scholarly connections has grown, one that dates back to my youth.

2 The School of Milan

In addressing the subject of this paragraph, I must refer to certain autobiographical elements stemming from my own academic formation in Milan and from having experienced, so to speak, "firsthand," the key moments in the life of its School. What follows, therefore, is the perspective of someone who studied at the Università Statale of Milan, where I enrolled in 1975 and attended courses throughout the second half of the 1970s. To establish a timeline: I followed Paci's (1911–1976) final and remarkable course – on Leibniz and Husserl, focusing on monadology and relationism – and I had the opportunity to attend lectures by Remo Cantoni, Ludovico Geymonat, and Mario Dal Pra. That university – my university – no longer exists in the form I experienced it. I also attended the very first course taught by Carlo Sini (b. 1933), who had just arrived in Milan and was lecturing on Nietzsche. In my view, Sini would go on to become the last true philosophical voice of the Milan School.

In reconnecting with the reflections presented in the previous paragraph, we can now raise a legitimate question: why were all these archives related to the Milan School not gathered in Milan itself, but rather in Varese, at the University of Insubria? In truth, something important has indeed begun to take shape there, perhaps in part as a result of the creation of our Centre. The Library of Philosophy at the Università Statale has recently begun new archival work, and has already acquired several important collections. Even prior to that, it already housed the archives of Giuseppe Rensi (1871–1941) and Giovanni Vailati (1815–1890), a partial photocopy of Preti's archive (which I collected while completing my degree), and more recently, the complete archive of Mario Dal Pra. It now also includes the collections of Guido Neri (1935–2001), Rodolfo Mondolfo (1877–1976), and a group of particularly valuable documents from Antonio Banfi.

But there is a deeper reason that explains Milan's failure to preserve the materials and legacy of this School – something that took place after my time at the Statale. I am speculating here – since I was no longer directly involved – but this is the impression I formed: it was not a matter of chance. In my view, the principal responsibility lies with the very same professors. The decline of the Statale di Milano, in my opinion, stems from certain serious mistakes made by those masters themselves. In particular, I refer to the academic and institutional decisions made by two figures I worked closely with: Geymonat and Dal Pra. Their choices in faculty appointments and succession planning were, to put it plainly, lacking in foresight – and ultimately led to the self-destruction of this intellectual tradition.

From a cultural standpoint, we are now confronted with a process of erasure – what might be called a removal – of the Milan School in relation to the extraordinary events that unfolded during the 1930s. When I reflect on what took place in Milan during that decade, I remain astonished by the sequence of events and cannot help but ask: why did they happen? One key transition, in my view, has not been sufficiently examined. When Banfi arrived in Milan, it was upon the recommendation of Piero Martinetti. Martinetti was the only philosopher – among twelve in a total of roughly 1,200 university professors – who refused to swear allegiance to Fascism. He and Banfi shared the same civil stance. In fact, Banfi had initially been reluctant to accept the position in Milan, precisely because he did not want to swear loyalty to the Fascist regime. It was Martinetti who ultimately persuaded him to do so. But Banfi was a remarkable teacher. And beyond his pedagogical skill, he was also extraordinarily fortunate. Anyone who teaches knows this well: to encounter an outstanding student is a rare and precious gift. Banfi was exceptionally lucky in this regard – within a few short years, he found himself surrounded by a group of students, each more brilliant than the last.

So, if I consider the "School of Milan" – which may also be more strictly defined as the "Banfian School" – from this perspective, I see it as a system of epicycles: at its core lies the *hard core* made up of philosophers, especially the brilliant students of the 1930s, who form its nucleus. Around this core, however, orbit various epicycles that extend into aesthetics, literature, poetry, musicology, epistemology, the history of science, cinema, and more. Banfi's good fortune, then, was not only to be a great teacher, but also to have such students. I believe the key to this fruitful outcome lies in Banfi's phenomenological orientation, which allowed each of his students the freedom to pursue their own intellectual "demons."

This, of course, raises a historiographical question: what is the relationship between the intellectual development of Banfi's prominent students from the 1930s and his formative influence on them? Current historiography – limited though it may be – has crystallized around two main positions. On the one hand, some argue that the School of Milan – represented by figures such as Cantoni, Paci, Preti, Rognoni, Formaggio (1914–2008), and others – gradually drifted away from Banfi's original critical stance. This thesis is supported, for example, in a notable book by Franco Cambi (1983; see also Coniglione 2011). On the other hand, a different interpretation is advanced by other scholars – including myself – who contend that Banfi's students carried forward his critical spirit, developing it along highly original and creative lines. There was *something* in Banfi's pedagogical and philosophical approach that proved especially fertile, allowing each student to pursue their own trajectory while remaining rooted in a shared critical framework.

What was this "something"? Why was the Milan School so fruitful and intellectually compelling? Was there something distinctive in the theoretical formation that Banfi offered his students? Did he provide them with a particular advantage – one that could be seen not only in his student philosophers, but also in his student poets, writers, and other intellectuals? I believe so. And if I were to summarize briefly, I would say that the critical stance which characterized the School's intellectual edge – though culturally problematic, as noted in Sini's recent reflections on Paci (Sini 2015; 2013, pp. 105–115) – was rooted in Banfi's highly original way of structuring the School, itself a product of his own unique philosophical trajectory. It is clear that Banfi's work must be approached through a proper periodization. The activity he carried out in Milan during the 1930s is firmly rooted in his earlier masterpiece, *I principi della teoria della ragione* (1926). But what, from a theoretical standpoint, was the philosophical operation Banfi undertook in that work?

To answer this question, we can turn to an essay by Mario Dal Pra – an excellent piece, later published in *Filosofi del Novecento* (Dal Pra 1990²; see also Coniglione 2018). In this text, Dal Pra poses the question: What is the *core* of the tradition of critical rationalism upon which Banfi drew? His answer is unequivocal: the *critical rationalism* of Kant and his *transcendentalism*. Within this tradition, *the very soul of reason is the transcendental*. But how did Banfi interpret the transcendental? *He sought to reinterpret Kant's transcendentalism through a Hegelian lens – without, however, falling into a full-fledged philosophy of history*. In Banfi's reading, the challenge lay in holding together two distinct poles: on the one hand, the *formal* structure of reason; on the other, its *historical genesis*. His philosophical operation, then, was to preserve the tension between these two dimensions, seeing reason as both structurally formal and historically dynamic. As Dal Pra notes, according to Banfi, "reason must be thought historically. If, from a Kantian perspective, reason is form, it is also, from a Hegelian one, a structure constructed in time." (Dal Pra [1984] 1990, p. 68).

These, I believe, are the two theoretical poles that constitute the core of Banfi's innovative contribution. Even in the 1930s, one can rediscover this dialectic throughout his published lectures, especially in the well-known courses collected by Formaggio and other students. One notable instance appears in his 1937 course on Plato. In the final section, where Banfi reflects on the generative power of Plato's teaching, it becomes evident that he is, in fact, speaking about himself. His analysis of the relationship between Plato and his students is, implicitly, a reflection on his own relationship with his pupils. This issue – central to Banfi's model of education – was also the object of discussion within the Milan School itself. If it is true that the Milan School developed within the framework of critical rationalism, then it is only natural that internal tensions would eventually emerge. As I see it, one of these tensions became particularly evident to figures such as Remo Cantoni (1914–1978), Enzo Paci, Giulio Preti, and even Antonia Pozzi (Minazzi 2015d; 2019b), each of whom responded to it through their own intellectual formation and personal journey.

In this regard, the position that seems to me the clearest is that of Giulio Preti. In a posthumous essay entitled *In principio era la carne* (Preti 1983), originally written in 1948, Preti identified a profound theoretical affinity between Banfi's thought and that of Giovanni Gentile (1875–1944). It would be possible to pursue a much longer discussion here, as Enzo Paci had made similar remarks, emphasizing the significance of Gentile's influence – something also evident from the fact that some of Banfi's students had direct contact with the two major neo-idealists, Croce and Gentile. While it is known that some of Paci's letters to Croce have been published, it is perhaps less well known that Preti himself visited Gentile. We have published two of Preti's letters to Gentile, which make this intellectual relationship abundantly clear (Minazzi 2015a, vol. II, pp. 1229–1233). But what did it mean for a student of Banfi to engage with Gentile in the 1930s? At that time, Gentile was referred to as "His Excellency," and not just in philosophical circles. There exists a particularly revealing letter by Preti, sent to Gentile – which we first published in *Il Protagora* – in which he candidly reconstructs the condition of a young philosopher in fascist Italy. Preti had the rare merit of believing that, in theoretical life and in philosophy, one must always pursue the truth head-on – without resorting to any form of mediation, not even social or political ones – even if this meant facing serious consequences.

In this letter, Preti explained that he had submitted an essay for publication in the *Giornale Critico della Filosofia Italiana* – a piece of Hegelian historiography in which he explored the relationship between *continuity* and *discontinuity* in the history of philosophical thought. This theme, deeply aligned with Banfian concerns, is particularly significant (Preti 1983; Coniglione 1993). As he had to introduce himself to Gentile, Preti wrote: "You cannot consider me one of your students, because I am not; nor am I an idealist. But neither can you consider me – as I am not – a representative of the Catholic intellectual tradition," nor, he added, of any of the practical alliances that formed within idealism after the Lateran Concordat (this was in 1929) which Garin describes so well in his *Cronache di filosofia* (Garin 1975, vol. II, pp. 345–488). He concluded: "The only one who listens to me is Banfi, who gave me the opportunity to publish and with whom I now work. And yet, I feel that I have only one destiny, and that is to follow My Lady Philosophy." It is noteworthy that students like Paci and Preti – though aligned with Banfi – nonetheless established personal and intellectual relationships with the major neo-Hegelian figures of their time, Croce and Gentile. This aspect of their formation remains particularly intriguing.

I believe that Preti, particularly at a certain point in his development, came to recognize that the problem he encountered in Banfi's thought was essentially the same theoretical challenge found in Gentile's philosophy. And what was that problem? It was the emergence of a divide between the highly formal – yet often hollow – structures of rationality on one side, and the richness of life and lived experience on the other. However, what is missing is a critical mediation between these two poles – the pole of rational formalism and the pole of experience, that is, that living experience (*Erlebnisse*) which would come to represent the central concern of Husserl's later thought. And it must be kept in mind that what applied to all of Banfi's students also held true for Preti: for them, Banfi stood for Husserl, he embodied phenomenology.

Preti completed his degree with a thesis – now lost, as he intentionally destroyed it – on the historical significance of Husserl. Portions of the thesis were published in *Archivio di filosofia* (1934) and in *Sophia* (1935) (Minazzi 1984, pp. 113–114), but the full manuscript was never preserved. In fact, Preti went so far as to retrieve the original copy from the University of Pavia archive and made sure it disappeared. Although his entire personal file is still held in the archive, the thesis itself is conspicuously missing. He graduated in Pavia under the supervision of Guido Villa (1867–1949), and later – thanks to Enzo Paci – he met Banfi. Immediately after graduation, Preti moved with his family to Milan, settling on the now-famous *Via dei Cinquecento* (Pozzi 2015b, p. 439), where Dino Formaggio (1914–2008) also lived.

From the archival documents, one striking fact emerges: the lasting intellectual and personal harmony between Formaggio and Preti. What explains this strong bond? I believe it lies in their shared social background. Banfi's circle included many who came from modest origins: Formaggio came from a working-class family and had himself been a laborer, while Preti, born into extreme poverty, was raised by a mother who at one point even worked as a laundress in Pavia. I remember his wife, Daria Menicanti, telling me that, at his mother's insistence, Preti completed a four-year teaching diploma at the *Istituto Magistrale* and then enrolled at the university. However, what he truly wanted was to study philosophy. The extra year he had "gained" from his earlier diploma, he spent studying on his own, as an autodidact, in preparation for the university entrance exam, which he took as a private candidate. He was the same age as the two sons of Pietro Vaccari (1880–1976), who served as *podestà* of Pavia and was a distinguished professor at the University of Pavia; Preti was a friend of one of the two, who was his contemporary and attended the Liceo Classico "Foscolo" in Pavia. To prepare for the exam, Preti would visit the Vaccari household, where the brothers would share their class notes with him, especially from their Greek lessons. There is an amusing anecdote in this regard: eager to follow the Greek classes more closely, Preti arranged with the Vaccari brothers that, whenever a Greek lesson was being held, they would open the window of their first-floor classroom. He would then position himself underneath the window and try to eavesdrop on the teacher's explanations.

This anecdote sheds light on a revealing episode recorded in Zambelloni's 1971 testimony (Minazzi 2015a, vol. II, pp. 1105–1106), during a speech delivered by Preti here at the Università Statale. It was 1971, and we were discussing university education amid the intensity of the student movement and broader cultural upheavals. At a certain point in the debate, Mario Dal Pra, who was seated at the presidential table, noticed Preti among the audience and invited him to speak. After touching on the larger picture, Preti offered a simple yet striking remark: "I believe that education should form people with butts as tough as leather." Behind these words lies the lived experience of someone who valued culture so deeply that, as a young man, he would hide beneath a window to listen in on Greek lessons. One should also add that, due to his poverty, Preti would go to the Vaccari household not only to study, but also because, as the sons of the *podestà* and of the Dean of the Faculty of Law at the University of Pavia, the brothers always had some tea and food to offer – something that was often missing from Preti's own home.

3 Banfi's Great Pupils of the 1930s

The problem of *formal reason* lies in its inability to grasp the richness of lived experience. This failure produces a dichotomy: on one side, an irrationalism that fully reflects the depth of experience, but remains ungoverned by reason; on the other, a form of reason that is *abstract, formal, empty* and ultimately devoid of content.

Faced with this tension, what was Preti's theoretical response? He sought to bridge this divide by turning to epistemology – particularly to the developments within neo-positivism in its various phases – as a space for constructing the *intellectual structures* (*Verstand*, in Kantian terms) capable of mediating knowledge. In this framework, reason operates not as an empty formality, but as a tool that *constructs objectivity through a theoretical framework grounded in experimental control.* But I believe this same effort can be found in the other key figures of the Milan School: Paci turned to existentialism, then to relationalism and phenomenology; Cantoni developed an anthropological approach; Formaggio pursued it through artistic practice. Each, in their own way, responded to the same critical limitation inherent in Banfi's original project. From a broader cultural perspective, Antonia Pozzi (1912–1938) also confronted this challenge with great sensitivity, though in a more difficult and personal way. Her relationship with Banfi, as reflected in her *Diary* (Pozzi 1988), was intellectually intense but marked by conflict. And yet, what survives in that published version is only a fragment of the original diary, most of which was destroyed. Still, from what remains, it is evident that for Pozzi, poetry became a space in which to work through her realization that something within Banfi's theoretical system was no longer convincing (Minazzi 2007).

I should also note – and here the human dimension naturally comes to the forefront – that in Banfi's case, these tensions never escalated, perhaps because he had the kind of personality that could hold everything together, much like Goethe. His optimistic disposition, and even his smile, contributed to this balance. Banfi embodied a kind of inner harmony that was notably absent among his students (Minazzi 2013b). That harmony was lacking, I believe, because his students had lived through the darkest years of modern Italian history: the 1930s. They were acutely aware of what Formaggio would later describe so vividly – the *leaden weight* pressing down on their future. This, of course, had a political dimension as well: many of Banfi's students would later become active participants in the Resistance. But this rupture became most evident in the case of Giulio Preti, who, as early as the spring of 1946, chose not to renew his membership in the Italian Communist Party. His decision reflected a divergence from the Party line, but also, significantly, a growing divergence from Banfi himself (Minazzi 2011b).

But if you look closely, you will see that this principled stance came at a cost for Preti. He made this decision before securing a university chair and therefore paid a real price for it – unlike other Banfian students, who made similar choices only after obtaining academic positions and thus incurred far fewer consequences. In this respect, I find Preti especially admirable: if he had something to say, he said it. His wife, who was more pragmatic, once told me that she warned him: "Don't say these things to Banfi in public, in front of everyone! You'll ruin your academic career!" But Preti wouldn't listen. He had an independent spirit, and he was unwilling to compromise his convictions.

Now, in the volume I edited on the Milan School (Minazzi 2019), I noted that reinterpreting Banfi's legacy has posed challenges even for Preti. But here I would like to emphasize two specific points. The first is this: we often speak of the "School of Milan," but I believe we should do so in broad and inclusive terms. Looking back from the today's perspective, it is essential to take into account all the impulses and contributions at work within that tradition. In my view, the person who most effectively reconstructed the unity underlying the School's orientation toward critical rationalism was Mario Dal Pra. In his excellent essay on critical rationalism in Milan, published in the 1985 volume *La filosofia italiana dal dopoguerra a oggi* (Laterza), he succeeded in showing how all these thinkers ultimately responded to a shared philosophical horizon. And what was the novelty of that horizon? I believe it lay precisely in its transcendental dimension (Dal Pra 1985; Coniglione 2015; 2018). That, in my view, is the distinctive contribution that came from Banfi.

In this context, the connection with another voice becomes clear – a voice that, although following a different trajectory, must nonetheless be considered part of the broader landscape of the Milan School: the voice of the Geymonat School. If I ask myself: what kind of image of rationality do we find in Geymonat's work? The answer is that Geymonat was a neo-rationalist, a neo-Enlightenment thinker. All of the authors we are discussing broadly identify with this neo-Enlightenment orientation. However, the conception of rationality we find in Preti – who also belongs to this tradition – is markedly different from that of Geymonat. The distinction is profound: Geymonat's notion of rationality *lacks the transcendental dimension*. And this absence is not incidental; it is a consistent and defining feature of his thought.

I must say that on this point, a kind of cultural deafness has persisted. At stake here is not merely a divergence of opinion, but a deeper difference in philosophical lineage: Geymonat remained firmly rooted in the great tradition of positivism, while Preti's critique to him emerged from a different conceptual horizon, one rooted in the necessity of a transcendental dimension within the structure of reason. His central objection was precisely that Geymonat failed to acknowledge the role of the transcendental dimension in the functioning of reason itself. There is, however, an interesting development in Geymonat's intellectual trajectory. When he wrote the small volume on scientific thought for the Garzanti "Saper tutto" series (Geymonat 1954), he offered insights that would later prove foundational – not only because this book marked the beginning of his scholastic *Manuale di filosofia*, but also because that *Manuale* would eventually become the basis for his much larger *History of Philosophy*, the seven-volume work published between 1970 and 1976 (Geymonat 1970–1976). In that early volume, when Geymonat reflects on the concept of reason – on critical rationality – he includes a particularly striking passage. Writing about Galileo, he develops a remarkably insightful discourse on rationality, in which he asserts: "We must never confuse the structures of rationality with those of reality. We cannot short-circuit them, or else we risk falling into a metaphysical position grounded in a philosophy of identity." And yet, when Geymonat later wrote what is arguably his most metaphysical work, *Scienza e realismo* (1977), he returned to the very position he had criticized in the 1950s: the idea that there is no identity between the structures of rationality and the structures of reality was now reinterpreted in a dialectical key. This reversal is particularly telling. For reasons that are perhaps still not entirely clear – possibly in defense of a renewed form of realism – Geymonat ultimately returned to a stance that, paradoxically, reintroduced a fundamentally metaphysical position.

This brings us to a fundamental question: does the tradition of critical rationalism within the Milan School have its own intrinsic significance, its broader value? To answer this, we must expand our inquiry. A deeper reflection reveals a foundational – yet often overlooked – source within the School's genealogy: Piero Martinetti (1872–1943). Recognizing this lineage allows us to better understand the relationship between Martinetti and Banfi – and, to be fully accurate, we should also include another important figure: Giovanni Emanuele Barié (1894–1956). In Banfi's essay on his three masters – *Husserl, Simmel, and Martinetti* (Banfi 1946) – what does he say about Martinetti? He uses a phrase that is particularly relevant for anyone reflecting on the Milan School's conception of reason: "Martinetti represented the commitment of reason in his life." This legacy of Martinetti's rationalism, I believe, is what Cantoni later referred to in a fine essay published in *Studi filosofici* (first series), where he spoke of the "mythology of reason" – a concept he identified also in Banfi (Cantoni 1943). When Banfi's great students of the 1930s criticized the lack of critical mediation between the formal structures of rationality and the richness of experience, what they were confronting was precisely this mythologized aspect of reason – an inheritance transmitted through Banfi from Martinetti.

This leads us to a key observation: the broader Milan School – comprising Banfi's legacy, the remnants of the neo-Enlightenment debates of the 1950s (including Geymonat), and the Martinettian tradition – constructed a philosophical current that was both intellectually fertile and culturally significant in the history of Italian thought. Yet this tradition has been defeated. It has been systematically marginalized, erased, ignored, and in some cases even scorned. This also explains why so many of the archives ended up in Varese, at our *Centro Internazionale Insubrico*. If you consult the standard reference works – for instance, the twelve-volume *Enciclopedia filosofica* (Centro di Studi Filosofici di Gallarate, Bompiani 1968–69), or even a widely used dictionary like Abbagnano's (1971) – you will find under the entry "Critical Rationalism" only references to Popper. The Milan School's conception of critical rationalism is entirely absent. And this is not just a historiographical omission – it is a theoretical problem. Why is the Milan tradition not mentioned? First, because it was defeated. And secondly, what link exists between the critical rationalism of the Milan School and that of Popper? There is none. In fact, anyone who reads Popper will immediately perceive a radically different world. Where does this major difference lie? It lies in the image of reason itself. And here Preti was particularly insightful: in his analyses of the various stages of the Vienna Circle and logical empiricism (Preti [1954] 1976), he clearly identified the limitations of the neopositivist conception of rationality, without considering Popper's falsifiability criterion a viable remedy for the difficulties encountered with the verification principle.

So, speaking more broadly – and now I'm referring to the history of European critical rationalism within the Western tradition (so forgive the "barbarism" of the comparison I'm about to make) – I would say this: there exists a line of rationalism that begins with Socrates and runs, with many variations, through the Western tradition in a kind of subterranean, karstic flow. This rationalism is rich, complex, and enduring. Its major turning point – its deepest theoretical complication – comes with the Kantian Copernican revolution. And the rationalism of the Milan School, with its Husserlian and phenomenological inflections, belongs to this second trajectory: the post-Kantian line. What distinguishes it? Once again, the concept of the transcendental. If I were to offer a synthesis of this point, I would turn to a passage in Paci's *Diario fenomenologico*, a truly extraordinary book (Paci [1961] 1973, pp. 84–88). There is one passage that has always stayed with me. Paci recounts: "Today I went to see Banfi and asked him, 'What is phenomenology?' Instead of answering, he said: 'Describe the flowers on the table.'" Antonia Pozzi, too, devotes a page of her *Diary (1935–1938)* to that same moment, referring to what she calls that "horrid" vase of artificial flowers (Pozzi 2018, p. 173).

Nevertheless, at a certain juncture, Paci makes a statement that resonates deeply. He writes: "The transcendental is man." I believe that in this formulation, we find the core of what constitutes the problem, the value, and the limitation of Banfi's philosophical legacy. To say that "the problem is man" means that the central question has been recognized: *How do I grasp the object of my comprehension of the world?* I grasp it through the operation that intelligence performs upon the world – an operation that inevitably impoverishes the world by constructing an *objectivity* that I know to be *relative*. Preti then poses the key question: "Relative to what?" His answer: relative to the *regional ontology* within which the object has been *constructed*, and to the *practices* through which this ontology is tested and exercised. This issue is central because it touches on the *status of objectivity* – a theme that became increasingly prominent in the 1950s. What makes Preti particularly interesting is that, in the 1960s, when he once again took up this question from a theoretical perspective, he returned to Banfi. The curvature of that return is clearly visible in *Retorica e logica* (Preti 1968; 2018), a book that is unmistakably "Banfian" in both its theoretical chapters and its historiographical reflections.

But this also opens up a more interesting and complex question. If we interpret the Milan School through this lens (Minazzi 2019; Coniglione 1993; 2011), we see that what fundamentally characterizes it is its sustained engagement with the Kantian concept of the *transcendental*. And to invoke the Kantian transcendental means, at the very least, recognizing the importance of the *structures of intentionality* and their relation to *historical temporality*. Preti followed this trajectory to its most mature expression. At one point, through his studies in medieval logic (Preti 1976, vol. II, pp. 3–194), he returned to the debates of late Scholasticism, where he identified the early emergence of what he called "logical neo-realism" or "phenomenological realism." This was the foundation upon which he sought to construct a *critical ontology* – a project that remains in the unpublished part of his work. Preti died in July 1972, just as he was preparing a collection of essays on precisely this topic (Preti 1969–1970; Minazzi 2011a). The terminology may have changed – "logical neo-realism," "phenomenological realism," "critical ontology" – but the core issue remained the same: the attempt to think *intentionality* through the lens of the *transcendental*. I will stop here. It seems to me that one of the crucial tasks in revisiting the Milan School today is not only to understand its theoretical contributions, but also to grasp the reasons for its removal from the dominant narratives of twentieth-century philosophy.

I will close with a brief, but telling, personal anecdote. I completed my degree here in Milan. My thesis advisor was Ludovico Geymonat, and Mario Dal Pra was to serve as my co-examiner. This was in 1978. Shortly thereafter, Geymonat retired and was replaced by one of his former students, who then became my new advisor. My thesis was on Preti's historical epistemology, and of course it began with Banfi's critical rationalism and the broader framework of *problematicism*. I share this episode because it stayed with me – and because it reveals something essential about the cultural silence that surrounds this tradition. After reading just one chapter of my thesis, the new advisor looked at me and said: "*Problematicism*? I don't know *problematicism*. I only know problems." That remark, I believe, encapsulates the removal of the Milan School from contemporary philosophical discourse.

4 Conclusion: the Decline of the School and the Enduring Relevance of Its Enlightenment Legacy

I fully agree that, through their teaching, Paci and the other philosophers of the Milan School educated their students to think critically. This was, in fact, the core contribution of its *problematicism* – namely, the adoption of an epistemological and methodological orientation that rejects any closed or dogmatic system in favor of an open, dynamic, and questioning conception of thought, standing in opposition to both positivism and systematic idealism. That said, there was more at play than just individual trajectories – there was something deeper. Today, the Archives are beginning to reveal this, for example in the correspondence between Geymonat and Dal Pra (Minazzi 2010a, pp. 263–292), in which they discussed philosophical questions in depth. Among other things, there is a letter from Geymonat, written in 1978, in which he reflects on the history of the Milan School: "I am retiring; I can finally devote myself full-time to philosophy," he wrote. Dal Pra, in response, expressed his dismay at the changes taking place in the philosophy department at the Statale. Geymonat reassured him, citing Vico: "Don't worry – it's a matter of recurring cycles, of historical recurrences." He then invoked the figure he considered his ideal teacher, Piero Martinetti, adding: "Of course, we are no longer in the glorious age of Martinetti," suggesting that even the era of the Milan School to which he had belonged already felt like a phase of decline: "Now we are entering a period of deep decadence – but we must remain optimistic about the future." This exchange is telling: they were reasoning in terms of *schools* – not in the institutional sense of the Frankfurt School, but as a *tradition of thought* in a broader sense, where each thinker followed his own *demon*, with different methods, questions, and sensibilities. Yet they still recognized themselves as the heirs of a common tradition: that of *critical rationalism*.

I would add that it was not only an *education in thinking*. In 1975, when I attended the lectures of these philosophers, I could feel – behind certain words – the weight and force of an *entire life*'s commitment. Let me try to explain this. The operation Banfi had undertaken in the 1930s was a significant one. It helps us understand the repression that followed, of which the problematicism I mentioned earlier is an unwitting epigone. Banfi's position was this: "We are living in a time of profound crisis, even though fascism appears triumphant. But it is not enough to simply oppose fascism or declare oneself against it; we must fight it by *building a new culture*." This, I believe, was the ethical and intellectual commitment of that generation of philosophers, whose participation in the Resistance was of the highest caliber. They fought fascism, as Preti put it, because it represented a tradition that was regressive and out of step with the cultural debates taking place in Europe and America. Today, philosophy may no longer aspire to be a "grand narrative." But if it confines itself to the narrowly defined problems of analytic philosophy, it risks becoming, as Evandro Agazzi once put it (metaphorically), *like someone taking up a shotgun to go butterfly hunting* – when what is really needed is a butterfly net.

Philosophy lives insofar as it has a social presence. This is the lesson of Socrates – he was a philosopher because he engaged with and influenced the society of his time. In the case of the Milan School, its thinkers did not occupy a marginal place within the intellectual tradition of critical rationalism. On the contrary, they all identified with it – albeit through distinct and sometimes even conflicting positions. Dal Pra, in his 1985 essay, highlighted precisely this: the capacity to bring together diverse thinkers, each with their own autonomous theoretical concerns, within a shared tradition, enabling them to participate in a *common dialogue*, even if their voices at times stood in tension with one another (Dal Pra 1985). I would also like to challenge a widely repeated claim: that these thinkers are no longer read. That is simply not true. They are, in fact, read abroad. Take for example *Per un nuovo illuminismo* by Jean Petitot (Petitot 2009), a philosopher of mathematics and epistemologist steeped in French intellectual culture. His testimony is emblematic: he states that the only body of thought that truly supports his epistemological concerns is that of the Milan School. He is astonished that it is not more widely acknowledged in Italy, calling it "a jewel." He reads Banfi and Preti with deep appreciation, just as Gereon Wolters (2014, p. 284) has also done in relation to Preti, "to whom we owe among other things a fascinating pragmatist embedding of philosophy of science". For Petitot, the philosophical dialogue continues only with the Milan School, because the central epistemological question they raised in the 1960s is the same one that international philosophical culture only later came to recognize: How can we relate the *formal structure* of the transcendental – the a priori – with the *historical dimension of conceptual development within scientific thought*? This, indeed, is the core of Preti's project.

A useful counterproof can be found in the case of Paolo Rossi Monti (1923–2012), a thinker who passed through Milan but later developed a conflicted relationship with the Milan School. If one examines his autobiography (Rossi 2012) or earlier essays dedicated to members of the School – written in a polemical tone that often goes too far – it becomes clear that many of his criticisms are not well-founded. For example, in his 1986 work (pp. 244–247), Rossi reconstructs Geymonat's *Filosofia e filosofia della scienza* (1960), yet entirely overlooks the fact that the Turin epistemologist explicitly addresses the issue of conceptual transformations within scientific theory. In this regard, there is a well-known page in Geymonat's text that directly echoes reflections already developed more rigorously by Giulio Preti in his *Storia del pensiero scientifico* (Preti 1957). Preti, not by chance, drew upon a position rooted in Gentile's idealism, while Rossi instead links the problem to Kuhn – an author who is in fact completely irrelevant in this context. Kuhn's theory centers on the incommensurability of scientific paradigms, whereas Geymonat's concern was quite the opposite: "How can we construct *continuity* through *discontinuity*?

In effect, Paolo Rossi Monti's work does not help us to understand the Milan School; rather, it contributes to its erasure. What emerges here is a crucial feature of the School's project: its intention to construct a *positive philosophy* in opposition to a centuries-old Italian tradition – one that dates back to the seventeenth century and is deeply marked by the legacy of the *Counter-Reformation*. Moreover, their work carried within it a call to think on behalf of those who could not think – a commitment rooted in the partisan experience of the Resistance. It was a philosophy meant to serve society, and in this sense, it aligned profoundly with the Enlightenment spirit (Luzzini 2018). And indeed, much like the Enlightenment thinkers of the eighteenth century, the philosophers of the Milan School conceived of reason as a critical instrument of emancipation, not as a pure or neutral faculty. Yet unlike the classical Enlightenment, they also recognized its historical, situated, and problematic limits. This is what makes theirs a *neo-Enlightenment* project. As a consequence, one of the implicit (and at times explicit) targets of their philosophical endeavor was the Counter-Reformation and idealist tradition that had dominated Italian philosophy for centuries. Just as the Enlightenment opposed the absolute authority of religion and monarchy, these thinkers resisted authoritarian uses of culture, advocating instead for a conception of knowledge as critical, dialogical, and open-ended construction.

If, by contrast, we reduce philosophy to a set of purely technical problems – as Diego Marconi suggests (Marconi 2014) – we risk extinguishing philosophy itself. For philosophy has value precisely to the extent that it unsettles us, just as Socrates did with his questions. Consider Banfi's *Socrate* (1943) – an extraordinary book, later reviewed by Preti in 1946 in *Studi Filosofici*. Banfi argued that Socratic philosophy was inseparable from the conduct of life, relying on the testimony of Plato in the *Theaetetus* and Aristotle in the first book of *Metaphysics*. Preti, however, offered a different reading: in Socrates, he saw not only an ethical figure, but also a *theoretical mind*, one who raised the problem of the concept itself.

Beyond this, ours was a School that did not define itself through a manifesto. It was not the Frankfurt School. Rather, it was a school in the sense of a shared tradition – a tradition of critical rationalism that stood in open conflict with what Preti once called "the dead weight of the Italian tradition." As he wrote: "There has always been a minority in Italy that looked toward culture wherever it was most alive, internationally; a minority that did not identify with the dominant tradition and sought instead to construct a European Italy" (Preti 1994 pp. 348–349). This, too, remains our problem today. The fact that, from a historiographical standpoint, little space has been devoted to the rationalism of the Milan School does not stem solely from any internal weaknesses it may have had, but above all from the fact that the orientation emerging from the Kantian Copernican revolution never took shape as a movement with a manifesto, a journal, a formal school, or an organized structure for dissemination. By its very nature, it consists of individuals who, in principle, think for themselves. That is why we find, scattered across Europe – in Poland, France, Switzerland, Italy – epistemologists who followed their own paths. And yet, if we take a broader view of these voices – Gonseth, Piaget, Bachelard, the Lvov-Warsaw School, the Milan group, Enriques – we discover a heterogeneous but intellectually rich tradition. And in a sense, this was also its weakness: it never became a philosophical party (Papi and Minazzi 1995; Coniglione 1993; 2014).

That said, I also see this as the sign of a broader cultural defeat – one that, from this perspective, bears a name: Federico Enriques (1871–1946). Enriques, for example, argued that philosophy should have a transversal role across university curricula. Instead, Benedetto Croce (1866–1952) prevailed, and to this day, philosophy remains bound to the domain of literature and the humanities. As for art, beyond the group around the magazine *Corrente di vita giovanile* (1938–1940) (Fagone 1978), there was constant osmosis between philosophy and artistic production. Preti, for instance, maintained a close relationship with a younger artist like Piero Manzoni (1933–1963); they wrote about the *Galleria Il Milione*, attended exhibitions, and engaged in debates about contemporary art. This, too, was part of Banfi's legacy – his *Lectures on Aesthetics* emphasized the importance of not judging life a priori, but instead understanding the inner logic of what was emerging in each field. It is no coincidence that Preti published more essays on art than on philosophy in *La Fiera Letteraria* (Preti 1970), or that the *Minuziano* series edited by Rognoni explored aesthetic and literary themes. Banfi's students engaged deeply with aesthetics, writing on Hume's *Essay on Taste*, Fracastoro, Spalletti, Schelling, and many others. It was a philosophy that did not close itself off, but listened and responded to the world in motion.

In the end, we arrive at the question of *schematism*. I believe this is a misleading focus; rather than clarifying the issue, it confirms the deeper problem within the Milan School. This is especially apparent when Preti equates Gentile and Banfi on a crucial point: both, he claims, arrive at the "nullification of the intellect" – that is, the dissolution of the intermediate level between *empty formal structures* and the *richness of reality*, which must be accounted for. Let me cite Preti directly: "What is epistemology? Epistemology is the study of the *functional structures* of the intellect, analyzing the *construction of objects* in light of those very structures" (*In principio era la carne*, 1948, in Preti 1983). If I were to translate this indication, I would say: philosophy must locate itself within a meta-reflection. It no longer possesses its own predefined purpose or content, but must instead go out and learn from other disciplines. This is also what Geymonat, in very different terms, meant when he said: "I go to learn, not to teach." There is no longer a "grand narrative"; rather, I place myself within the problem – not as a technician of that specific problem, but as a philosopher in the Socratic sense, guided by the key of *critical rationalism*.

From this perspective, there is a document we intend to publish – a manuscript dating back to 1945–46 – concerning a series of seminars held at Banfi's home. Participants included Enzo Paci, Remo Cantoni, Giulio Preti, Giovanni Maria Bertin, and Dino Formaggio. Each session featured a designated speaker and a note-taker (Angela Colantoni Stevani), with minutes later reviewed and corrected by Cantoni. It is in this form that the records have come down to us. The discussions, which continued for two years, focused on the problem of reason and the concept of the a priori. The contributions are remarkable and reflect many of the theoretical movements addressed in this essay. The participants rarely agreed. Banfi intervened infrequently, and only to stimulate further discussion. What emerges clearly from this is that the central question was precisely this: How can we relate the structures of rationality to experience? Banfi emphasized that the system inherited from phenomenology is grounded in the idea of correlation – between subjectivity and objectivity – a correlation that, when translated into the *Leben* of the Milan School, meant the living relationship between *Geist* and *Leben*.

This relationship – between life and thought – originates with Simmel: *life – more life – more than life*. And this, in a sense, was the underlying problem. Paci, Preti, and Formaggio were acutely aware that thought arises from life, as an emergence rooted in vitality, in embodiment, in the materiality of the body itself. There is no concept of death without life – it is life that grounds even its own limit. This sensibility is reflected not only in their philosophical writings, but in their biographies and their deep engagement with the artistic world. The artist is seen through their *praxis*, which opens up a discourse on practice, on objects, on techniques. It is a deeply relevant orientation today, as it anticipates the relational dimension of practice that Carlo Sini has more recently explored – though the terms and concerns were already present in their work. As for Hegel, this aspect does not trouble me. They did not inherit the burden of Hegelian metaphysics. Rather, theirs is a Hegel analytically disassembled – one who brings forward the richness of *Phenomenology*, not the system of the *Encyclopaedia*. In this sense, dialectics is not a dogma, but a *heuristic instrument*, a way to rationally trace the polarity between *subject* and *object*. This, for example, is the very lesson that Mario Dal Pra drew from his study of Marx's thought (Coniglione 2018, p. 311).

•••

If we were to summarize the trajectory traced thus far, we could say that the Milan School's conception of critical rationalism is a *progetto sui generis* – a distinctive intellectual undertaking that cannot be assimilated to other forms of critical rationalism, such as Popper's. It does not rest on the falsifiability of hypotheses or the external critique of systems; rather, its concept of rationality acknowledges its own transcendental conditions while remaining open to the lived complexity of experience. Thought, for these philosophers, is not an abstract operation, but an emergence from life (*Leben*), always shaped by the dialectic between subjectivity and objectivity. This critical rationalism is not a method imposed from outside, but a form of meta-reflection – Socratic in spirit – that disturbs, interrogates, and transforms. Its strength lies in its refusal to become a philosophical party or system. Its weakness, perhaps, is that this very refusal led to its erasure. And yet, precisely for this reason, it remains one of the most intellectually fertile and ethically urgent traditions in twentieth-century European philosophy.

Moreover, the critical rationalism of the Milan School differs fundamentally from Popper's version due to the *central role it assigns to the transcendental dimension*, as has been repeatedly emphasized in the preceding pages. For Banfi, Preti, Cantoni, Formaggio, Dal Pra and Paci – each in his own way – reason is not an external apparatus that judges from above, but a historically situated activity that constructs objects of knowledge from within functional structures (as in Preti's idea that each science builds its own regional ontology) (Preti [1954] 1975[3], p. 163), or intentional horizons (as in Paci), always in connection with life (*Leben*) and with praxis. This understanding of the transcendental is clearly rooted in a post-Kantian tradition, not dogmatic or systematic, but mediated through Husserlian phenomenology and an open, non-metaphysical reading of Hegel. It is not the transcendental conceived as an a priori foundation of validity (in the neo-Kantian sense), but rather as an operative structure of knowing, always intertwined with historicity, embodiment, language, and the social dimension. From this point of view, the contrast with Popper and the Vienna Circle becomes stark. In Popper's critical-falsificationist method, the transcendental is either removed or reduced. Rationality is framed as the external logical control of propositions, and no reflection is offered on the conditions of possibility of knowledge, the intentionality of consciousness, or the internal relationship between subject and object. Everything the Milan School explores under the names of "form," "regional a priori," or "construction" is considered by Popper either irrelevant, metaphysical, or psychological in a pejorative sense. Also, classical neo-positivism (Carnap, Schlick, Neurath), even when acknowledging the linguistic structure of knowledge, never addresses the historicity of rational forms nor their connection to concrete praxis, something that Preti in particular foregrounds, especially in his notion of "critical ontology."

In this light, Milanese transcendentalism emerges as a form of rationalism that is internal to life, not external; constructive rather than reductionist; reflective rather than merely methodological. It conceives reason as a self-problematic device, developing within the tension between form and experience, between theoretical construction and socio-historical embeddedness. In this sense, it not only resists assimilation into Popper's model, but constitutes an alternative tradition, one that aligns more closely with continental philosophy (Kant, Husserl, Hegel), yet remains firmly anchored in the clarity and accountability of scientific discourse.

To put it briefly: the Milan School's critical rationalism is not a method for judging knowledge from the outside, but a living reflection on the conditions through which reason, historically and practically situated, constructs meaning within experience.

References

AA. VV. (1968–1969). *Enciclopedia Filosofica*, IV Vols. Florence: Sansoni (new edition Rome: Lucarini, 8 Vols., last and new revised and expanded edition Milan: Fondazione Centro Studi Filosofici di Gallarate-Bompiani, 12 Vols.).

Abbagnano, N. (1971). *Dizionario di filosofia*. Turin: Utet.

Agazzi, E. (1961). *Introduzione ai problemi dell'assiomatica*. Milan: Vita e Pensiero.

Banfi, A. (1943). *Socrate*. Milan: Garzanti.

Banfi, A. (1946), Tre Maestri (Simmel, Husserl, Martinetti). *L'Illustrazione Italiana*, LXXIII, 44, 284–285 (reissued in Banfi 1970, pp. 245–251).

Banfi, A. (1970). *Scritti letterari*, edited by Carlo Cordiè. Rome: Editori Riuniti.

Cambi, F. (1983). *Razionalità e prassi a Milano (1945–1954)*. Milan: Cisalpino-Goliardica.

Cantoni, R. (1943). L'illuminismo religioso di Piero Martinetti. *Studi Filosofici*, IV, 2–4, 216–233.

Coniglione, F. (1993). Scetticismo e scienza in Giulio Preti in G. Preti *Scetticismo e conoscenza*, a cura di F. Coniglione. Catania: CUECM, pp. VII–XXXVII.

Coniglione, F. (1999). *L'idea di filosofia scientifica. Per una storia dell'epistemologia del Novecento*. Catania: CUECM.

Coniglione, F. (2002). Russell e la nascita dell'idea di filosofia scientifica. In: G. Bentivegna, S. Burgio, G. Magnano San Lio (eds.), *Filosofia, scienze, cultura*. Soveria Mannelli (CZ): Rubbettino, pp. 181–218.

Coniglione, F. (2005). Per la storia della filosofia scientifica. Il Circolo di Vienna e la Scuola di Leopoli-Varsavia. In: G. Gembillo (ed.), *Filosofia e scienze. Studi in onore di Girolamo Cotroneo*. Soveria Mannelli: Rubbettino, pp. 109–141.

Coniglione (2007). The Place of Polish Scientific Philosophy in the European Context. *Polish Journal of Philosophy*, 1, pp. 7–27.

Coniglione (2008a). Il pensiero infermo. Origine e destino della filosofia scientifica. In: B. Bonghi, F. Minazzi (eds.) *Sulla filosofia italiana del Novecento*. Milan: Franco Angeli, pp. 151–174.

Coniglione (2008b). Filosofia scientifica europea e positivismo italiano, In: F. Coniglione, G. Bentivegna e G. Magnano San Lio (eds.), *Il positivismo italiano: una questione chiusa?*, Atti del Congresso tenutosi a Catania, 11–14 settembre 2007. Bonanno Editore: Acireale-Rome, pp. 39–67.

Coniglione F. (2011). Giulio Preti, Ludovico Geymonat e la filosofia scientifica in F. Cambi and G. Mari (a cura di), *Giulio Preti intellettuale e critico attuale*. Florence FUP, pp. 51–83.

Coniglione F. (2014a). Positivismo, filosofia scientifica e scienze umane nell'Italia d'inizio secolo. In: G. Cimino and G.P. Lombardo (a cura di), *La nascita delle "scienze umane" nell'Italia post-unitaria*. Milan: Franco Angeli, pp. 29–45.

Coniglione F. (2014b). La fondazione metodologica della psicologia in Roberto Ardigò. In: R.L. Cardullo, D. Iozzia, ΚΑΛΛΟΣ ΚΑΙ ΑΡΕΤΗ. BELLEZZA E VIRTU'. Studi in onore di Maria Barbanti. Acireale-Rome: Bonanno, pp. 759–774.

Coniglione F. (2015). La funzione dello scetticismo nell'epistemologia di Giulio Preti. In: F. Minazzi (a cura di), *Sul bios theorètikos di G. Preti*, vol. I. Milan-Udine: Mimesis, pp. 259–275.

Coniglione F. (2018). Mario Dal Pra e la filosofia scientifica del Novecento. In: F. Minazzi (a cura di), *Mario Dal Pra nella "scuola di Milano"*. Milan-Udine: Mimesis, pp. 307–323.

Dal Pra, M. ([1984] 1990^2). Kantismo ed hegelismo in Banfi. In: M. Dal Pra 1990^2, pp. 65–76.

Dal Pra, M. (1985). *Il razionalismo critico*, in AA. VV. *La filosofia italiana dal dopoguerra ad oggi*, Rome-Bari: Laterza, 31–92.

Dal Pra, M. (1990^2). *Filosofi del Novecento*. Milan: Franco Angeli.

Dal Pra, M., Formaggio, D. and Rossi, P. (1984). *Antonio Banfi: le vie della ragione*. Milan: Edizioni Unicopli.

Fagone, V. (a cura di). (1978). *Corrente (reprint gennaio 1938–maggio 1940)*. Pollenza (Macerata): La Nuova Foglio.

Garin, E. (1975). *Cronache della filosofia italiana 1900–1943. In Appendice Quindici anni dopo 1945–1960*, 2 Vols. Rome-Bari: Laterza.

Geymonat, L. (1954). *Il pensiero scientifico*. Milan: Garzanti.

Geymonat, L. (1960). *Filosofia e filosofia della scienza*. Milan: Feltrinelli.

Geymonat, L. (a cura di) (1970–1976). *Storia del pensiero filosofico e scientifico*, 7 Vols. Milan: Garzanti.

Geymonat, L. (1977). *Scienza e realismo*. Milan: Feltrinelli.

Luzzini, F. (2018). Bibliographical Distortions, Distortive Habits: Contextualizing Italian Publications in the History of Science. *Isis*, 109, SI, 3–13 – Italian Version: Luzzini, F. (2019). Distorsioni bibliografiche e storture etiche. Le pubblicazioni di storia della scienza nel contesto italiano. *Il Protagora*, XLVI, 31–32, 327–340.

Marconi, D. (2014). *Il mestiere di pensare*. Turin: Einaudi.

Minazzi, F. (1984). *Giulio Preti. Bibliografia*. Milan: Franco Angeli.

Minazzi, F. (1994), *L'onesto mestiere del filosofare. Studi sul pensiero di Giulio Preti*. Milan: Franco Angeli.

Minazzi, F. (2007). La pura teoreticità banfiana e il pensiero scientifico. In: S. Chiodo and G. Scaramuzza (a cura di), *Ad Antonio Banfi cinquant'anni dopo*. Milan: Edizioni Unicopli, pp. 257–269.

Minazzi, F. (2010). *Geymonat epistemologo. Con documenti inediti e rari (un inedito del 1936, il carteggio con Moritz Schlick, lettere con Antonio Banfi e Madio Dal Pra)*. Milan-Udine: Mimesis.

Minazzi, F. (2011a). *Suppositio pro significato non ultimato. Giulio Preti neorealista logico*. Milan-Udine: Mimesis.

Minazzi, F. (2011b). *Giulio Preti: le opere e i giorni. Una vita più che vita per la filosofia quale onesto mestiere*. Milan-Udine: Mimesis.

Minazzi, F. (2013a) (ed.). *Nel sorriso banfiano. Scritti, cartolettere e foto inedite per Alba Binda*. Milan-Udine: Mimesis.

Minazzi, F. (2013b). Nel sorriso banfiano. In: Minazzi (2013a), pp. 19–147.

Minazzi, F. (2015a) (ed.). *Sul bios theoretikós di Giulio Preti*, 2 Vols., Milan-Udine: Mimesis.

Minazzi, F. (2015b). L. Geymonat. Scegliere la ragione contro la barbarie nazifascista. *Humanitas*, LXX, 1, 83–94.

Minazzi, F. (2018) (ed.). *Mario Dal Pra nella "scuola di Milano"*. Milan-Udine: Mimesis.

Minazzi, F. (2019) (ed.). *Sulla scuola di Milano*, Florence: Giunti.

Paci, E. ([1961] 1973). *Diario fenomenologico*. Milan: Bompiani.

Peruzzi, F. (2011). *L'estetica di Ermanno Migliorini. La militanza e la misura*. Florence: Leo S. Olschki.

Petitot, J. (2009). *Per un nuovo illuminismo*, edited and transl. from French by F. Minazzi. Milan: Bompiani.

Popper, K.R. ([1930–1933] 2009). *The Two Fundamental Problems of the Theory of Knowledge*. London and New York: Routledge.

Pozzi, A. (2015). *Parole. Tutte le poesie*, edited by G. Bernabò and O. Dino. Milan: Centro Nazionale Insubrico.

Pozzi, A. (2018). *Mi sento in un destino. Diari e altri scritti*, ed. by Graziella Bernabò e Onorina Dino. Milan: Ancora (ed. ePub).

Preti, G. ([1954] 1975). Le tre fasi dell'empirismo logico. In: Preti 1976, pp. 295–313.

Preti, G. (1957). *Storia del pensiero scientifico*. Milan: Mondadori.

Preti, G. ([1957] 1975^3). *Praxis ed empirismo*. Turin: Einaudi, 1957.

Preti, G. (1968). *Retorica e logica*. Turin: Einaudi.

Preti, G. (1969–1970). *Il neorealismo logico. Saggi di ontologia filosofica*. Unpublished autograph of 33 pages dating from 1969–1970 kept at "Centro Internazionale Insubrico" of Università degli Studi di Varese (Italy).

Preti, G. (1970). *Que serà, serà*. Florence: Il Fiorino.

Preti, G. (1976). *Saggi filosofici*, 2 Vols. Florence: La Nuova Italia (English selected transl. Preti 2011).

Preti, G. (1983). *In principio era la carne. Saggi filosofici inediti (1948–1970)*, edited by M. Dal Pra. Milan: Franco Angeli.

Preti, G. (1994). *La cultura nella società italiana* (*A proposito di E. Garin, La cultura e la scuola nella società italiana*. Turin: Einaudi 1960) in Minazzi 1994, pp. 345–349.

Preti, G. (2011). *Philosophical Essays. Critic Rationalism Historical-objective Transcendentalism*, edited by F. Minazzi, transl. by R. Sadleir. Bruxelles et al.: Peter Lang.

Preti, G. (2018). *Retorica e logica*, new amended and enriched edition, edited with an introduction and notes by F. Minazzi. Milan: Bompiani.

Rossi Monti, P. (1986). *I ragni e le formiche. Un'apologia della storia della scienza*. Bologna: il Mulino.

Rossi Monti, P. (2012). *Un breve viaggio e altre storie. Le guerre, gli uomini, la memoria*. Milan: Raffaello Cortina Editore.

Sini, C. (2013). *Incontri. Vie dell'errore, vie della verità*. Milan: Jaca Book.

Sini, C. (2015). *Enzo Paci*. Milan: Feltrinelli.

CHAPTER 2

Science and Philosophy between "Open Society" and "Depersonalized Society": from Claude Bernard to Karl Popper

Riccardo Roni

Abstract

In this article I focus on the *ante litteram* popperism of the great French physiologist Claude Bernard, developing an important observation that Francesco Coniglione reports in his 1978 book, *La scienza impossibile. Dal popperismo alla critica del razionalismo*. Based on Bernard's theories – who emphasizes the dialectic between facts and theories according to what I define "weak falsificationism" – the relationship between science and philosophy within the "open society" is addressed, starting from a comparison between Henri Bergson, Karl Popper and Walter Lippmann. The article closes with a theoretical assessment of the function of dialogic rationality within the open society, understood as an antidote to what Popper calls an "abstract and depersonalised society".

Keywords

Claude Bernard – Karl Popper – Falsificationism – Science – Open Society.

1 About Claude Bernard's "*Ante Litteram* Popperism"

In a footnote to his 1978 book, *La scienza impossibile. Dal popperismo alla critica del razionalismo*, Francesco Coniglione quotes a passage by Hans Reichenbach from *The Rise of Scientific Philosophy* (1951) on the "creative function of genius", together with another observation by Ernest Nagel reported in *The Structure of Science* (1961), according to which "there are no rules of discovery and invention in science" (Coniglione 1978, p. 13, footnote 5). Reichenbach's observation allows us to recall the epistemological perspective which – precisely in the name of the creative function of genius – redefines a certain conception of induction understood as a simple passage from facts to theories. I'm referring

 | HTTPS://DOI.ORG/10.1163/9789004756496-004

to the case of Claude Bernard (1813–1878), the father of experimental medicine, quoted by Coniglione in the same footnote to suggest the thesis of his "popperism *ante litteram*", namely, that "there are no rules for giving birth in the brain, as a result of observation, to good and fruitful ideas that pave the way for successful research"; because it is "only after having had an idea" that one "can know how to submit it to certain rules and precise logical orders from which no researcher can deviate". But we must not forget, Bernard goes on, that the appearance of such an idea is "completely spontaneous and its nature depends on the individual who formulated it. It is a special feeling, a *quid specialis* that characterizes the originality, the imagination, the genius of the researcher" (Bernard 1865, p. 59; quoted in Coniglione 1978, p. 13, footnote 5).

Based on what Coniglione suggests, I want to dwell for a moment on the theses of Bernard, famous physiologist and professor of medicine at the Collège de France, as explained in his *Introduction à l'étude de la médicine expérimentale* (1865), in order to connect them to those of Bergson and Popper regarding the "open society". With the aim of developing, to conclude, a theoretical proposal on the function – within the open society – of dialogic rationality, interpreted as the right antidote to a depersonalized society.

In developing these issues, I import a different sense of falsification than the usual one applied to Popper regarding the boundary between "science" and "pseudo-science." If falsification has been understood as the idea that legitimate science is a discipline that enquires into problems where one's answers may turn out to be false, according to my approach, falsification becomes the view that false results or facts are as important as those that are established through experimentation and observation. I therefore modify Popper's usual definition of the boundary between science and pseudoscience to highlight the heuristic function of error for science in its historical development, as Bernard clearly shows when he states that true progress always consists in changing an old theory that contains fewer facts with a new one that contains more.

According to Bernard, what distinguishes observation from *expérience* can be identified in the difference between the mere observation of a fact (which can never constitute a science) and the "control of an idea through a fact" (Bernard 1865, p. 29). Every experience presupposes both a reasoning applied to the observational fact, a comparison between facts and a judgment through other facts that control the first experiences. In this respect, an "experimental science" corresponds to a science derived from experiences, from facts obtained under conditions determined by the experimenter. The latter, being able to assert his *puissance d'agir* over phenomena, manifests a pragmatic vocation, which is summed up in his ability to create artificial conditions for the phenomena he is going to study. The simple *observateur*, therefore, becomes *expérimentateur* when he imagines investigative procedures that penetrate bodies by varying the conditions of their phenomena (Bernard 1865, p. 30). Within this epistemological framework, one can speak of "theory" only when facts, understood as necessary materials, are "actualized" by the experimental reasoning that constitutes science (Bernard 1865, p. 47). However, we mustn't overlook an anthropological fact. According to Bernard, man has a natural and sentimental inclination towards metaphysics. We read further in the *Introduction*:

> Man is naturally metaphysical and proud; he could believe that the ideal creations of his mind, which correspond to his feelings, also represented reality. It follows that the experimental method is not primitive and natural for man, and that only after wandering for a long time in theological and scholastic discussions, did he end up recognising the sterility of his efforts in this direction.
>
> BERNARD 1865, p. 48

The point is that we must learn to exploit the potential implicit in a priori ideas (or intuitions) by defining them in terms of hypotheses, whose implications must be subjected to the scrutiny of an experimental criterion (which must in any case include feeling as well as reason and experience): "in fact", writes Bernard, "the greatest truths are, after all, only a feeling in our minds" (Bernard 1865, p. 51).

But this is not enough, a further, decisive step is needed. Bernard clearly emphasizes the need to "overturn" theories, as is the duty of any good scientist who is attentive to the facts, putting forward a proposal that we can define as "weak falsificationism":

> it follows that when we have an idea or theory in science, we should not aim to preserve it by looking for everything that can support it and discarding everything that can invalidate it. On the contrary, we must examine most carefully the facts that seem to overthrow it (*renverser*), because true progress always consists in changing an old theory that contains fewer facts with a new one that contains more.
>
> BERNARD 1865, p. 72

Coniglione valorizes this reading, as can be seen from the passage in his book in which, following Feyerabend and Popper, he asserts that experience refers "to a theory that is the result of a historical-psychological stratification of meanings" (Coniglione 1978, p. 47). This last statement is very relevant to our approach. If we start from this premise in the search for real scientific progress, we are led to value the invention, writes Coniglione, "even gratuitous, of alternatives to existing scientific theories, which are capable on the one hand of producing new 'facts' in their support and at the same time in refutation of existing theories, and on the other hand to decode the experience consolidated over millennia of human practice and make it compatible with a new way of seeing the world" (Coniglione 1978, p. 47). The invention alone, however, is not enough, because a theory, precisely because it gives order to the world, must be put to the test by examining its durability over time. And in this case Coniglione reminds us, through Popper, of the importance of a certain dogmatic attitude *lato sensu*, which consists precisely in being able to adhere to a theory "for as long as possible" (Coniglione 1978, p. 50), in the very terms in which Bernard stated that "one must have firm faith and not believe" (Bernard 1865, p. 295).

As a corollary to the previous remarks, Bernard emphasizes that in the presence of a new fact that contradicts the theory, one should not abandon the fact to preserve the theory at all costs:

> *When the fact one encounters is in opposition to a prevailing theory, one must accept the fact and abandon the theory, even if the latter, supported by big names, is generally adopted.*
>
> BERNARD 1865, pp. 287–288 (Italics by BERNARD)

The most advanced theories, therefore, are those that include more facts. The creative function of genius in this case consists of opening a new path, certainly exploiting the contribution of old theories, but without falling into the error of absolutizing their verification (Bernard 1865, p. 290). We should remember that one of Bernard's most important insights in his research on the formation of sugar in animals had as its starting point an apparently trivial experimental circumstance. He discovered that sugar is formed in animals both after death and during life, through a mechanism quite similar to that which occurs in plants, since it is produced in the liver by means of a diastatic substance that reacts with a starchy substance called "glycogen."

From these experimental results we can see that if we had in mind only the "confirmation" of a theory, "we probably would not have set out on the trail of the new facts we have encountered" (Bernard 1865, p. 290). We therefore talked about a weak falsificationism in Bernard, because it does not exclude verification but emphasizes the dialectic between facts and theories, understood as the two levels of investigation that must be kept very distinct in experimental criticism. Every theory has a provisional value that must be based on the "absolute determinism" of the facts, established with precision in their conditions of existence. So, while on the one hand science requires that we agree on fact as the foundation for reasoning, on the other hand, when it comes to interpretations and ideas, "it is even a good thing for them to be discussed, because these discussions lead to further research and new experiences" (Bernard 1865, p. 332). From this methodological framework, Bernard puts forward some interpretative insights regarding the "vital force" that are very useful for the theoretical purposes of this contribution. I am referring to what he writes in ch. IV (§ 2: *L'ignorance scientifique et certaines illusions de l'esprit médical sont un obstacle au développement de la médecine expérimentale*), when he talks about the properties of living matter, distinguishing between *physical-chemical* properties (known and determined) and *vital* properties (unknown and indeterminate). To avoid the risk of a reductionist physicalist approach – and this is the point – Bernard doesn't exclusively value the point of view of determinism, but also that of indeterminism, precisely in cases where theory is called upon to illuminate practice. In this last point of view lies the intuition of the scientist who, placing himself in the perspective of the conservation of life, cannot limit himself exclusively to acting in accordance with empirically recorded natural tendencies. To guarantee life, Bernard significantly emphasizes, "it is often necessary to act against the tendencies of nature" (Bernard 1865, p. 372), because "experimental medicine is nothing more than the combination of expectation and empiricism illuminated by reasoning and experimentation" (Bernard 1865, p. 373). At least in this respect, it doesn't make much sense to talk about materialism as opposed to spiritualism, because there are increasingly complex phenomena whose conditions must be determined by science.

2 What Science of Man in the "Open Society"?

2.1 *From Bernard to Popper, through Bergson*

Still focusing on Bernard, it is worth emphasizing another relevant fact, which is the corollary of what has just been said. Bernard showed that the most advanced theories are always perfectible and therefore essentially "mobile", outside the logic of "systems", and therefore cannot represent exclusively personal achievements, but the current state of science that includes the *impersonal* result of the efforts of a large community (Bernard 1865, p. 383).

The experimental method, while based on the experimental verification of a scientific hypothesis, also requires the opposite method, that of falsification, as we have seen. If we also recognize that invention consists precisely in the creation of a hypothesis, which reveals the feeling or genius of the scientist who created it, we must not neglect "that philosophic spirit (*esprit philosophique*) which, without being anywhere, is everywhere, and which, without belonging to any system, must reign not only over all the sciences, but over all human knowledge" (Bernard 1865, p. 387). Philosophy remains the guardian of the "indeterminate":

> Philosophy, by constantly stimulating the inexhaustible mass of unresolved questions, stimulates and maintains this salutary movement in the sciences. Indeed, in the limited sense in which I consider philosophy here, only the indeterminate (*indéterminé*) belongs to it, while the determinate necessarily falls within the scientific domain
>
> BERNARD 1865, p. 390

Further on we read:

> For this reason, while recognizing the superiority of great men, I believe that, in the particular or general influence they exert on the sciences, they are always and necessarily more or less a function of their time. The same is true of philosophers; they can only follow the course of the human mind (*esprit humain*) and contribute to its progress only by opening more widely (*ouvrant plus largement*) to all the way of progress that many would perhaps not perceive.
>
> BERNARD 1865, p. 393

Still focusing on the French epistemological debate, it is impossible to ignore what Bergson expounds in his speech given at the ceremony for Claude Bernard's centenary at the Collège de France on December 30, 1913 (Bergson [1938] 2013b, pp. 229–237), which is very close to the theses of the *Introduction*, a work that is compared by Bergson to Descartes' *Discours de la méthode*. What strikes Bergson about Bernard was not only the strength of the brilliant physiologist and the great experimenter who made great discoveries, but the spirit of invention of which Bernard was a great witness, a spirit that, in science, unites teacher and student in the common search for truth outside of systems, according to a theoretical line that leads Bergson to consider the pragmatism of William James (*Sur le pragmatisme de William James. Vérité et réalité*, in Bergson [1938] 2013b, pp. 239–251).

As a "philosopher and master of contemporary thought", Bergson argues, Bernard was able to develop a philosophy "capable of following concrete reality in all its folds" (Bergson [1938] 2013b, p. 237), and as a critic of "systems" he presented three ideas: 1) the idea of a single philosophy out of the dialectical conflict between different schools; 2) that of a philosophy realized in collaboration; and finally, 3) the prospect of a collaboration between philosophy and science. Through Bernard, Bergson can thus express the full radicality of his conception of knowledge, in which the subject no longer stands inertly in front of the object – in a "mirror-like" relationship – but is the constituent part of reality interpreted as a "whole". On this basis, the concept of intuition that Bergson places within the creative process of reality refers to a conception of ideas as "instruments", and not as simple "mirror-reflections" of reality. Instruments whose function is precisely that of "expanding" reality, starting from the intelligence that is part of life, to reach a more complete coincidence with the "whole."

At this point in the discussion, it is necessary to raise the question of the role of science and philosophy in the *société ouverte*, an expression introduced and discussed by Bergson in his mature work, *Les deux sources de la morale et de la religion* (1932), and later taken up by Popper in *The Open Society and Its Enemies* (1945), albeit with a different interpretation. When Bergson addresses the question of the *closed* and the *ouvert*, he refers rather to morality than to science. The first occurrence of the notion of a *closed society* is in the first chapter of the *Deux sources* dedicated to moral obligation, in which Bergson explains that small social groups, to which we are instinctively tied, set us against the greater whole represented by humanity. However, the point that we aim to discuss here concerns in particular the ethical-political implications that derive from this Bergsonian assumption. Because the transition from primitive instinct, which imposes a defensive attitude towards our fellow men and even justifies war, to the feeling of sympathy, which instead unites us to humanity by making us recognize in the other person the object of the utmost respect, is not as abrupt as one might imagine. In this case the difference is more of degree than of nature. Moreover, it should not be forgotten that "our civilized societies (*civilisées*), however different from the society to which we were immediately destined by nature, are fundamentally similar to it" (Bergson [1932] 2013a, p. 25). And it is at this level that science and knowledge come into play, without neglecting also the experience of mystics, who, in Bergson's view, go directly from society to humanity to God. Bergson clearly explains that the first element that distinguishes the primitive from the "civilized" subject is to be found precisely in the "enormous mass of knowledge and habits" that he drew "after the first awakening of his consciousness, from the social environment in which they are preserved" (Bergson [1932] 2013a, p. 24). In fact, according to Bergson – critical of Durkheim's moral conception [1925], in which family, homeland and humanity represent the three main stages of our social and moral "evolution" –, the passage from "nation" to "humanity" implies a break, that is, the passage from the closed to the open that only an "enlarged" moral conscience can guarantee (according to a holistic vision).

This is the case of the man of genius, previously encountered in Bernard's epistemological reflection. The point now is that exceptional men, conceived as true *distractions heureuses* of nature, can become themselves and assert themselves on condition that there is a society predisposed to receive their transformative influence. Consider, for example, the emblematic case of St. Francis of Assisi, the *homme d'élite* faithful to the original impulse, as described by Félix Ravaisson in his *Testament philosophique* (1933, p. 163). St. Francis is the "outcast" who appeals to the law of the heart to bring the pure spirit into the "profane" world of men (cf. Coniglione 2022). All these recurring cases in the history of humanity demonstrate that in order to loosen the cohesion of a closed society (in which the individual and society merge and become confused), thus allowing individuality to sprout, a sufficiently advanced level of *civilization* is required, that can create an environment (*milieu*) favorable to the genesis and development of genius. This cultural environment is guaranteed by the *société ouverte*, which in Bergson's words "can in principle embrace the whole of humanity", and which "dreamed up, from time to time, by chosen souls, each time realizes something of itself in a series of creations, each of which, through a more or less profound transformation of man, allows to overcome hitherto insurmountable difficulties" (Bergson [1932] 2013a, p. 284).

As for Popper, after the publication in Vienna of *The Logic of Scientific Discovery* (1934), he gave a series of lectures in London criticizing political projects based on scientific forecasts, which were collected in *Poverty of Historicism* (1944). After the Second World War, Popper closely confronted the Bergsonian distinction between closed and open society, giving it a different, somewhat much more radical interpretation (see Boyer 2008, p. 194).

In a footnote to the *Introduction* of *The Open Society and Its Enemies* Popper argues:

> The terms "open society" and "closed society" were first used, to my knowledge, by Henri Bergson, in *Two Sources of Morality and Religion* (Engl. ed., 1935). In spite of a considerable difference (due to a fundamentally different approach to nearly every problem of philosophy) between Bergson's way of using these terms and mine, there is a certain similarity also, which I wish to acknowledge. (Cp. Bergson's characterization of the closed society, *op. cit.*, p. 229, as "human society fresh from the hands of nature".) The main difference, however, is this. My terms indicate, as it were, a rationalist distinction; the closed society is characterized by the belief in magical taboos, while the open society is one in which men have

> learned to be to some extent critical of taboos, and to base decisions on the authority of their own intelligence (after discussion). Bergson, on the other hand, has a kind of religious distinction in mind. This explains why he can look upon his open society as the product of a mystical intuition, while I suggest (in chapters 10 and 24) that mysticism may be interpreted as an expression of the longing for the lost unity of the closed society, and therefore as a reaction against the rationalism of the open society. From the way my term "The Open Society" is used in chapter 10, it may be seen that there is some resemblance to Graham Wallas' term "The Great Society";[1] but my term may cover a "small society" too, as it were, like that of Periclean Athens, while it is perhaps conceivable that a "Great Society" may be arrested and thereby closed. There is also, perhaps, a similarity between my "open society" and the term used by Walter Lippmann as the title of his most admirable book, The Good Society (1937)
>
> POPPER [1945] 2013, pp. 512–513

As can be seen from the passage quoted above, Popper identifies holistic societies, such as those of ancient Greece, with closed societies organized according to collective beliefs, distinguishing them from those with an individualistic imprint – whose precursor is Socrates – and which are based on the criticism and rejection of scientific assumptions.

> The magical or tribal or collectivist society will also be called the *closed society*, and the society in which individuals are confronted with personal decisions, the *open society*
>
> POPPER [1945] 2013, p. 165

While a closed society "is still a concrete group of concrete individuals, related to one another not merely by such abstract social relationships as division of labor and exchange of commodities, but by concrete physical relationships such as touch, smell, and sight", in the open one, "many members strive to rise

---

1 Cf. Wallas 1914, p. 3: "During the last hundred years the external conditions of civilised life have been transformed by a series of inventions which have abolished the old limits to the creation of mechanical force, the carriage of men and goods, and communication by written and spoken words. One effect of this transformation is a general change of social scale. Men find themselves working and thinking and feeling in relation to an environment, which, both in its world-wide extension and its intimate connection with all sides of human existence, is without precedent in the history of the world. Economists have invented the term The Great Industry for the special aspect of this change which is dealt with by their science, and sociologists may conveniently call the whole result The Great Society".

socially, and to take the places of other members. This may lead, for example, to such an important social phenomenon as class struggle. We cannot find anything like class struggle in an organism" (Popper [1945] 2013, p. 165).

Within this framework, if Socrates becomes the forerunner of modern individualism, Plato is seen by Popper as the founder of an organicist political philosophy that we find again in Hegel, who is presented – very reductively – as the philosopher who not only "developed the historical and totalitarian theory of nationalism", but that "also clearly foresaw the psychological possibilities of nationalism" (Popper [1945] 2013, p. 275). Now, leaving aside this polemical confrontation with Hegel (as well as with Karl Marx), which would require much more space to be discussed analytically, Popper's greatest point of distance from Bergson's position concerns mystical intuition, which Popper can no longer admit as the basis of open society. The mystical approach is traced back to the practices of a closed society, and not as a vehicle for opening to the new, and therefore as an expansion of the perspective on the human in a universalistic sense, a characteristic of the man of genius already mentioned by Bernard. Popper's greatest distance from Bergson concerns his position on rationality, his "critical rationalism", "which recognizes the fact that the fundamental rationalist attitude results from an (at least tentative) act of faith – from faith in reason" (Popper [1945] 2013, p. 437).

However, Popper places a moral requirement at the basis of his "critical rationalism": "For the question whether we adopt some more or less radical form of irrationalism, or whether we adopt that minimum concession to irrationalism which I have termed 'critical rationalism', will deeply affect our whole attitude towards other men, and towards the problems of social life. It has already been said that rationalism is closely connected with the belief in the unity of mankind" (Popper [1945] 2013, p. 437).[2] Moving from these assumptions, Bergson is traced back by Popper to an alleged evolutionary biologism of Hegel (an author, the latter, of whom Bergson knew little, as he had personally admitted to Benedetto Croce at the Philosophy Congress of Bologna in 1911).

Popper retorts:

> The transubstantiation of Hegelianism into racialism or of Spirit into Blood does not greatly alter the main tendency of Hegelianism. It only gives it a tinge of biology and of modern evolutionism. The outcome is a

---

2 He then clarifies as follows: "I too believe that our Western civilization owes its rationalism, its faith in the rational unity of man and in the open society, and especially its scientific outlook, to the ancient Socratic and Christian belief in the brotherhood of all men, and in intellectual honesty and responsibility" (Popper [1945] 2013, p. 448).

> materialistic and at the same time mystical religion of a self-developing biological essence, very closely reminiscent of the religion of creative evolution (whose prophet was the Hegelian Bergson).
>
> POPPER [1945] 2013, p. 273

In another passage, Bergson is presented, in line with most "German philosophers and intellectuals", as a theorist of "oracular irrationalism":

> An oracular irrationalism has established (especially with Bergson and the majority of German philosophers and intellectuals) the habit of ignoring or at best deploring the existence of such an inferior being as a rationalist. To them the rationalists – or the "materialists", as they often say – and especially, the rationalist scientist, are the poor in spirit, pursuing soulless and largely mechanical activities, and completely unaware of the deeper problems of human destiny and of its philosophy.
>
> POPPER [1945] 2013, p. 435

From the last two passages mentioned above, it can be seen that Popper is interested in emphasizing a specific aspect, morally characterized as a clear choice between different "beliefs": that in reason and in human individuals, on the one hand, and that in the mystical faculties of man that unite him to a collectivity, on the other. Thus, the ultimate choice remains "between an attitude that recognizes the unity of mankind and an attitude that divides men into friends and foes, into masters and slaves" (Popper [1945] 2013, p. 450). But that's not the point. If we completely exclude from the analysis some of the more original characteristics that Bergson attributes to the open society[3] – which Popper instead convincingly attributes to the closed society –, we witness the degeneration of the open society into a "depersonalized society", whose configuration Popper describes as follows:

> As a consequence of its loss of organic character, an open society may become, by degrees, what I should like to term an "abstract society". It may, to a considerable extent, lose the character of a concrete or real group of men, or of a system of such real groups. This point which has been rarely understood may be explained by way of an exaggeration. We

3 Such as a certain mysticism and its characteristic language (the inner word), the positive role of myth in relation to *logos* (cf. Coniglione 2022b), a certain emotional atmosphere within which *fraternité* between men is built, the intuition of the charismatic, generous and brilliant subject, capable of representing the "destiny" of humanity.

> could conceive of a society in which men practically never meet face to face – in which all business is conducted by individuals in isolation who communicate by typed letters or by telegrams, and who go about in closed motor-cars. (Artificial insemination would allow even propagation without a personal element.) Such a fictitious society might be called a "completely abstract or depersonalized society".
>
> POPPER [1945] 2013, p. 166

Now we need to understand if a rationality that is not anonymous and impersonal, but dialogic and open to the symbolic dimension (cf. Cardullo-Coniglione 2017), can really be a valid antidote to the degeneration of open society into an abstract and depersonalized society.

3 A New Dialogic Rationality for the "Depersonalized Society"

Also in *The Open Society*, Popper had lucidly admitted that "although society has become abstract, the biological make-up of man has not changed much; men have social needs which they cannot satisfy in an abstract society" (Popper [1945] 2013, p. 166). Before him, Bergson had emphasized that morality "is of biological essence" (Bergson [1932] 2013a, p. 103), giving the word biology a "much fuller meaning" (Bergson [1932] 2013a, p. 103). For social life cannot be conceived as a mere fact (*fait*), but it is necessary to keep in mind both that even the individual "can judge society and obtain from it a moral transformation", and that life itself can communicate "a new impulse to privileged individualities", capable of taking society even further forward (Bergson [1932] 2013a, p. 103). In other words: intuition must be put back at the center of man's ethical-political life, something that Popper rejects. As the liberal Walter Lippmann observed in the same years, we must not forget that every government is composed of mortal men and that there must be limits to that superior intuition. The social order must be modelled on the principle of justice and sympathy, already theorized by Adam Smith in the *Theory of Moral Sentiments* ([1759] 2002). So Lippmann:

> There must be a strong desire to be just. There must be a growing capacity to be just. There must be discernment and sympathy in estimating the particular claims of divergent interests. There must be moral standards which discourage the quest of privilege and the exercise of arbitrary power. There must be resolution and valour to resist oppression and

> tyranny. There must be patience and tolerance and kindness in hearing claims, in argument, in negotiation, and in reconciliation.
>
> LIPPMANN [1937] 2017, p. 363

On this basis, if we think of a hypothetical open society that is truly reconciled – and therefore that is also a new society, modelled on the principle of *fraternité* (see Coniglione 2021) – we cannot once again ignore the moral dimension, remaining aware with Lippmann that "the virtues that are required for the overhead administration of a civilization are superhuman". But there is another very problematic passage in Lippmann's text, which brings us back to what Bergson said about the mystical-intuitive subject which opens society:

> The supreme architect who begins as a visionary, becomes a fanatic, and ends as a despot. For no one can be the supreme architect of society without employing a supreme despot to execute the design. So, if men are to seek freedom from the arbitrary dominion of men over men, they must not entertain fantasies of the future in which they play at being the dictators of civilization. It is the bad habit of an undisciplined imagination.
>
> CONIGLIONE 2021, p. 365

The issues raised here by Lippmann are many and all extremely important. In some way, he brings us back to the theme of essentialism introduced by Popper when he wrote *The Poverty of Historicism* (see Coniglione 2002, pp. 247–281),[4] alluding to the need to find definitive explanations, not only within the perimeter of scientific investigation in the strict sense, as Lippmann's quoted passage clearly shows. The fact remains that the right of human beings as such to recognize themselves as supreme architects by virtue of a superior intuition depends first and foremost on the correct and reasonable exercise of virtue, on a form of "generosity" (both towards oneself and towards the world in general) that Aristotle in the *Nicomachean Ethics* already identified not so much in the war, nor in the typical acts of temperance or even in judicial decisions, but

4 Coniglione asks himself in this regard: "But how is it possible to admit, even as a goal never to be reached, the existence of an *absolute* truth in a coherently *anti-essentialist* position, for which there are no *ultimate* explanations? It only makes sense to talk about absolute truth on the condition that one accepts an ultimate level of reality and therefore the possibility of an ultimate, fundamental explanation; but the existence of this has been denied by Popper. Not only, therefore, does the concept of truth (in Popper's sense) lack any criteria that would make it operational, but it is in contrast with what Popper argued in his critique of essentialism" (Ibid., pp. 257–258).

in the capacity to "give" (Aristotle 1999, 1, 119b, p. 127). Therefore, recognizing the "complexity" of reality – as opposed to the idea that it can be exhausted within a purely theoretical model – leads to a more nuanced understanding of knowledge itself:

> Thought is therefore complex insofar as it needs to be articulated in a plurality of cognitive models, none of which is exhaustive of the totality of reality; it is simple because each of these models requires a filtering of reality that reduces its logical openness and allows for its logical-rational treatment. To think, instead, that thought must become complex in order not to mutilate the complexity of reality is to make thought the mirror of this reality, a reduplication of it, and thus to prevent it from developing what is its most proper function, the symbolic and modelling capacity that presupposes simplification.
>
> CONIGLIONE 2022a, p. 70

At least from this point of view, recognizing oneself as the *supreme architect of society* means not only thinking of oneself in an Aristotelian way as worthy of great and beautiful things (think of important associations such as Freemasonry that also make use of a rich symbolic and ritual apparatus), but to always feel ready to confer freedom and rational richness to a world that otherwise would remain just a rough stone. It also means to assert the Pascalian "reasons of the heart". In this regard, Coniglione writes:

> And this language is that of reason, which all men have in common. And this is not because reason is in itself a guarantee of arriving at a unique, universal, identical for all men and therefore requiring a uniformity that binds everyone, but rather in the sense that only by resorting to it, and therefore to dialogue in which it finds its highest realization, can we reach that consensus and that common perspective that have nothing constrictive or violent about them, based on conviction and acceptance.
>
> CONIGLIONE 2021, p. 186

Therefore, considering the different stages of our investigation, we feel the need to define a model of dialogic rationality for an open and supportive society (understood more in Bergson's than Popper's way, for the reasons we have explained here), capable of rediscovering – albeit intuitively but at the same time beyond any biological automatism – a higher reality of common brotherhood in which one learns the original meanings of things. The moral man, who is a communitarian man of elite, thanks to such cognitive experiences, can rediscover the value of "inner work", of "personal journey", of "maturation of the individual that can only take place in close harmony with a community" (Coniglione 2021, p. 194).

References

Aristotele (1999). *Etica Nicomachea*, edited by C. Natali. Rome-Bari: Laterza.

Bergson, H. ([1932] 2013a). *Les deux sources de la morale et de la religion*. Paris: PUF «Quadrige».

Bergson, H. ([1938] 2013b). *La pensée et le mouvant*. Paris: PUF «Quadrige».

Bernard, C. (1865). *Introduction à l'étude de la médicine expérimentale*. Paris: G. Baillière.

Boyer, A. (2008). Popper, Bergson: l'intuition et l'ouvert. *Revue philosophique de la France et de l'étranger*, 2, 133, 187–203.

Cardullo, R.L., Coniglione, F. (Eds.) (2017). *Reason and no-reason from ancient philosophy to neurosciences*. Sankt Augustin: Academia Verlag.

Coniglione, F. (1978). *La scienza impossibile. Dal popperismo alla critica del razionalismo*. Bologna: Il Mulino.

Coniglione, F. (1980). Scacco a Popper. *Spirali*, 6, 10–11.

Coniglione, F. (1996). Humanistic interpretation between Hempel and Popper. *Poznań Studies in the Philosophy of the Sciences and the Humanities*, 47, 283–301.

Coniglione, F. (2002). *La parola liberatrice. Momenti storici del rapporto tra filosofia e scienza*. Catania: CUECM.

Coniglione, F. (2003). *Introduzione alla filosofia della scienza. Un approccio storico*. Acireale-Rome: Bonanno Editore.

Coniglione, F. (2008). *Popper addio. Dalla crisi della filosofia della scienza alla fine del* logos *occidentale*. Acireale-Rome: Bonanno Editore.

Coniglione, F. (2021). Etica laica ed etica come fraternità. In: P. Amato *et al.* (eds.), *Metafisica dell'immanenza. Scritti per Eugenio Mazzarella, vol. II: Etica e religione*, pp. 185–199. Milan: Mimesis.

Coniglione, F. (2022a). Complessità e pluralismo cognitivo. *Complessità*, 2, 32–72.

Coniglione, F. (2022b). Dal mito alla scienza e ritorno. Per una visione non imperialistica della conoscenza. In: E. Coco (ed.), *L'invenzione della realtà. Scienza, mito e immaginario nel dialogo tra psiche e mondo oggettivo. Una prospettiva filosofica*, pp. 11–30. Pisa: Edizioni ETS.

Coniglione, F. (2022c). *L'uomo venuto da un altro mondo. Francesco d'Assisi*. Acireale-Rome: Bonanno Editore.

Coniglione, F. (2022d). Visione olistica e visione riduzionistica della natura nell'epistemologia contemporanea. In: R.L. Cardullo (Eds.), *Cura di sé, cura del mondo. L'impatto della crisi ambientale sul fisico* (sôma) *e sul morale* (psychê) *dell'uomo.*

Atti della Giornata di Studi, Catania 30 giugno–1° luglio 2021, pp. 118–137. Milan: Mondadori Education.

Coniglione, F. (2025). *Lontano da Popper. L'epistemologia post-positivista e le metamorfosi della razionalità scientifica*. Pisa: Edizioni ETS.

Durkheim, É. (1925). *L'éducation morale*. Paris: Alcan.

Lippmann, W. ([1937] 2017). *The Good Society* (*with an introduction by G. Dean Best*). London-New York: Routledge.

Mossé-Bastide, R.M. (1955). *Bergson éducateur*. Paris: PUF.

Nagel, E. (1961). *The Structure of Science. Problems in the Logic of Scientific Explanation*. New York et al.: Harcourt, Brace & World, Inc.

Popper, K. ([1934] 2002). *The Logic of Scientific Discovery*. London-New York: Routledge.

Popper, K. ([1945] 2013). *The Open Society and Its Enemies* (*with a new introduction by A. Ryan and an essay by E.H. Gombrich*). Princeton: Princeton University Press.

Popper, K. ([1957] 1964). *The Poverty of Historicism*. New York: Harper & Row.

Popper, K., Eccles, J.C. (1977). *The Self and Its Brain*. New York: Springer.

Ravaisson, F. (1933). *Testament philosophique et fragments* (edited by Ch. Devivaise). Paris: Boivin et C[ie].

Reichenbach, H. (1951). *The Rise of Scientific Philosophy*. Berkeley-Los Angeles: University of California Press.

Roni, R. (2015). *La visione di Bergson. Tempo ed esperienza del limite*. Milan: Mimesis.

Smith, A. ([1759] 2002). *The Theory of Moral Sentiments*. Cambridge: Cambridge University Press.

Wallas, G. (1914). *The Great Society. A Psychological Analysis*. London: Macmillan and Co.

CHAPTER 3

Francesco Coniglione on Ernst Cassirer's Theory of Concept-Formation: Some Critical Remarks

Giacomo Borbone

Abstract

In Francesco Coniglione's extensive scientific production there are many references to the works of Ernst Cassirer, especially regarding the formation of scientific concepts. As the scholars of Coniglione's books and essays are aware, the latter, in the international scene, was among the first to valorize the idealizational approach to science elaborated by Leszek Nowak and the Poznań School of Methodology. However, Coniglione has been able to extend these theoretical acquisitions to other disciplinary fields and, above all, has employed the scientific concept of idealization as a key for interpreting many authors and currents of thought. It is in the light of the concept of idealization that Coniglione has analyzed the epistemological thought of Ernst Cassirer, thus emphasizing the fruitfulness of the Breslau philosopher's functionalist approach. In the paragraphs of my contribution, I will discuss just Coniglione's interpretation of the difference between abstraction and function present in Ernst Cassirer, while in the concluding parts I will mention some critical points of his hermeneutical reading.

Keywords

Cassirer – function – idealization – science – substance

1 Introduction

> In his epistemological masterpiece *Substanzbegriff und Funktionsbegriff*, Cassirer clearly articulated the distinction between abstraction as understood by empiricism – whose origins he rightly traced back to Aristotle – and the specific mode of conceptualization characteristic of modern science in the context of the "theory of concept formation."
>
> F. CONIGLIONE, *Between Abstraction and Idealization*

 | HTTPS://DOI.ORG/10.1163/9789004756496-005

> Any attempt to interpret the concepts of natural science as mere aggregates of facts of perception must necessarily fail. No scientific theory is directly related to these facts, but is related to the ideal limits, which we substitute for them intellectually.
>
> E. CASSIRER, *Substance and Function*

In his long and still ongoing philosophical journey Francesco Coniglione has often addressed intricate epistemological questions: the main-trends of western epistemology – neo-positivism, Popper's critical rationalism, and post-positivist reflections on science (Coniglione 1978; 1992a, 1996a; 2008b, 2025) – but also more "heterodox" thinkers, such as Ludwig Fleck and his concept of "thought-style" (Coniglione 2019; 2020) culminating in a systematic study of twentieth-century Polish scientific philosophy (Coniglione 1992b; 1996b; 2007). This is particularly true with regard to the conceptualization of the so-called "Poznań School," which originated with the works of Jerzy Topolski, Jerzy Kmita, and primarily Leszek Nowak (Coniglione 1983; 2010[2], 2023). Coniglione not only translated, edited, and published their writings in Italian (see, most notably, Nowak 1987), but was also the first to promote their idealizational approach at the international level, dedicating a comprehensive monograph to it well ahead of its wider recognition (Coniglione 1990a; 2010[2]). More, Coniglione has been able to extend these theoretical acquisitions to other disciplinary fields and, above all, has employed the scientific concept of idealization as a key for interpreting many authors and currents of thought (Coniglione 1990b; 1994; 1997; 2012; 2016).

It was precisely through his study and in-depth analysis of the core themes of this school – especially the central concept of "idealization" – that Coniglione had the opportunity to rediscover and further explore the thought of Cassirer, a philosopher he had already known since his formative years, having read his major epistemological works: *Substance and Function*, *Determinism and Indeterminism in Modern Physics*, and *Einstein's Theory of Relativity* (Cassirer [1910] 1953; [1937] 1956). This reflection unfolds in close connection with his analysis of Leszek Nowak's theories, starting from his early articles, which already contain significant references to Cassirer's work (Coniglione 1990b; 2004), culminating in a systematic treatment of the Breslau philosopher's functionalist approach in an essay specifically dedicated to him (Coniglione 2008a).

Coniglione's interest in Cassirer's work is primarily driven by his theory of concept formation (*Begriffsbildung*), with the aim of highlighting not only its difference from the conceptions developed within classical empiricism and later adopted by neo-positivism, but also its enduring epistemological fertility. For Coniglione, it is not only about highlighting the transition operated by Cassirer from the concept of substance to that of function (Cassirer [1910] 1953) – an aspect frequently noted by the most meticulous scholars of Cassirer's epistemology, and often assimilated, though not always with due awareness, in the form of his concept of function – but rather about emphasizing Cassirer's critique of the concept of abstraction in the first chapter of his 1910 masterpiece, which has been almost completely neglected by the main epistemological currents of the 20th century. To this end, Coniglione focuses his attention on a critical examination of the relationship between abstraction and function, seeking to clarify the reasons behind its neglect and, in turn, the relevance it has gained in the contemporary debate on models, particularly within post-positivist philosophies of science.

In the following paragraphs of this contribution, I will discuss Coniglione's role in the rediscovery of Cassirer's concept of abstraction, interpreting it through the lens of idealization and highlighting its divergence from the conception embraced by the dominant strand of neo-positivism. In the concluding section, I will briefly address some critical aspects of his hermeneutical interpretation.

2 Abstraction and Function in Cassirer: Coniglione's Interpretation

> Any physical assertion involves certain logico-mathematical constants. If we desire to bring all of these constants into a short formula, we can point out the concept of number, the concept of space, and the concept of function as the fundamental elements, which enter as presuppositions into every question which physics can raise.
>
> E. CASSIRER, *Substance and Function*

In his 1910 theoretical work, Cassirer, as is well known, outlines the fundamental features of his functionalist conception of science. The theoretical need to distinguish – on the theoretical level – the concept of substance (as in Aristotle) from the mathematical concept of function enables Cassirer to insightfully grasp the formation of the concepts on which modern science is based. Much like Plato (Borbone 2023a), Cassirer maintains that the data of perception – being subject to the Heraclitean flux – offer no "Archimedean point" to thought. Instead, such a point must be sought in the act of determination and differentiation that thought itself imposes upon perceptual data, thereby granting experience the unity necessary to subsume each particular element. Through this functionalist framework, the philosopher of symbolic forms not only distances himself from reductionism and naïve realism, but also offers a valuable instrument for engaging with the complexity of reality.

However, in order to properly comprehend Cassirer's concept of function, a close examination of his critique of abstraction in the first chapter of *Substanzbegriff und Funktionsbegriff* is essential. It is precisely this aspect that Coniglione focuses on in his analysis of how Hegel – and subsequently Marx – understood abstraction. In doing so, he highlights the distinctions from the classical empiricist view, showing how, from the English empiricism of Locke and Hume to the more developed version proposed by Mill, the concept of abstraction was conceived in a manner analogous to Aristotle's critique of Plato, namely, as the isolation of properties common to multiple entities. However, this procedure, Coniglione asserts – following Cassirer ([1910] 1953, pp. 3–9) – "leads to the useless, empty Aristotelian concept of substance, which is absolutely fruitless from a scientific point of view" (Coniglione 1990b, p. 66). In this regard, Coniglione states that Aristotle:

> conceived of the universal, the result of the process of abstraction, as that which is common to many individuals; as such – i.e. as primary substances – individuals cannot be fully captured in terms of universals. However, although real individuals elude our thought, we can conceive of the qualities, as real as the individuals themselves, which constitute what is shared by a certain number of individuals. It is just because individuals possess these common qualities that we can group them together and classify them as belonging to species and genera.
>
> CONIGLIONE 2004, p. 60

In his early engagement with Cassirer's work, Coniglione already emphasizes how Cassirer's position sharply opposes the scientific concept of schematic representation of types (in the Aristotelian sense), and how a parallel can be drawn with Marx's method of abstraction, reinterpreted through the lens of the idealizational conception:

> So, it is evident that for Cassirer abstraction "doesn't leave out the distinctive features of the contents" since his departure from concrete experience is only a "methodical step" preceding a "return to earth", the phenomenological plane, through the process of concretization. It shows that the particular contents are "necessary" since they derive from the law once one introduces corrective parameters which take into account formerly omitted factors, otherwise the law is falsified. This is the same content Marx speaks of in his "advancing from the abstract to

> the concrete". The scientific concepts reached through idealization (or scientific, Marxist abstraction) both "deny" and "preserve" empirical particulars and so are fruitful. The "abstract" concept, on the other hand, is fruitless in that it does not allow reacquisition of experience by explaining it particulars are "denied" but not "preserved".
>
> CONIGLIONE 1990b, p. 87

Subsequently, Coniglione returns to the analysis of the concept of abstraction in a comprehensive essay, where he examines its most significant formulations throughout the history of philosophy, along with its explicit theorization by empiricist currents and their inductive approach (Coniglione 2004, pp. 60–69, *"The empiricist mainstream"*). He contrasts these with both actual scientific practice – which often does not adhere to the features of inductivist philosophy, despite being strongly influenced by it (as in the case of Newton) (Coniglione 2004, pp. 70–80, *"The abstraction of the scientists"*) – and the theories of philosophers who position themselves outside the mainstream of contemporary epistemology and who have, at times, displayed openly anti-scientific tendencies (Coniglione 2004, pp. 80–88, *"The philosophical outsiders"*).

As for the philosophical approaches – this is where we now turn our attention – Coniglione observes that Edmund Husserl had already devoted a substantial portion of his *Logical Investigations* (Husserl [1900–1901] 1968) to analyzing the "modern theories of abstraction" (especially in the *Second Investigation*), highlighting their internal contradictions and difficulties, and opposing empirical abstraction (Coniglione 2004, p. 86). Yet these acute and profound reflections had little impact on 20th-century epistemological discourse, largely because the prevailing empiricism "does not grasp the 'ideal' nature of scientific concepts, which can be arrived at only by what Husserl calls 'idealizing abstraction'" (Coniglione 2004, p. 86). As a result, it fails to account for those ideal meanings that are embodied in typical scientific laws, meanings that go beyond mere empirical generalizations. This leads Husserl to reject "a purely extensionalistic concept of scientific concepts" (Coniglione 2004, p. 87). However, Coniglione notes, Husserl, especially in his later writings and in his last work ([1954], 1970),

> is no longer interested in evaluating this typical mode of acquiring scientific knowledge so much as proposing as an alternative – on account of its artificial nature – re-discovery of the "world of life" as the only resource that can help European science to get over its "crisis". Husserl's thought thus belongs to a broad current of contemporary philosophy that sees scientific rationality as an enemy of philosophical speculation and

> therefore wishes to confine it within narrower limits, depriving it of any authentic cognitive value.
>
> CONIGLIONE 2004, p. 88

And this is just one of the main reasons for the limited resonance of his thought in the field of epistemology, which is why in this respect "Husserl remains a misunderstood philosopher" (Coniglione 2004, p. 88).

What was said for Husserl can be partly reiterated for Cassirer as well. But in this case, there is no anti-scientific stance; rather, there is a sympathetic attitude toward the rationality of science, which is examined in terms of its underlying structures and characteristics. As a result, Cassirer's epistemological thought has, in recent times, attracted renewed attention (Friedman 2011; Borbone 2016; 2019; 2022). Francesco Coniglione has also contributed to this revival of interest, particularly by drawing attention to Ernst Cassirer's critique of the concept of abstraction. Coniglione notes that the traditional theoretical procedure of abstraction is nothing more than the simple generalization of empirical facts. Yet in the history of science, abstraction has often remained enveloped in a certain ambiguity, making it unclear what is actually meant by abstraction when one looks at the concrete practices of scientists. It is precisely here that the originality of Cassirer's epistemological perspective becomes evident, something that Coniglione has insightfully grasped. He emphasizes how Cassirer sharply criticizes the sterility, and therefore the inadequacy, of empiricist abstraction to meet the demands of scientific methodology. Mature science, according to Cassirer, does not operate primarily through inductively derived abstractions, but rather through *idealizations* that serve to overcome the constraints of matter. For this reason, the philosopher of symbolic forms, with his gaze turned toward the most recent developments in logic and mathematics within modern science, shifts the core of his epistemological inquiry to the fundamental mathematical concept of function.

For example, as Coniglione points out, in his analysis of the modern concept of energy Cassirer shows that it becomes unintelligible if approached through the traditional lens of abstraction, as this would lead to a process of substantialization that ultimately conceals its functional nature. This implies that the fundamental concepts of natural science continuously transcend empirical data: even laws that seem closely tied to experience can only be properly formulated by referring to ideal constructs, which conceptually replace the raw data of sensory perception. As Coniglione writes, "even laws which appear to be close to experience can be reasonably expressed only with reference to ideal extreme figures with which we replace the empirical data of sensorial perception by means of a concept. […] Direct contact between scientific concepts and perceived facts is lost [...]" (Coniglione 2004, p. 84). Thus, there emerges a complete detachment – a total separation – between concept and reality. None of the concepts of science are rooted in sensorial perception, as we cannot dispense with the use of limiting concepts. As Cassirer himself says in a quote reported by Coniglione:

> The fundamental theoretical laws of physics throughout speak of cases that are never given in experience nor can be given in it; for in the formula of the law the real object of perception is replaced by its ideal limit. [...] The insight gained through them never issues from consideration of the real alone, but from the possible conditions and circumstances; it includes not only the actual, but also the "virtual" process.
>
> CASSIRER [1910] 1953, p. 174

In effect, according to Cassirer, modern science reveals a gradual process of the de-substantialization of reality, which appears not so much as a compendium of sensible things or qualities, but rather as a system of abstract symbols of thought. Consequently, idealized constructs never find a perfect correspondence with the world of concrete physical objects; no Aristotelian-Scholastic *adaequatio rei et intellectus* is to be found. Indeed, when we speak of the physical nature of the link between phenomena, we do not mean to affirm the corporeality or materiality of that link. Rather, a norm is established, one that governs the entire material system under analysis and is always expressed in purely mathematical terms. If the concepts forged by modern science possess an ideal and therefore counterfactual character, it follows that it is impossible, through mere generalization, to conceive of entities such as the ideal gas or a perfectly elastic body. It is precisely at this point that the theoretical procedure of abstraction reveals its full inadequacy and sterility, since – if applied – it would render the conception of such entities impossible, as Cassirer asserts in *Substanzbegriff und Funktionsbegriff*:

> No scientific theory is directly related to these facts, but is related to the ideal limits, which we substitute for them intellectually. We investigate the impact of bodies by regarding the masses, which affect each other, as perfectly elastic or inelastic; we establish the law of propagation of pressure in fluids by grasping the concept of a condition of perfect fluidity; we investigate the relations between the pressure, temperature and volume of gas by proceeding from an "ideal" gas and comparing a hypothetically evolved model to the direct data of sensation.
>
> CASSIRER [1910] 1953, p. 130

Coniglione emphasizes that the *modus operandi* described in several sections of Cassirer's 1910 work is characterized by a specific theoretical procedure which, rather than simply mirroring empirical data, consists in disregarding those factors considered secondary in order to operationalize, in a functional manner, those deemed essential. In scientific inquiry, objectivity is thus provided by the *ideal*, not by things as they are given in their mere materiality and *empirical* immediacy. It is precisely in this respect that Coniglione identifies the originality of Cassirer's epistemological perspective, namely, in having rethought the formation of scientific concepts in a novel way, albeit without the benefit of a full awareness of twentieth-century epistemology. As Coniglione afterwards notes, Cassirer "did not have the time to witness it, and he was also much more concerned with the works of physicists and scientists (from Galilei and Newton to Helmholtz, Hertz, Mach, Duhem, Pearson, and many others, with an expertise worthy of admiration) than with those of philosophers of science" (Coniglione 2008a, p. 178). Cassirer's functionalist approach – which he would later extend to the various symbolic forms through which human beings confer meaning upon reality – has its undeniable roots in scientific inquiry, where the emblematic figure is, significantly, that of Galileo (Coniglione 2016). And in this regard, we may further support Coniglione's argument by citing what Cassirer states in his posthumous *Lectures on Ancient Philosophy*:

> Galileo rejects, in express words, the saying of Aristotle that physical problems never can be treated in the same way as mathematical ones and that the ideal of exactness, maintained in Geometry, cannot be transferred to physical investigations. [...] as Galileo declares, nature is not to be defined as an aggregate of things, it must be understood as a system, as a concatenation of causes and effects. And between these causes and effects there always exists an invariable relation, a necessary connexion. [...] It is by this conception of causality that Galileo is led back to the Platonic concept of knowledge and truth. He adopts and acknowledges in its full sense the Platonic concept of truth; he repeats in his *Dialogues* the saying of Plato that true knowledge is attainable only of such objects as are in themselves firm and steadfast depending on unchangeable and eternal causes.
>
> CASSIRER 2022, p. 278

For Cassirer, the relationship between idea and reality must therefore be conceived in entirely new terms, not in terms of opposition, but of mutual connection, or, as Cassirer puts it, in terms of the *condition* and the *conditioned* (Cassirer [1916] 1975, pp. 162–163). This functionalist conception of science includes an additional dimension that Coniglione captures particularly well in his analysis of Cassirer's epistemology. Indeed, it is precisely within the framework of this functionalist approach that Cassirer, as Coniglione observes, interprets "all science in a new way compared to what coeval epistemology did" (Coniglione 2008a, p. 181). Scientific concepts, in fact, are not the result of mere generalization of empirical facts; rather, they are ideal constructs that lack any immediate empirical referent.

3 Towards the Concept of Idealization

> The existing of the ideal, which can alone be critically affirmed and advocated, means nothing more than the objective logical necessity of idealization.
>
> E. CASSIRER, *Substance and Function*

What Coniglione has stated so far finds a more complete and systematic treatment in an essay entirely devoted to Cassirer (Coniglione 2008a). In this work, the arguments already presented in previous publications are revisited, but now developed through a more rigorous comparison with the concept of abstraction, first in classical empiricism, beginning with Aristotle, and then in contemporary epistemology. Coniglione highlights how the same symbolic formulation of the function $\varphi(x)$ assumes two epistemologically distinct meanings in the empiricist tradition and in Cassirer's Neo-Kantian approach: in the former, it signifies inductive generalization; in the latter, idealization. As Coniglione argues:

> The whole of 20th-century epistemology – relying on Frege's and Russell's concept of function – chose the first interpretation. Berkeley's conception, indeed, was later taken as a point of no return in the debate about general objects, and as a basis for building the standard contemporary theory of universals and predication. This was the work of Frege: reviving Berkeley's thesis that only individual objects exist, he based the logic of generality on quantifier-variable notation. Due to his refusal to accept the existence of indefinite (or general) objects, a statement like "a whale is a mammal" is for Frege equivalent to saying "all whales are mammals". So he also shares the view that concrete universals (the idea of a "whale in general") do not exist, and the function of the abstract universal ("mammality") is to act as predicate in the corresponding statement.
>
> CONIGLIONE 2008a, p. 176

This perspective – further developed through Russell's notion of "definition by abstraction" and the reformulation of the abstract universal in set-theoretical terms – led to the understanding of general concepts as equivalence classes. As Carnap succinctly states: "Such a definition of a family of properties by way of the equivalence class of an equivalence relation is often called *definition by abstraction*" (Carnap 1958, p. 137). This procedure is seen as the refinement and culmination of the empiricist discussion on concept formation and the nature of abstraction. As Coniglione emphasizes by quoting a prominent Italian mathematician: "The passage from set A to the quotient set A/ε (where ε is an equivalence relation) schematizes and precisely defines the process of concept formation from objects and, more generally, the ordinary process of abstraction, which consists in the identification of different elements, yes, but all sharing a common 'property'" (Lombardo-Radice 1982, p. 27).

This is not how Cassirer understands scientific concepts. As we have seen, Coniglione emphasizes that the meaning attributed to them by the thinker from Breslau highlights the creative character of the human mind, which "constructs" them on the basis of experience, yet in a way that departs from its immediate givenness. In essence, Cassirer "puts forward what is a typical demand of his Kantianism: the need for a 'genetic account' of the concepts of science, which must be performed by the intellect in an ongoing and never-ending process" (Coniglione 2008a, p. 179). Thus, for example, in order for the physical body of mechanics to be integrated into the laws of physics, it must relinquish any determination that is directly derived from experience and acquire a new nature (Coniglione 2008a, p. 180).

According to Coniglione (2008a, p. 181), what has been said so far is already sufficient to recognize how, in many respects, Cassirer anticipated the criticisms that, in the second half of the twentieth century, would be raised by the New Philosophy of Science against the classical formulations of neo-positivism. At the core of Cassirer's philosophy lies the conviction that there is no such thing as empirical data unmediated by theory, that is, there can be no "pure experience" free from conceptual presuppositions. Experience, Cassirer argues, cannot be treated as part of an *experimentum crucis* in the Baconian sense, where data would serve as a decisive empirical criterion to discriminate between competing theories. Indeed, for Cassirer, even the simplest measurement relies on theoretical principles, and therefore on general functions of connection, formation, and coordination of concepts. What is commonly referred to as a "datum" of perception must already be transcended and replaced by a conceptual symbol.

This critical orientation could not, as Coniglione notes, fail to be associated with a rejection of Mach's reductionism and the positivist descriptivism which held that the concepts of physics are nothing more than mere symbols for complexes of impressions, useful only for the economy of thought. According to this view, the task of a philosophy of physics would be to reduce each of its concepts to the sum of the perceptions it synthesizes. In Cassirer, by contrast, Duhem's lesson is clearly present, a lesson that was later rediscovered following Quine's reflections in twentieth-century epistemology (Coniglione 2008a, p. 182). However, the primacy of the theoretical framework does not take on a dogmatic character; it does not lead to the loss of any discriminating role of experience, as happens in certain strands of twentieth-century epistemology that made the predominance of theory their banner, through the notion of *theory-ladenness*, as seen for instance in Kuhn and Feyerabend (Coniglione 2025). The critique of Baconian empiricism and its notion of the *experimentum crucis* is in fact always accompanied by an emphasis on the reciprocal interplay between theory and facts. What defines the "true objectivity" of physical concepts, according to Cassirer, are the physical consequences they generate – consequences that can be intersubjectively tested. This objectivity relies less on verification and more on what Popper would later describe as the decisive function of falsification.

However, Cassirer's thought can be more fully appreciated when viewed against the background of the theory of idealization developed by Nowak and the Poznań School. It was precisely this perspective that allowed Coniglione to revisit Cassirer's epistemology with new eyes and to discern within it an anticipation – albeit an unconscious one – of a model-based conception of science. This view was first advanced by the Poznań School (which Coniglione studied and explored in depth) and later taken up by other prominent scholars who frequently referred to it (Borbone 2023b). The connection is identified precisely in the concept of idealization, which lies at the core of Cassirer's philosophy and, at the same time, constitutes the epistemological heart of the Poznań School.

In the latter, use is made of the concept of an *idealizing assumption*, which is defined by the following propositional function: $p(x) = 0$, if and only if 0 denotes the minimum value of the magnitude p, and moreover, for every real object a, $p(a) \neq 0$. From this perspective, then, idealizing assertions are not satisfied by any real object because – much like Platonic ideas – they are not empirical in nature. Although an idealizing assumption describes an ideal rather than a real condition, it is nonetheless possible to move from idealized conditions to less idealized, and thus more realistic, ones through the procedure known as *concretization* (Nowak 1980; Borbone 2020, pp. 189–217), that is, the way in which the individual is reintroduced at the end of the cognitive process, descending from the ideality of concepts back to the concreteness of experience. Cassirer illustrates this by referring to the progression from Boyle–Mariotte's law to Gay-Lussac's law and finally to Van der Waals' equation of state, through the introduction of constants that allow for a better "approximation" to reality (Cassirer [1937] 1956, pp. 85–6). This is, in essence, the very same example employed within the idealizational conception of science to explain the procedure of concretization (Coniglione 2010², pp. 255–256).

Cassirer, who also possessed a strong sense of historical awareness, identifies the true scientific revolution brought about by Galileo and his experimental method precisely in the concept of idealization; in this regard, one can also discern a substantial convergence of perspectives with the theoretical framework developed by the Poznań School, which likewise emphasized the revolutionary import of Galileo's contribution (Nowak 1995). Coniglione effectively underscores this central feature of Cassirer's epistemological reflection, in which historical and systematic dimensions are inextricably intertwined: "Not surprisingly, Galilei is seen as the first to have understood such truth of knowledge when he had to imagine a completely isolated body, which is not a real object and which could never be found in nature" (Coniglione 2008a, p. 182). It is on this crucial point that a methodological bifurcation arises, giving way to two distinct thought-styles: on the one hand, the Aristotelian-Ptolemaic tradition, which asserts the inapplicability of the geometric-mathematical method to the realm of nature; on the other, the tradition already foreshadowed by Plato, based on the necessity of mathematizing the sensible world by framing reality within ideal models derived from ideas (Coniglione 2016).

The fact that, in Cassirer's conception, reality cannot simply be "mirrored" or synthesized by the subject, but is constituted through our theories, highlights the possibility of multiple "points of view" or perspectives, depending on the specific universe of discourse to which the disciplines engaged in the knowledge process belong. This idea receives one of its clearest conceptual illustrations in Cassirer's well-known example of the idol in a shrine, presented in his important epistemological work *Determinism and Indeterminism in Modern Physics* (1937). Cassirer writes:

> An idol revered in a shrine can be described according to purely scientific principles, and can be represented according to the concepts and categories of science. In this way it becomes a "piece of nature," subject like any other to physicochemical laws. Yet we know that with all these determinations we do not penetrate to its full meaning. The latter is not exhausted by the mere enumeration of scientific data. It demands other criteria, different in principle. No matter from how many viewpoints we may observe and analyze the marble as a natural object, the result

> will never divulge anything about its form and the beauty of this form, or about its significance in religious worship as an object of religious reverence.
>
> CASSIRER [1937] 1956, p. 206

Cassirer's functionalist approach also demonstrates its full potential when applied to the problem of complexity, as it allows for the coexistence of different theories which – each from its own standpoint and within its own symbolic system – offer a distinct perspectival view of the same phenomenon. This perspectivist insight has been further developed by Coniglione in his recent work on contemporary epistemology (Coniglione 2022a; 2022b; 2025) and is one he has also emphasized in his interpretation of Cassirer (Coniglione 2020b, pp. 15–16). Man, actively impresses upon external impressions a determinate form: it is precisely this *Wille zur Gestaltung* that thinkers such as Herder and von Humboldt identify in the essence of language, Schiller in art, and Kant in the structure of the theory of knowledge. In short, there are specific modes of production that refer to a kind of productivity which constitutes the distinctive feature of the human being. For these reasons, transcendental idealism, from the perspective of Cassirer's functionalism, "does not begin with any assertion about the nature and essence of the objects themselves; it begins with a critical investigation of the various modes of cognition by which the different classes of objects – the objects of experience, the objects of science, the objects of religious or metaphysical thought – are made accessible to us" (Cassirer [1936] 1979, p. 69). As Coniglione notes, "for Cassirer, even the simplest measurement presupposes theoretical principles and thus general functions of connection, formation, and coordination of concepts, so that the so-called 'datum' of perception must already be overcome and replaced with a conceptual symbol" (Coniglione 2008a, p. 183). Knowledge of concrete reality, therefore, is grounded in the cooperation of what Cassirer, in the concluding pages of his *Determinism and Indeterminism in Modern Physics*, refers to as "functions of the spirit" (*Funktionen des Geistes*); and it is only through the totality of these functions that reality can be comprehensively apprehended, certainly not through the absolutization of any single point of view.

This human capacity to transcend the sensible datum has found one of its most fruitful applications, according to Coniglione, in the domain of science, and in particular in the idealizational approach to scientific knowledge. Coniglione has insightfully identified this orientation within Cassirer's epistemological reflection, and by drawing upon the categorical framework of the Polish philosopher Leszek Nowak, he has succeeded in articulating the originality of Cassirer's position with remarkable clarity. Indeed, as

Coniglione notes in the conclusion of his essay on abstraction and function in Cassirer, the reappraisal of the modeling and idealizing dimension of empirical science, advanced by Nowak, as well as by epistemologists such as Ronald Giere, Nancy Cartwright, Craig Dilworth, and Evandro Agazzi, "has basically agreed with what Cassirer expressed so efficaciously, albeit in a style so profoundly continental that it proved hostile to the analytical mentality of most twentieth-century philosophers of science" (Coniglione 2008a, p. 188).

4 Conclusion: Some Critical Remarks

While I fully agree with the reappraisal of the idealizational approach to science as found in Cassirer's functionalist conception, I would nonetheless like to raise some doubts concerning the reasons identified by Coniglione for the lack of attention that epistemology has paid to Cassirer's functionalism. While Coniglione is certainly right in attributing this neglect, at least in part, to the different understanding of abstraction and concept formation that characterized twentieth-century epistemology – especially within the empiricist tradition – it does not seem to me that this *ignoratio* can be ascribed primarily to a general mistrust of Kantianism (cf. Mormann 2014). In fact, the philosophy of science that emerged in the early twentieth century – particularly under the influence of neo-positivism – did not entirely disregard Cassirer's reflections. As Jeremy Heis points out, "Cassirer's commitment to a philosophy of science that engaged with cutting edge science ran deep and was widely known. For example, as a letter from Hans Reichenbach makes clear, Cassirer was the only philosopher to sign onto a petition, composed by Reichenbach in 1931 and co-signed by Hilbert and Einstein, petitioning the German government to create a chair in the philosophy of science" (Heis 2014, p. 10).

It is also sufficient to recall the ongoing dialogue between the Breslau thinker and leading neo-positivists such as Moritz Schlick and Hans Reichenbach, a dialogue well documented in their correspondence with Cassirer himself (Ferrari 2015, pp. 261–284). Of course, Schlick, Reichenbach, and even Carnap, while recognizing and appreciating Cassirer's extensive scientific knowledge, firmly rejected those aspects of his philosophy related to Kant's synthetic *a priori*. Nevertheless, this does not justify speaking of a general *ignoratio* of Cassirer's epistemological contributions on the part of early twentieth-century philosophy of science. Consider, for instance, Carnap's *The Logical Structure of the World* (1928), in which Cassirer is mentioned several times, especially his *Substanzbegriff und Funktionsbegriff*. Carnap even expresses agreement with Cassirer on a central point: that the basic elements of a system of constitution, which positivism sought in the raw data of lived experience, are not sufficient on their own. As Carnap states, "order concepts, our basic relations, must be added" (Carnap [1928] 2003, p. 122). Moreover, Moritz Schlick, with whom Cassirer also maintained correspondence, wrote a rather extensive review of Cassirer's 1921 volume on Einstein (Schlick [1921] 1979, pp. 322–333). In short, despite the obvious differences, it cannot be argued that Cassirer's epistemology has been ignored by the protagonists of neo-positivism.

References

Borbone, G. (2016). The Concept of Idealization in Ernst Cassirer's Theory of Knowledge. *Analysis and Metaphysics*. 15, 88–109.

Borbone, G. (2019). *Pensieri al limite. Sostanza, funzione e idealizzazione in Cassirer e Husserl*. Naples: Diogene Edizioni.

Borbone, G. 2020. Lokale und globale Idealisierungen. Das Wissenschaftsmodell von Ernst Cassirer. *Deutsche Zeitschrift für Philosophie*, 68, 2, 189–217.

Borbone, G. (2022). La statua in un santuario. Ernst Cassirer e l'approccio funzionalista. In: E. Coco (ed.), *L'invenzione della realtà. Scienza, mito e immaginario nel dialogo tra psiche e mondo oggettivo. Una prospettiva filosofica*, pp. 71–82. Pisa: ETS.

Borbone, G. (2023a). *Forma e idea. L'interpretazione di Platone nella Scuola di Marburgo e in Ernst Cassirer*. Milan-Udine: Mimesis.

Borbone, G. (2023b). The Originality of Leszek Nowak's Philosophical and Epistemological Thought. *Organon F*, 30, 2, pp. 171–181.

Carnap, R. ([1928] 2003). *The Logical Structure of the World*, English Translation by R.A. George. Chicago and La Salle: Open Court.

Carnap, R. (1958). *Introduction to Symbolic Logic and its Applications*. New York: Dover.

Cassirer, E. ([1910] 1953). *Substance and Function*, English translation by W.C. Swabey and M.C. Swabey. In: Cassirer, E. 2003. *Substance and Function & Einstein's Theory of Relativity*. Mineola-New York: Dover Publications, Inc.

Cassirer, E. ([1916] 1975). *Freiheit und Form. Studien zur Deutschen Geistesgeschichte*. Darmstadt: Wissenschaftliche Buchgesellschaft.

Cassirer, E. ([1936] 1979). Critical Idealism as a Philosophy of Culture. In: Cassirer, E., *Symbol, Myth, and Culture. Essays and Lectures 1935–1945*, edited by D.P. Verene, pp. 64–91. New Haven and London: Yale University Press.

Cassirer, E. ([1937] 1956). *Determinism and Indeterminism in Modern Physics*, Translated by Theodor Benfey. London-Oxford: Yale University Press.

Cassirer, E. (2022). *Lectures on Ancient Philosophy*. Herausgegeben von G. Borbone. Hamburg: Meiner Verlag.

Coniglione, F. (1978). *La scienza impossibile. Dal popperismo alla critica del razionalismo*. Bologna: il Mulino.
Coniglione, F. (1983). Da Treviri a Poznan. Leszek Nowak ed il marxismo come scienza rigorosa. *Laboratorio*, 10/11, 20–76.
Coniglione, F. (1990a). *Realtà ed astrazione. Scuola polacca ed epistemologia post-positivista*. Catania: CUECM.
Coniglione, F. (1990b). Abstraction and Idealization in Hegel and Marx. In: J. Brzezinski, F. Coniglione, L. Nowak, and T.A.F. Kuipers, eds., *Idealization I: General Problems*, pp. 61–88. Amsterdam-Atlanta: Rodopi.
Coniglione, F. (1992a). Irracjonalizm epistemologiczny Paula Feyerabenda [Paul Feyerabend's Epistemological Irrationalism]. *Poznańskie Studia z Filozofii Nauki*. 12, 111–132.
Coniglione, F. (1992b). Scientific Philosophy and Marxism in Poland. In: F. Coniglione-R. Poli-J. Woleński (eds.), *Polish Scientific Philosophy. Poznań Studies in The Philosophy of the Sciences and the Humanities*, vol. 27, pp. 67–112. Amsterdam-Atlanta: Rodopi.
Coniglione, F. (1994). Filosofia e scienza in Jan Łukasiewicz. *Epistemologia*, 17/1, 73–100.
Coniglione, F. (1996a). Humanistic Interpretation between Hempel and Popper. In: A. Ziedler-Janizewska (ed.), *Epistemology and History. Poznań Studies in The Philosophy of the Sciences and the Humanities*, vol. 47, pp. 283–301. Amsterdam-Atlanta: Rodopi.
Coniglione, F. (1996b). *Nel segno della scienza. La filosofia polacca del Novecento*. Milan: Franco Angeli.
Coniglione, F. (1997). Astrazione e modelli di scientificità in Croce e Gentile. *Synaxis*, XV/2, 677–727.
Coniglione, F. (2004). Between Abstraction and Idealization: Scientific Practice and Philosophical Awareness. In: F. Coniglione-R. Poli-R. Rollinger (eds.). *Idealization XI: Historical Studies on Abstraction and Idealization. Poznań Studies in The Philosophy of the Sciences and the Humanities*, vol. 82, pp. 59–110. Amsterdam-Atlanta: Rodopi.
Coniglione, F. (2007). The Place of Polish Scientific Philosophy in the European Context. *Polish Journal of Philosophy*, I, 1, 7–27.
Coniglione, F. (2008a). Astrazione e funzione in Ernst Cassirer. In: M. Castellana et al. (2008). *Filosofia e storiografia. Studi in onore di Giovanni Papuli*. Vol. III.1. *L'età contemporanea*, pp. 165–188. Lecce: Congedo Editore.
Coniglione (2008b). *Popper addio. Dalla crisi dell'epistemologia alla fine del logos occidentale*. Acireale-Rome: Bonanno.
Coniglione, F. (2010). *Realtà e astrazione. Scuola polacca ed epistemologia post-positivista*. Acireale-Rome: Bonanno.
Coniglione, F. (2012). Il rapporto tra Marx ed Hegel alla luce del concetto di idealizzazione. In: G. Borbone (a cura di), *Scienza, marxismo e metafisica. Leszek Nowak e la scuola metodologica di Poznań*, pp. 23–74. Villasanta: Limina Mentis.

Coniglione, F. (2016). Galileo and Contemporary Epistemology. Do we still have something to learn from Galileo's 'methodological revolution'? *Przegląd Filozoficzny. Nowa Seria*, 25, 1, 123–139.

Coniglione, F. (2019). Lontano da Vienna, lontano da Leopoli. Ludwik Fleck: l'uomo giusto nel posto e nel tempo sbagliati. In: L. Fleck, *Stili di pensiero. La conoscenza scientifica come creazione sociale*, edited by F. Coniglione, pp. 7–93. Milan: Mimesis.

Coniglione, F. (2020). Conoscenza tacita: Ludwik Fleck nostro contemporaneo. In: A. Czajka-G. Cunico-E. Colagrossi (eds.), *Cento anni di filosofia e di cultura polacca*, pp. 235–251. Milan: Mimesis.

Coniglione, F. (2020b). Prefazione. In: E. Cassirer, *Determinismo e indeterminismo nella fisica moderna. Studi storici e sistematici sul problema della causalità*, edited by G. Borbone, pp. 7–17. Milan-Udine: Mimesis.

Coniglione, F. (2022a). Dal mito alla scienza e ritorno. Per una visione non imperialistica della conoscenza. In: E. Coco (Ed.), *L'invenzione della realtà. Scienza, mito e immaginario nel dialogo tra psiche e mondo oggettivo. Una prospettiva filosofica*, pp. 11–30. Pisa: Edizioni ETS.

Coniglione, F. (2022b). Complessità e pluralismo cognitivo. *Complessità*, n. 2, luglio–dicembre, 32–72.

Coniglione, F. (2023). Leszek Nowak, a Neglected Thinker. In: *Organon F*, 30/2. Special Issue. Author Meets Critics Symposium Discussion on *The Relevance of Models*, ed. by Krzysztof Brzechczyn, 130–136.

Coniglione, F. (2025). *Lontano da Popper. L'epistemologia post-positivista e le metamorfosi della razionalità scientifica*. Pisa: ETS.

Ferrari, M. (2015). Cassirer and the philosophy of science. In: N. De Warren-A. Staiti (Eds.), *New Approaches in Neo-Kantianism*, pp. 261–284. Cambridge: Cambridge University Press.

Friedman, M. (2011). *A Parting of the Ways. Carnap, Cassirer, and Heidegger*. Chicago-La Salle: Open Court.

Heis, J. (2014). Realism, functions, and the a priori: Ernst Cassirer's philosophy of science. *Studies in History and Philosophy of Science*, 48, 10–19.

Husserl, E. ([1900–1901] 1968). *Logische Untersuchungen*, Vol. 1. Tübingen: Niemeyer.

Husserl, E. ([1954], 1970). *The Crisis of European Sciences and Transcendental Phenomenology: An Introduction to Phenomenological Philosophy*. Evanston: Northwestern University Press.

Lombardo-Radice, L. (1982). *Istituzioni di algebra astratta*. Milan: Feltrinelli.

Mormann, T. (2014). Critical Idealism Revisited: Recent Work on Cassirer's Philosophy of Science. *Vienna Circle Institute Yearbook*, 6, 295–306.

Nowak, L. (1980). *The Structure of Idealization*. Dordrecht: Reidel.

Nowak, L. (1987). *Oltre Marx. Per un materialismo storico non-marxiano*, ed. by F. Coniglione. Rome: Armando.

Nowak, L. (1995). Remarks on the Nature of Galileo's Methodological Revolution. In: M. Kuokkanen, eds., *Idealization, Structuralism and Approximation. Poznań Studies in the Philosophy of the Sciences and the Humanities*, vol. 42, pp. 111–126. Amsterdam/Atlanta: Rodopi.

Schlick, M. ([1921] 1979). Critical or Empiricist Interpretation of Modern Physics. Remarks on Ernst Cassirer's *Einstein's Theory of Relativity*. In: Schlick, M. (1979). *Philosophical Papers*, vol. 1 (1919–1922), edited by H.L. Mulder and B.F.B. van der Velde-Schlick, pp. 322–333. Dordrecht: Reidel.

CHAPTER 4

The Development of the Theory of Revolution in Comparative Historical Sociology

An Attempt at a Methodological Analysis of Skocpol and Goldstone's Approaches to Revolution

Krzysztof Brzechczyn

Abstract

The subject matter of this article is an analysis of the relations between two theories of revolution – Theda Skocpol's and Jack A. Goldstone's – formed within the framework of comparative historical sociology (hereinafter, CHS). For that purpose, in the first section, I will briefly characterize that subdiscipline of science by discussing its basic research methods. In the second section I reconstruct of Skocpol's theory of revolution. In third section, I will outline the main directions of the developments of Skocpol's theory, including a detailed examination of Goldstone's theory of revolution. Finally, I will reconstruct the relations between these two theories in the notion apparatus of an idealizational theory of science.

Keywords

comparative historical sociology – idealization – modelling – revolution

1 Comparative Historical Sociology: Theoretical Assumptions and Explanatory Strategies

Comparative historical sociology (CHS) is an interdisciplinary enterprise intended to change, on the one hand, historians' idiographic approach and, on the other hand, sociologists' ahistorical approach to the studied social phenomena.[1] CHS begins from posing a problem, research questions, theories, an interpretive scheme, and then looks for material, not necessarily

1 With regard to the history of comparative historical sociology, see, for example: Smith (1991), Collins (1999). Smith dates that field back to the 18th century and mentions Montesquieu, Hume, Tocqueville, Marx, Durkheim, and Weber as its precursors.

 | HTTPS://DOI.ORG/10.1163/9789004756496-006

source material, which will provide the answer. In the research, a comparative method is used so that it is possible to propose and verify hypotheses. Dennis Smith characterizes that scientific discipline as follows:

> To oversimplify, historical sociology is the study of the past to find out how societies work and change. Some sociologists are "non-historical": empirically, they neglect the past; conceptually, they consider neither the time dimension of social life nor the historicity of the social structure. Similarly, some historians are "non-sociological": empirically, they neglect the way processes and structures vary between societies, and conceptually, they consider neither the general properties of processes and structures, nor their relationships to acts and events. By contrast, historical sociology is carried out by historians and sociologists who investigate the mutual interpenetration of past and present, events and processes, acting and structuration. They try to marry conceptual clarification, comparative generalization and empirical explorations.
>
> SMITH 1991, p. 3

In the light of the Smith's interpretation CHS combines some research methods used in sociology and history. These research methods can be divided into theoretical assumptions and explanatory strategies. Theoretical assumptions consist some general vision of human being and society and way of their explanation. These assumptions steer particular explanatory strategies that is a general way of constructing a theory of the studied phenomena. Within the framework of CHS, a full reconstruction of the theoretical assumptions can be difficult because there are many researchers working in the field and they assume only partially overlapping sets of research directives. This will be done in the case these historical sociologists who work in theory of revolution. Nevertheless, it seems that this set of theoretical assumptions can be characterized in the following way:

(i) a structural (and not voluntaristic) approach to social life which leads to interest in mass phenomena;
(ii) the approach of autonomism (and not social reductionism), which presupposes the mutual autonomy of the distinguished areas of social life: politics, economy, and culture, the economy being understood in materialist terms, politics – institutionalist ones, and culture – idealistic ones;
(iii) an approach in which one of the sources of the autonomy of the state organization with respect to the social-economic structure is deemed to be the international political system of which the given state is a part;
(iv) a globalist approach, where the relations between the internal and external factors for the given society are taken into consideration (and not

only the historical and social processes, with the assumption that the society is isolated).

The explanatory strategy in CHS can be reconstructed as follows:

(i) look for the purpose of explaining, for a sufficiently 'theory-genic' social phenomenon (for example, a revolution) in a selected – with respect to some criterion – set of societies;

(ii) with the use of the comparative method (its two basic varieties are looking for similarities and looking for contrasts), build a model (theory) of a social phenomenon which occurs in selected societies, such that it will only apply to the distinguished cases;[2]

(iii) try to transform that model – having reformulated it appropriately – into a model with a universal scope, which applies to analogous phenomena (for example, a revolution) which also occur in other societies which do not fulfill the original criterion of selection;

(iv) if you cannot use the constructed model directly for explaining phenomena, modify it by taking into account the factors which have been initially omitted from it

The purpose of this paper is to interpret relations between Skocpol and Goldstone's theories of revolution in the notion apparatus of the idealizational theory of science.[3] In the first step, I will reconstruct Skocpol's theory of revolution and present the some of its extensions. In the next step I will present Goldstone' theory of revolution. The theoretical puzzle is relation between mentioned above two theories: do they silently assume one general model of revolution or Goldstone's contribution is development of Skocpol's theory? By the use of the notion apparatus of idealizational theory of science I analyze relations between these two theories and answer this question. Results of my analysis will be enriched by an application of Franco Coniglione's typology of scientific attitudes.

2 Theda Skocpol's Theory of Social Revolution: an Attempt at a Reconstruction

According to Theda Skocpol, revolutions are sudden transformations of political and economic structures of societies, accompanied by general social

2 For a reconstruction of these methods within the notion apparatus of idealizational theory of science, see: Brzechczyn (2016).

3 The first approach to the method of idealization was presented in L. Nowak (1970; 1971); for a mature reconstruction of that concept, see Nowak 1974; 1977, in English: Nowak (1980) and Nowak and Nowakowa (2000), on Nowak's theory and his school see: Coniglione (2010).

mobilization and grassroots mass movements. She considers revolutions in France (1789) Russia (1919) and China (1948). Revolutions in these countries were results of political crisis caused by geopolitical pressure and peasants revolts strengthened by the support of oppositional political elites (Skocpol 1979, p. 4; detail analysis of Skocpol's work, see Brzechczyn 2022).

2.1 *Modernization in the International System of States*

Modernization is understood as a supranational process of economic development, characterized by industrialization, dissemination of wage labor and trade by means of market mechanisms. Those changes, in turn, lead to changes in other spheres of social life, especially in political and legal institutions, and they contribute to the creation of a modern, bureaucratic state. It led to the creation of an international economic market which included capitalist countries as well as those which, for various reasons, remained outside of the scope of such modernizing processes.

Such a modernization was taking place in an international system of competing countries. Europe was specific in that no imperial state controlled the whole area of Europe and its overseas territories. The competition between England and France led to the construction of the first colonial empire by the former country. England owed its advantageous position in the international rivalry to its earlier start on the path of capitalist development. Finally, in the 19th century, the competition among the European countries led to their conquest of the whole world. The mechanism of international competition overlapped with the modernization processes although it was independent from them.

2.2 *The State Power in Agrarian Bureaucracy Societies*

According to Skocpol autonomy of state stems from its roots in the international political system and in the social and economic social order (Skocpol 1979, p. 32). Normally, the state fulfills two basic functions: it competes with other states and maintains internal order. For those purposes, states usually support the existing class order in the economy. However, during a crisis, the state can also make decisions incongruent with the economic interests of the (economically) dominant classes and only motivated by its own political interest.

Skocpol considers the situation of disparate interests of the state and the economically dominant classes with the use of the example of agrarian bureaucracy societies. She describes that social system as follows:

(i) the semi-bureaucratic state and the landowners' class cooperate;
(ii) the landowners' class exerts limited administrative power over the peasantry, on behalf of the state;

(iii) the central authority collects taxes from the peasants and avails itself of the prerogative to demand free labor from them, via the owners' class and directly;
(iv) for the landowners' class, the support of the state which has at its disposal the means of coercion is necessary for collecting rent and/or deriving other benefits from peasant labor without resistance on their part (Skocpol 1976, pp. 178–179).

In the agrarian bureaucratic system, the state was a proto-bureaucratic structure. That meant that administrative and military power functioned within the institutional framework of an absolute monarchy. In that kind of a state, offices were partially specialized and clerks were subjected to hierarchical control which, however, did not extend to all their functions. The functions of, on the one hand, holding an office and fulfilling the related duties and, on the other hand, managing one's assets became only partially independent from each other. The bureaucratization of that type of social system was incomplete, so the state organization could not directly control the social relations in the agrarian sector of the economy. The agrarian bureaucratic state performed best in the tasks set for it: waging wars abroad, supervising the society to maintain the general order, appropriating social and economic resources by recruiting to the army, and taxing the peasants or businesses (Skocpol 1979, p. 48).

A market economy often developed in that type of society, but capitalist relations of production did not dominate the economy. Consequently, the emerging trade and bourgeoisie classes were interrelated on many levels with the landowners' class and the proto-bureaucratic state with which they lived – in Skocpol's words – in a symbiosis.

Let us now turn to the relations between the landowners' class and the state. On the one hand, the state and the landowners' class were partners in the exploitation of the peasantry. On the other hand, they competed with each other. Skocpol describes it as follows:

> at the local level, the dominant classes could not defend themselves against peasant revolts – they had to depend, although in varying degrees, on centralized monarchical states which supported their class position and prerogatives.
>
> SKOCPOL 1979, p. 49

However, according to her:

> The imperial states and the landed classes [...] were also competitors in controlling the manpower of the peasantry and in appropriating surpluses

> from the agrarian-commercial economies. Monarchs were interested in appropriating increased resources from society and channeling them efficiently into military aggrandizement or state-sponsored and centrally controlled economic development. Thus, the economic interests of the landed upper classes were in part obstacles to be overcome; for the landed classes were primarily interested either in preventing increased state appropriation or in using state offices to siphon off revenues in ways that would reinforce the domestic socio-economic status quo.
>
> SKOCPOL 1979, p. 49

In agrarian bureaucracy societies representatives of the owners held lower offices in the state administration what causes that the state did not exercise its right to interfere in the life of local peasant communities – it passed the administrative functions over to the landowners, into their "private capacities," (Skocpol 1976, p. 184) who then used that political power to increase their private assets (Skocpol 1979, p. 49). Such a state of affairs had the following consequences:

> to the extent the dominant-class members gained the capacity for self-conscious collective organization within the higher levels of the existing imperial state structures, they could obstruct the monarchical undertakings which threatened their income. Such obstruction could culminate in unconscious challenges to the autocratic political authority – and at the same time, it could have the quite unintended effect of destroying the administrative and military integrity of the imperial state itself.
>
> SKOCPOL 1979, p. 49

That caused constant conflicts between the state elite, which strove for greater centralization of power, and the owners' classes, which wanted to maintain the *status quo* with its guarantee of privileges.

In agrarian bureaucratic societies, the peasantry was the basic layer of population. Peasants were exploited by the state, the landed gentry, and the burgeoning bourgeoisie-trade classes. In such societies, peasant communities always had reasons for economic dissatisfaction, and the threat of an outbreak of a revolt was endemic. (Skocpol 1979, pp. 48, p. 115). The peasants' ability to rebel depended on three factors: (1) the degree to which their communities were organized and the manner of their organization; (2) the degree of their autonomy from the direct, everyday supervision by the lords and their apparatus; and (3) the weakening of state sanctions against peasant revolts (Skocpol 1979, p. 115).

As we can see, a revolution could break out given a certain level of organization of peasant communities. The first two conditions depended on the structure of agricultural ownership – when it obstructs organization of peasant communities, it prevented an outbreak of a revolt. That happened on great estates where the owners supervised the peasants directly or when the owners' class controlled the administrative and executive power at the local level. The third condition crucial for an outbreak of a revolution was determined by the relationship between the state and the landed gentry. When the international situation of the state was stable, the authorities and the landed gentry were partners in the exploitation of the peasantry. A crisis only began when:

> the old regime states became unable to meet the challenges of evolving international situation. Monarchical authorities were subjected to new threats or to intensified competition from more economically developed powers abroad. And they were constrained or checked in their responses by the institutionalized relationship of the autocratic state organization to the landed upper classes and the agrarian economies. Caught in cross-pressures between domestic class structures and international exigencies, the autocracies and their centralized administrations and armies broke apart, opening the way for socio-revolutionary transformations spearheaded by revolts from below.
>
> SKOCPOL 1979, p. 47

2.3 *The Mechanism of a Victorious Revolution*

In the international domain, agrarian bureaucracies encounter challenges which they can only rise up to when they begin to modernize. That means building an effective taxation system, spending more on weapons, and modernizing the economy so that it can achieve the goals of the state. However, because of the landowners' resistance, central power has limited ability to carry out internal, peaceful reforms. In particular, it cannot rationalize the taxation system. Because of the political structure of agrarian bureaucracies, the authorities also have restricted access to financial and human resources, which are controlled by the landowners' classes. Moreover, the government has a limited ability to mobilize the society because such an action would be against the interests of the landowners' class, which acts as both the state and an employer with regard to the peasantry.

Meeting the challenges of international competition requires an effective economy capable of financing the costs of the maintenance of the army, administration, and diplomatic service, so the state solves the crisis by modernizing the economic and political structures. The reforms weaken the relationships between the state structure and the owners' class. The state no longer protects the interests of the aristocracy. This, in turn, contributes to the possibility of outbreaks of peasant revolts because:

> Agricultural regimes featuring large estates worked by serfs or laborers tend to be inimical to peasant rebellion – witness the East Elbian Junker regime – but the reason is not that serfs and landless laborers are economically poor, rather that they are subject to close and constant supervision and discipline by landlords or their agents.
>
> SKOCPOL 1976, p. 193

Finally, the fourth factor was the influence of the radical political elites which existed on the margins of societies. According to Skocpol:

> Although peasant insurrections played a decisive role in each of the great historical social revolutions, nevertheless an exclusive focus on peasants – or on the peasant situation in agrarian bureaucracies – cannot provide a complete explanation for the occurrence of social revolutions.
>
> SKOCPOL 1976, p. 201

Such a complete explanation can only be obtained by taking into account the influence of marginal political elites. Generally speaking, members of those social groups came from the intelligentsia. They had the knowledge and skills which qualified them for service in the state apparatus, but they did not have traditional sources of prestige (such as noble birth) which would give them the right to independently wield power. That is why it was natural for the elites of the opposition to demand equality, abolition of status-group privileges, and introduction of political democracy. French marginal political elites originated chiefly from the third status group. The main leaders of national assemblies of the French Revolution belonged to a professional group called notables – they were lawyers and lower-rank officials. They were marginal in that they held lower offices in the administrative structure of the pre-revolutionary order or came from provincial towns. The marginality of members of political elites of the opposition in Russia was determined by their social origins or belonging to radical political groups. Bolsheviks were often graduates of universities, which prepared them for a career in the state. However, during their studies, they joined radical political movements, and they became professional revolutionists instead of professional officials. In China, the opposition against the system comprised educated members of the Kuomintang and the Communist Party of China.

The participation of marginal political elites in the revolution helped translate the practical postulates of mass peasant revolts into the language of political postulates which transformed the structure of the state and society; it also built the internal communication between the naturally dispersed and decentralized peasant communities. Moreover, those political elites which opposed the old order – Jacobins, communists – had at their disposal a universal ideology which encouraged the cooperation of people from various social environments and facilitated political mobilization of the masses. That ideology was a holistic world view which justified using unlimited (that is, immoral) means to realize limited (that is, moral in the light of a given ideology) aims. As noted by Skocpol, although revolutionary ideology was a necessary component of social revolutions, their course and results cannot be explained by referring to the content of the ideology – instead, one should look at its functions.

To sum up Skocpol's theory, we could say that for a victorious social revolution to take place, four analyzed factors must be at play: external modernizing pressure, a political crisis caused by the weakening of the relations between the authorities and the owners, peasant revolts, and the influence of marginal political elites. External modernizing pressure worsens the international position of agrarian bureaucracy states. State elites try to meet the challenges of the international situation by carrying out reforms which weaken the relationship between the political power and the owners' class. That, in turn, causes an internal political crisis which is aggravated by peasant rebellions made possible precisely by the reduction of the scope of the authorities' political control over the peasantry. When the political elites of the opposition form an alliance with grassroots peasant revolts, the actions of the peasantry can be coordinated on the national scale, and a separate vision of social order can be expressed. Consequently, the state authorities are overthrown, and new state order is built.

For Skocpol, a revolution only ends when the revolutionary elites have created an alternative political order. In her view, the significant features of that order are discernible in all three analyzed cases. Peasant revolutionary movements changed the agrarian structure of the states in which they took place. Centralized and completely bureaucratized nation states replaced autocratic and proto-bureaucratic monarchies. In particular societies, the pre-revolutionary class of great owners lost its privileged position to the benefit of lower classes and of the heretofore marginal political elites, who took over all power in post-revolutionary states.

3 Extensions of Skocpol's Theory: an Attempt at an Interpretation

Skocpol's approach belonged to so-called third generation theory of revolution (Goldstone, Grinin, Korotayev 2023, pp. 44–45; see also Chodak 2012). On the other hand, her mode of research could be viewed as an implementation of directives (i) and (ii) assumed in the explanatory strategy of CHS. Initially, the author maintained that her theory only applied to the three analyzed cases: Russia, France, and China. Later, however, she used the theory to explain the Iranian revolution. Since she could not apply the model directly, she had to modify the original model and, thereby, implement directive (iv) in the explanatory strategy. As she was pondering upon the phenomenon of the revolution in Iran, Skocpol expanded the set of explanatory factors by adding cultural and ideological variables to it, replaced the factor of international pressure with beneficial international conditions, and substituted the category of the rentier state for the category of the agrarian bureaucracy society (Skocpol 1982, pp. 265–283).

Other authors repeated those alterations as they developed Skocpol's initial approach. I will only mention sample developments here. As she reflected on the necessary conditions for effective political reforms, Elen K. Trimberger distinguished two types of agrarian bureaucracy societies. In the first, state power is dominated by the civil sector and in the second. by the military. Trimberger claims that top-down modernization has the greatest chance of success in the second type of agrarian bureaucracies (Trimberger 1978). Elsewhere, in analyzing the 1910 Mexican revolution, Walter Goldfrank substituted the factor of international pressure with the factor of "favorable geopolitical conditions" (Goldfrank 1979). By contrast, Rebecca S.K. Lee (2002) considered the influence of sudden market developments on the destabilization of state power.

A detailed analysis of the developments mentioned above, however, would require more space than is available in this article. I will thus restrict myself to an analysis of another development, that of Jack A. Goldstone (1991), whose approach belonged to the fourth generation of theory or revolution (Goldstone, Grinin, Korotayev 2023, pp. 46–47). Goldstone noted that state breakdowns were cyclical in early modernity – we could call them waves of state breakdowns. The first wave of political crises took place in 1550–1660, after which state structures remained stable for a hundred years until the next crisis of state structures, lasting from 1770 to 1850 (Goldstone 1991, p. 17).

The crisis of early modern statehood is a manifestation of a phenomenon called the "general crisis of the 17th century" in European historiography (Goldstone 1991, p. 17), but Goldstone views the answers given in traditional historiography as insufficient because they are too Eurocentric, and the phenomenon of state structure breakdowns occurred in the whole Eurasian region, not only in Europe. That prompts Goldstone to ask two questions:

> Why did the near simultaneous state breakdowns in major Asian empires resemble, in their origins, the state breakdowns in Europe? And to the extent that they had similar origins, why were the long-term outcomes so different, leading to that divergence in development commonly known as the rise of West'?
>
> GOLDSTONE 1991, p. 4

At the beginning of his analysis, Goldstone reviews Skocpol's model. He notes that both advanced capitalist states – like the Netherlands (1566–1648) and England (1640–1688) – and their less developed competitors underwent the same period of social and political disruptions. Moreover, revolutions broke out in countries which were not at war at the time (Goldstone 1991, pp. 20–21). On the other hand, Goldstone views Skocpol's theory as partly accurate because the preparation for international conflicts and participation in them can undermine the finances of a state. Still, Goldstone emphasizes that military conflicts alone are not a key factor in a financial crisis as they can be accidental, while the observed cyclical patterns of the state breakdowns in early modernity were not. According to Goldstone, the growing military costs of international rivalry resulted mainly from cyclical growths of populations (Goldstone 1991, p. 21).

4 Population Growth and State Crisis

Goldstone begins his reflections with the empirical contestation that since 1500, Eurasia experienced a noticeable population growth. That growth has not been properly accounted for yet. Some possible explanations include climate warming and the disappearance of epidemics. That trend remained more or less stable until 1650, and the population of the societies of particular Eurasian countries doubled in that time. About 1650, the population growth on the Eurasian scale stopped as diseases and epidemics returned. The next population growth happened in 1730–1860, in agrarian societies which, according to Goldstone: "were not equipped to deal with the impact of the steady growth of population that then began throughout northern Eurasia, eventually amounting to population increases in excess of the productivity gains of the land." (Goldstone 1991, p. 24). Those societies had the following characteristics:

(i) an agrarian economic base;
(ii) a hereditary ruler and administrative officials of the territorial state who were in a conflict of sorts with the local, religious, and intellectual elites;
(iii) an educated elite who remained in and out of the administrative apparatus and who had its own interests;
(iv) a people's class in the urban and rural areas which had overlapping loyalties and was exploited by the elites and the state;
(v) a relatively well-functioning internal market ensuring access to goods to various social actors (Goldstone 1991, p. 7).

In that type of society, we can distinguish between marginal social groups and groups which belong to the societal core. Marginal social groups "face some sort of boundary conditions, such as peasants who are seeking to gain new lands, or younger sons of elite families who are seeking new elite positions" (Goldstone 1991, p. 33). In the conditions of population growth, the number of marginal groups increases, and there is more internal competition among them. In the terms and conditions of stagnant economy, even a small population growth leads to a substantial increase of marginal social groups. Without a place in the social structure, those groups pose a serious threat to social stability. Goldstone describes the mechanism of the imbalanced population growth which leads to a revolution as follows. The increasing competition for resources (land) leads to higher prices of land and food and, consequently, to greater costs of the maintenance of the state apparatus, especially the army, which results in a financial crisis. The state responds by raising taxes, which, in turn, leads to a conflict with the social elites which are also feeling the pressures of the population growth because it increases the number of people desiring a position in the state system, while the financial crisis prevents the achievement of such aspirations. The rivalry inside the social elites becomes fiercer. The situation of the peasantry worsens, and people migrate from villages to cities. The real wages of hired workers in cities fall. The breakdown of the state is preceded by its crisis, which does not have to end with a revolution. Goldstone distinguishes the following variants of social development:

(i) if the elites control the state, the crisis of the state can end with peaceful reforms;
(ii) if the key elites are united and capable of achieving the desired changes without triggering social mobilization, the crisis of the state can be ended by means of a coup d'état;
(iii) if the elites are alienated from the state and have appropriate resources at their disposal, the crisis of the state will lead to a revolt of the elites and will exacerbate the conflicts among them; and

(iv) if the elites are alienated from the state, have at their disposal appropriate resources, and are capable of mobilizing the society, the crisis of the state will lead to a civil war.

Goldstone believes that a revolution will break out

> only when a society simultaneously experiences three kinds of difficulties: (1) a state financial crisis, brought on by a growing imbalance between the revenues a government can securely raise and the obligation and tasks it faces; (2) severe elite divisions, including both alienation from the state and intra-elite conflicts, brought on by increasing insecurity and competition for elite positions; and (3) a high potential for mobilizing popular groups, brought on by rising grievances (e.g. regarding high rents or law wages) and social patterns that assist or predispose popular group to action (e.g., large numbers of youth in the population, increasingly autonomous rural villages, growing concentration of workers in weakly administered cities.
>
> GOLDSTONE 1991, pp. xii–xiv

In that approach:

> The causes of revolutions and major rebellions therefore need not be sought solely among sudden events. Such events can be considered as "triggers" or "releasers," of pent-up social forces, but they are not the fundamental causes. Indeed, such "releasing" events are themselves generally the result of cumulating social pressures. A long-term cumulative factor, such as population increase, can therefore readily lead to a sudden event. What matters is whether the existing social and political institutions are flexible enough to move easily in response to such pressures.
>
> GOLDSTONE 1991, pp. 35–36

The first trend of population growth, in 1500–1650, led to a wave of crises of state organizations. The first of the series was the English Revolution in 1640–1688. Then, there were the Fronde in France and the Catalan revolts in Spain. In the first half of the 17th century, the Ottoman empire and Imperial China suffered similar political crises (overcome by a change of the dynasty).

The second population growth, in 1730–1860, caused a second wave of state crises, starting with the French Revolution of 1789. Later revolutionary periods were 1830–1834 (France, Belgium, Poland) and 1847–1852 (France, Germany, Italy, Austria, Poland, Hungary, and the Czech and Slovak lands). The first half of the 19th century was also a period of crisis for the Turkish and Chinese states, as exemplified by the separation of Greece from Turkey (1821) and the Taipei uprising in China.

5 The Development of the Theory of Revolution as a Case of (Methodologically) Unconscious Use of the Method of Idealization: Meta-Methodological Reflections

Thus, we have obtained two models of revolution. In Skocpol's model, social revolution (*SR*) was the result of the coincidence of four factors. External modernizing pressure (*Ip*) worsened the international position of agrarian bureaucracy states. State elites tried to meet the challenges of the international situation by carrying out reforms which weakened the relationship between the political power and the owners' class. The reforms brought about an internal political crisis (*Pc*), which facilitated the outbreak and spread of peasant revolts (*Pr*). The influence of the political elites of the opposition (*Oe*) on grassroots peasant revolts makes it possible to coordinate the actions of the peasantry on the national scale and to express a separate vision of social order. Consequently, the state authorities were overthrown, and new state order was built. That model can be illustrated with the use of the following scheme:

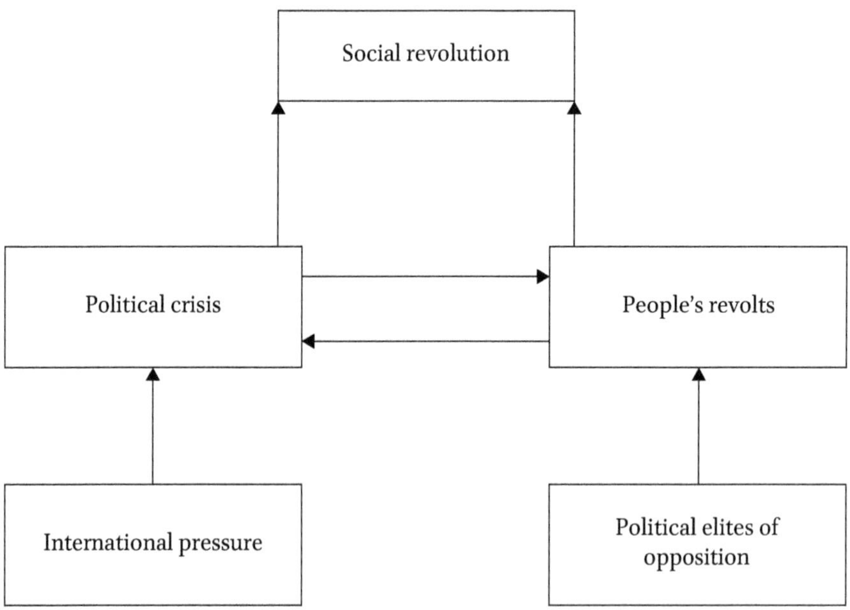

FIGURE 4.1 Skocpol's model of revolution. *SR* – social revolution, *Ip* – international pressure, *Pc* – political crisis, *Pr* – people's revolts, *Oe* – political elites of opposition, arrows symbolize influence.

In Goldstone's model, a revolution is also a result of four factors. Unbalanced population growth (*Gp*) leads to a political crisis of the state (*Pc*), with the following symptoms: the worsening of the economic situation of the lower class, increasing rivalry among the social elites, and decreasing resources at the disposal of the state. The state elites try to rise to the challenge by raising taxes and increasing other burdens on the population. The worsening of the economic situation of the people's masses in the urban and rural areas leads to outbreaks of people's rebellions (*Pr*). Those can be successful as the now fiercely competing social elites (*Ce*) take over leadership of the protests and turn local uprisings into a revolution capable of overthrowing state structures. Goldstone's model can be illustrated with the use of the following scheme:

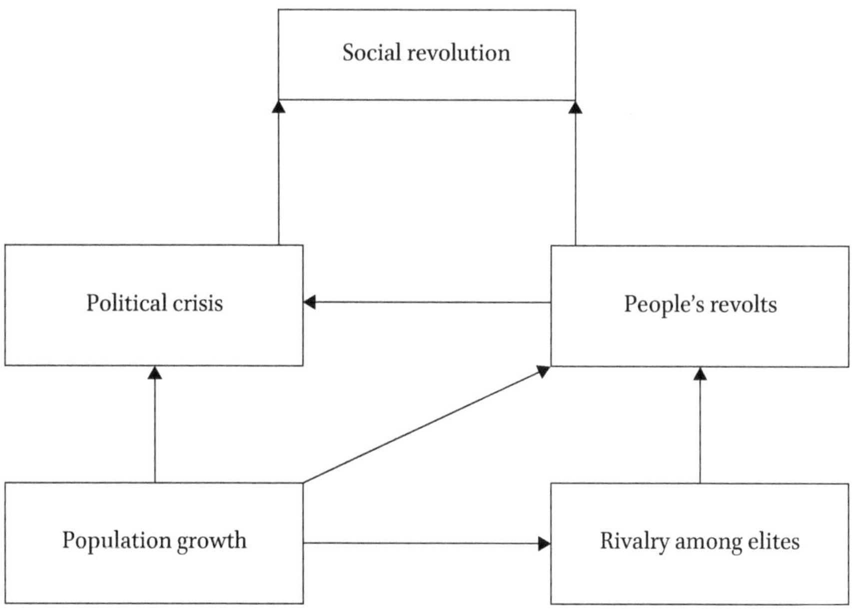

FIGURE 4.2 Goldstone's model of revolution. Explanations: *SR* – social revolution, *Pc* – political crisis, *Pr* – people's revolts, Gp – population growth, *Ce* – rivalry among elites, arrows symbolize influence.

The remaining question is what the nature of the relationship between the two models, Skocpol's and Goldstone's. The possible variants are:

(i) Goldstone's model is a concretization of Skocpol's more basic model;
(ii) Skocpol's model is a concretization of Goldstone's more basic model;
(iii) both Goldstone and Skocpol silently assume a third, the most idealized model.

Randall Collins was the first researcher who pointed out the similarities between the two approaches. He claimed that "The core of Goldstone's model is essentially the core of Skocpol's theory." (Collins 1999, p. 22). It would mean that Goldstone's proposal is a sort of a development of Skocpol's approach. Collins, then, would be a proponent of variant (i). An additional argument for that would be the biographic characteristic of Goldstone as Skocpol's intellectual student and a continuer of her work (Collins 1999, p. 21). Elsewhere, Collins clearly states that "Skocpol and Goldstone provide variants on the same model although disagree on some points." (Collins 1999, p. 23) and looks at "Goldstone's work as an extension of a core model of a state breakdown that can be activated in various ways" (Collins 1999, p. 25). On that basis, we can surmise that Collins espouses variant (iii) which he characterizes as follows in a different place:

> Skocpol and Jack Goldstone have shown that class conflict alone is insufficient for revolution and must be accompanied by a fiscal crisis of the state as well as a split between state elites and property owners over the repair of state finances. Skocpol's work marks the paradigm shift to what might be called the state-breakdown theory of revolution. Skocpol and Goldstone elaborate a common model of state breakdown into alternative chains of causes further back focusing respectively on geopolitical strains and population-induced price changes.
>
> COLLINS 1999, p. 3

The author, however, does not go beyond those initial distinctions and does not elaborate either on the status of the core model he postulates or on the methodological status of the "forking" of that more basic model into Skocpol's and Goldstone's alternative approaches.

For a more precise characterization of the relations between the two theories of revolution, one would need a methodological meta-language which would be more general than that of the authors and their interpreter. For that purpose – a reconstruction of those relations – I will use the conceptual apparatus of the idealizational theory of science.

Let us assume that there are the following factors in the essential structure of the phenomenon of social revolution (*SR*) assumed by both authors:

- political crisis (*Pc*),
- people's revolts (*Pr*),
- international pressure (*Ip*),
- political elites of the opposition (*Oe*),
- population growth (*Gp*),
- competition between elites (C*e*).

Consequently, we can represent Skocpol's theory with the use of the following formula:[4]

If *Ce* = 0 and *Gp* = 0, then SR = *Pc*, *Pr*, *Ip*, *Oe*

Goldstone's model can be represented as follows:

If *Oe* = 0 and *Ip* = 0, then SR = *Pc*, *Pr*, *Gp*, *Ce*

There are two factors in the consequent of both conditional statements: *Pc* and *Pr*. That makes it possible to reconstruct the first model of the theory silently assumed by both authors. It would look as follows:

If *Wp* = 0 and *Pm* = 0, and *Re* = 0, and *Oe* =0 and *Ip* = 0, and *Oe* = 0, then SR = *Pc*, *Pr*

In the most basic model, a revolution would be just a coincidence of two factors: a political crisis of the state and grassroots peasant (people's) revolts. Both Skocpol and Goldstone assume such an idealized basic model. The complete essential structure of the studied phenomenon of social revolution (SR) could be presented as follows:

SR: (2) *Pc*, *Pr*,
(1) *Pc*, *Pr*, *Ip*, *Gp*,
(0) *Pc*, *Pr*, *Ip*, *Gp*, *Ce*, *Oe*

At the deepest (2) essentiality level, there are two main factors: *Pc*, and *Pr*. Differences only appear at the level of concretization. According to Skocpol, the factor to be considered in the first order is *Ip*, and the second – *Oe*. According to Goldstone, the first factor to be considered is *Gp*, and the second – *Ce*.

It is worth noting that if we paraphrase Collins's reconstruction in the language of the idealizational theory of science, there are three factors at the deepest essentiality level: *Pr* – peasant revolts, *Pc* – the political crisis of the state caused by financial problems, and *Ce* – the rivalry of elites. However, at the deepest level in my reconstruction, there are two factors: *Pc* – the political

4 The formulas in this part of the article are only presented for demonstration purposes. In particular, I omit the distinction between direct and indirect factors leading to revolutions. Assigning value zero to the factor of population growth would not mean that the number of people in the given society does not increase at all but that that increase does not destabilize the social situation.

crisis and *Pr* – peasant revolts, while the rivalry of elites belongs to the surface (0) essentiality level, which means that it is a secondary factor.

Skocpol's development of the theory – which we can call, following Collins, *geopolitical* – can be presented as follows:

If $Ip = 0$ and $Oe = 0$, and $Gp = 0$ and $Ce = 0$, then $SR = Pc, Pr$
If $Ip \neq 0$ and $Oe = 0$, and $Gp = 0$ and $Ce = 0$, then $SR = Pc, Pr, Ip$
If $Ip \neq 0$ and $Oe \neq 0$, and $Gp = 0$ and $Ce = 0$, then $SR = Pc, Pr, Ip, Oe$

Goldstone's development of the theory – in Collins' interpretation that may be named *ecological* – can be presented in this way:

If $Ip = 0$ and $Oe = 0$, and $Gp = 0$ and $Ce = 0$, then $SR =Pc, Pr$
If $Ip = 0$ and $Oe = 0$, and $Gp \neq 0$ and $Ce = 0$, then $SR = Pc, Pr, Gp$
If $Ip = 0$ and $Oe = 0$, and $Gp \neq 0$ and $Ce \neq 0$, then $SR = Pc, Pr, Gp, Ce$

I would like to characterize the methodological self-consciousness of the authors of the said concepts. In the idealizational theory of science, the scheme of cognitive activity consists of the following elements (Nowak 1977, pp. 9–10, 53–54, 105–107):

(i) assuming a particular ontological perspective, one decides on the stratification rules which determine what kind of factors is significant (in particular, the main factors) for the studied phenomenon and what kind of dependencies there are between the distinguished main factors and the studied phenomenon;
(ii) the set of factors significant for the studied phenomenon – which constitute its essential structure – is determined;
(iii) a hierarchy is formed within the essential structure of the studied phenomenon, taking into account the main and secondary factors;
(iv) assumptions which eliminate the influence of factors considered to be secondary is introduced, and a hypothesis of how the studied phenomenon depends on the main factor(s) is formed;
(v) the assumptions are removed, and the initial hypothesis is modified to show how the studied phenomenon depends on the main and secondary factors.

It is hard to find the said directives applied in the research practice of the authors of the two discussed theories of revolution. First of all, Skocpol and Goldstone do not assume a uniform ontological perspective which would allow the stratification of the studied factors. For instance, Skocpol mentions the Marxian theory of class struggle and the theories of political conflict and globalization as sources of inspiration. Because of the theoretical eclecticism, the essential structure of the studied phenomenon (in this case, the set of factors leading to an outbreak of a revolution) is not clearly defined, and the influences of the factors are not hierarchized. The order of concretization is disturbed – it could be seen as arbitrary and accidental although it is possible to reconstruct a more idealized model of revolution to which both authors refer. Consequently, despite the fact that the authors use the term "model," one can hardly view the theory of revolution developed within the framework of CHS as an example of conscious use of the method of idealization. However, the theory of revolution, may be reconstructed within that framework. How could we explain that paradox? Francesco Coniglione (2004, p. 104) distinguished four basic scientific attitudes:

(i) scientists who do not use idealization in scientific practice and are not methodologically conscious of the possibility of using that method;
(ii) scientists who use idealization in scientific practice, but are not methodologically conscious of what they are doing (or they misinterpret that);
(iii) scientists who use idealization in scientific practice and are methodologically conscious of what they are doing;
(iv) scientists who do not use idealization in scientific practice, but are methodologically conscious of the application of that method (in other fields of knowledge, not in science).

It seems that the interpreted authors (in particular Skocpol) represent the second type scientific attitude. That is corroborated by Skocpol's methodological declarations that the theory she builds is to be an example of the application of John Stuart Mill's method of agreement and method of difference. Meanwhile, critics (Burawoy 1989, pp. 763, 771, Nichols, 1986, pp. 164, 177) of Skocpol have noticed that her methodological declarations are not congruent with her research practice, which Skocpol herself has indirectly admitted. (Skocpol 1986, pp. 188–190; see analysis of Skocpol's methodological standpoint: Brzechczyn 2016, pp. 187–191).

By contrast, Nowak distinguished two stages of the development of science: *empirical-gathering* and *theoretical-explanatory* (Nowak 1977, pp. 54–61). In the empirical-gathering phase, facts are systematically observed and generalizations are made based on those observations; the basic cognitive method is that of induction. In the second phase, theoretical-explanatory, the collected empirical material is explained. The basic cognitive method is idealization, which leads to the construction of very sophisticated theoretical models. In various disciplines of science, that methodological breakthrough took place at different times: in physics, it was at the beginning of the 17th century, because of Galileo, in the 19th century in biology because of Darwin, and in economy, because of Marx, in linguistics in 20th century, because of Chomsky. However, in Nowak's words:

> the transition has not taken place in all empirical sciences yet. Numerous humanities are still in the first phase of their development – they are collecting the empirical data which will make it possible to appropriately explain the areas of reality which are studied in those fields, and the explaining will only be possible with the use of a theory, the methodology of which will not be a set of empirical generalizations.
>
> NOWAK 1977, p. 61

This transition is marked by a change of methodological awareness of scientists. According to Coniglione:

> It is only with mature science, whose threshold was crossed by the work of Galileo, that a new scientific methodology based on idealizational procedures, and the systematic elaboration of ideal models was introduced; and it is only in the last few decades that epistemology has devised suitable conceptual tools for this change in scientific practice.
>
> CONIGLIONE 2004, p. 105

Be that as it may, it seems that the division of science into two phases is too coarse and that it should be complemented with a temporary *pre-theoretical-explanatory* stage in which science already has theoretical-explanatory goals but the basic cognitive method – declared but not practiced – is still the induction method.

While characterizing the stages mentioned above with the use of Coniglione's classification of scientific attitudes, one could say that during the empirical-gathering phase, the first type of scientific approach (scientists do not use idealization in scientific practice and are even not methodologically aware of the possibility of using that method) dominates. In the theoretical-explanatory phase, the third attitude: scientists use idealization in scientific practice and are methodologically aware of what they are doing) is the most common. However, in the transitional, *pre-theoretical-explanatory*, phase, the second type of a scientific attitude (scientists use idealization in scientific practice, but are not methodologically aware of what they are doing, with all the cognitive ramifications of that fact) prevails.

It is worth asking, then, if we could not also find such (unconscious) precursors of the use of idealization in other disciplines of science. In economy, they could be Adam Smith[5] and David Ricardo, in biology – Charles Darwin,[6] and in 19th-century linguistics – the creators of the theory of the existence of the Proto-Indo-European language. That topic merits a discussion in a separate article. It seems, then, that currently that interim, pre-theoretical-explanatory, stage of the development of science is represented by comparative historical sociology or, more precisely, the theory of revolution developed in that field.

References

Brzechczyn, K. (2016). *Strategies of Comparative Analysis in Comparative Historical Sociology. An Attempt at Explication within Notion Apparatus of Idealizational Theory of Science*. In: G. Borbone, K. Brzechczyn (eds.) *Idealization XIV: Models in Science. Poznań Studies in the Philosophy of the Sciences and the Humanities*, vol. 108, pp. 184–203. Leiden-Boston: Brill/ Rodopi.

Brzechczyn, K. (2022). *A Victorious Revolution and a Lost Modernization: An Attempt to Paraphrase Theda Skocpol's Theory of Social Revolution in the Conceptual Apparatus of Non-Marxian Historical Materialism*. In: K. Brzechczyn (ed.) *Non-Marxian Historical Materialism: Reconstructions and Comparisons. Poznań Studies in the Philosophy of the Sciences and the Humanities*, vol. 120, pp. 161–194. Leiden-Boston: Brill.

Burawoy, M. (1989) *Two Methods in Search of Science. Skocpol versus Trotsky*, "Theory and Society", 1989, no. 18, 759–805.

Chodak, J. (2012). *Teorie rewolucji w naukach społecznych* [Theories of Revolution in Social Science]. Lublin: UMCS.

Chodorowski, J. (1980). *Adam Smith*. Wrocław: Ossolineum.

Collins, R. (1999). *Macrohistory. Essays in Sociology of the Long Run*, Stanford: Stanford University Press.

Coniglione, F. (2004) *Between Abstraction and Idealization: Scientific Practice and Philosophical Awareness*. In: F. Coniglione, R. Poli, R. Rolinger (eds.), *Idealization XI: Historical Studies on Abstraction and Idealization. Poznań Studies in the Philosophy of the Sciences and the Humanities*, vol. 82, pp. 59–110. Amsterdam-Atlanta: Rodopi.

Coniglione, F. (2010). *Realtà astrazione. Scuola polacca ed epistemologia post-positivista*, Acireale-Rome: Bonanno.

---

5 See the chapter on the method in: Chodorowski (1980).

6 Krzysztof Łastowski's reconstruction can be interpreted in that way; see Łastowski 1987, pp. 24–27; for a different approach, see: Nowak 2004, 13–97.

Goldfrank, W.L. (1979). *Theories of Revolution and Revolution without Theory. The Case of Mexico*, "Theory and Society", vol. 7, 135–165.

Goldstone, J. (1991). *Revolution and Rebellion in the Early Modern World*, Berkeley: Routlegde.

Goldstone, J, L. Grinin, and A. Korotayev (2023). *The Phenomenon and Theories of Revolution*. In: J. Goldstone, L. Grinin, and A. Korotayev (eds.) *Handbook of Revolutions in the 21st Century The New Waves of Revolutions, and the Causes and Effects of Disruptive Political Change*, pp. 37–68. Cham: Springer.

Łastowski, K. (1987). *Rozwój teorii ewolucji* [The Development of the Theory of Evolution], Poznań: UAM Press.

Lee, R.S.K. (2002). *Alternative Route to State Breakdown: Toward an Integrated Model of Territorial Disintegration*, "Sociological Theory", vol. 20, 1–23.

Nichols, E. (1986) *Skocpol on Revolution: Comparative Analysis vs. Historical Conjuncture*, "Comparative Social Research", vol. 9, 163–186.

Nowak, L. (1971) *U podstaw Marksowskiej metodologii nauk* [The Foundations of Marxian Methodology of Science], Poznań: PWN.

Nowak, L. (1977). *Wstęp do idealizacyjnej teorii nauki* [The Introduction to the Idealizational Theory of Science], Warsaw: PWN.

Nowak, L. (1980). *The Structure of Idealization. Towards a Systematic Interpretation of the Marxian Idea of Science*. Dordrecht: Reidel.

Nowak, L. (2004). *O metodologii Karola Darwina* [On Charles Darwin's Methodology]. In: K. Łastowski (ed.), *Teoria i metoda w biologii ewolucyjnej*, pp. 13–97. Poznań. Poznań: Zysk i S-ka.

Nowak, L., I. Nowakowa (2000). *The Richness of Idealization. Poznań Studies in the Philosophy of the Sciences and the Humanities*, vol. 69. Amsterdam-New York: Rodopi.

Skocpol, T. (1976). *France, Russia, China: A Structural Analysis of Social Revolutions*, "Comparative Studies in Society and History," no. 2, 175–210.

Skocpol, T. (1979). *States and Social Revolutions*, Cambridge: Cambridge University Press.

Skocpol, T. (1982). *Rentier State and Shi'a Islam in the Iranian Revolution*, "Theory and Society", vol. 11, 265–283.

Skocpol, Th. (1986). *Analyzing Causal Configurations in History: A Rejoinder to Nicholas*, "Comparative Social Research", vol. 9, 187–194.

Smith, D. (1991). *The Rise of Historical Sociology*, Cambridge: Polity Press.

Trimberger, E.K. (1978). *Revolution from Above. Military Bureaucrats and Development in Japan, Turkey, Egypt and Peru*, New Brunswick: Transaction Books.

CHAPTER 5

The University "System" Has Long Been "Dead": Who Were the "Perpetrators" of Its Demise?

Francesco Alfieri

Abstract

In this rather unusual paper, I will be dealing both in personal terms and more generally with the state of universities in Italy, with the decline that has afflicted them and which they are still undergoing. I aim to look both at their internal dynamics and practices, such as how they go about recruiting, financing themselves, as well as their structural and external aspects which are related to their social and universal mission. My contribution is intended to be more than just a snapshot of how an entire "system" has imploded. I also hope it will help launch a discussion on how to bring about reform. I thought it would be fitting, to share these remarks with Professor Coniglione in his *Festschrift*, confident as I am in the power of debate, because I feel that we have a duty to "rebuild" or rather to "refound" our universities in an attempt to save them from the death they have been approaching for some time.

Keywords

system – university – professors – illegality – competitions

1 Addressing the "Issue"

When my friend and colleague Giacomo Borbone invited me to contribute to the *Festschrift* in honor of Professor Francesco Coniglione, I was eager to accept. I have long been familiar with the Professor's scholarly and academic work, and I found one of his books particularly enlightening (Coniglione 2011). This book offers a reflection on the evolution of the Italian university system, focusing on the reforms that have shaped it – reforms that can be traced back to those introduced in 1999 by University Minister Ortensio Zecchino, though they were largely conceived by Luigi Berlinguer, who served as Minister of University and Scientific Research from 1996 to 1998. The ideas expressed in

 | HTTPS://DOI.ORG/10.1163/9789004756496-007

that book resonated deeply with my own Heideggerian perspective on the university and have, over the years, led me to reflect on a thorny issue – one that many have come to take for granted. Indeed, there has been a marked increase in the level of dissatisfaction among scholars and university professionals with regard to reforms implemented over the past two decades. This is evidenced by the substantial body of studies and essays on the subject, as well as the lively critical activity of the website *Roars – Return on Academic Research*, to which Francesco Coniglione contributed numerous articles, at least until 2021. Across all these contributions, a nearly unanimous assessment emerges: rather than improving the quality and efficiency of the Italian university system, these reforms have lowered academic standards and increasingly entangled scholarly activity in bureaucratic procedures, diverting precious time away from both research and effective teaching. As a result, even those who played a role in shaping these reforms have come to perceive the university's decline – a decline they themselves helped bring about. Indeed, it often seems that the very people lamenting the current state of the university are the ones who led the institution to the edge of a precipice – and even pushed it over – through the unchecked power games they've been allowed to play for years.

What I aim to do here is share a few reflections on the Italian university system and the kinds of policies that have brought it to this point. I do so in a spirit of dialogue and collaboration, in the hope that a constructive debate might emerge – one in which each of us can contribute during the crucial "reconstruction" phase. In my view, this phase has become absolutely essential, given how much of the university system has already been eroded.

Rather than offering a mere critique, I believe we must honestly and clearly examine how we arrived at this situation in the first place.

2 The University Is "Dead"

Currently we find ourselves in the situation of wanting to "save appearances" so that all the knowledge produced within academic institutions seems merely to sustain the mechanisms of their own production and reproduction, without much interest in cultivating the search for foundational knowledge, which seems to have been banished from our university as if it were obsolete. This is particularly noticeable with regard to one of the issues that has been most debated in these years and has been the subject of multiple legislative interventions, namely that of the recruitment and advancement of university teachers. Indeed, to put it bluntly, how can we even aspire to fair and meaningful academic competitions when recruitment is primarily based on personal connections and kinship? How can we tolerate selection processes that completely disregard candidates' actual academic merit? More importantly, what future can we envision for our universities if the present is so compromised by considerations of personal power? Individuals are often forced to take part in a hollow performance, appealing to committees whose members have themselves been co-opted by the system. I would even go so far as to suggest that committee members are often selected precisely because they are known to comply with the established, and deeply flawed, recruitment practices. Such a "system" can never be expected to generate any kind of future as long as it continues to erode the present and sabotage any hope for meaningful change.

One might even be willing to tolerate these abuses – if only we had the courage to stop pretending that our academic competitions are "public," when in reality they are entirely "private" and devoid of merit. However, I would like to clarify exactly what I mean, to avoid any risk of being misunderstood – even though those familiar with the system likely already grasp my point. I will address this complex issue in the final part of this paper, as it deserves a dedicated and systematic analysis.

The foundational principle of our institutions of higher learning has traditionally been their "autonomy" – at least until the late 1980s. It is on this basis that universities were expected to shape their internal policies and manage the recruitment of faculty and staff. However, this notion of independence from external powers – particularly from political influence – has been completely overturned. The original spirit of the university, rooted in the pursuit of knowledge and the advancement of human civilization, has been subordinated to dynamics of clientelism and personal influence that are entirely incompatible with its mission. Tying the pursuit of free and independent research to a political agenda represents an even deeper hypocrisy, as it sustains the illusion that recruitment is based on academic merit – whether in research or teaching – when, in fact, it is not.

We must seriously reconsider what a university is meant to be, and liberate it from a "system" that has subordinated it to external priorities, stripping it of any genuine autonomy. Unfortunately, the commodification of academic careers – where professorships are acquired under the influence of entrenched university "barons" engaged for years in a system of mutual exchange – is a disheartening spectacle. Consider the countless so-called "open" competitions funded by public money (from the state budget), which are treated as if they were the personal property of whichever professor happens to control them. In reality, private cooptation has become the only selection mechanism in use. All of this deeply undermines the ideals of free research, open competition, and meritocracy – principles that now seem to have been shipwrecked on the reefs of internal power struggles, which ultimately decide who receives an academic post and when. Worse still, these corrupt practices are destroying the university system and the future of higher education itself. The very purpose of a public, competitive selection process is nullified when the committee fails to impartially select the most qualified candidate for the benefit of all. Furthermore, interference from political elites and economic maneuvering exerts a harmful influence on the final selection of candidates. The result is a near-total collapse: the university has not only lost its direction in the pursuit of knowledge, but also its higher calling – the "spiritual mission" of shaping minds and educating the individuals who will contribute to civil society. This shameful state of affairs – fueled in part by legislation that Coniglione (2019) aptly described as "criminogenic" in relation to the co-optation and recruitment mechanisms prevailing in academia – has contaminated nearly all public universities (with the exception of private institutions governed by different principles), fostering a widespread network of corruption involving nearly everyone involved in this system.

The university is obviously an institution structured from the outside, meaning it reflects the social and collective life in which we all participate. However, what is scarcely convincing is precisely this "structure" and the ways in which it is managed, which have little to do with true "collegiality" or even "legality" in its proper sense. To clarify what is meant here by corruption of the "university system," one can cite the typical case experienced by many researchers and scholars who, after years of work and research in institutions outside the Italian public university system – often spent in an exhausting condition of precariousness with temporary contracts and research grants – thought they could gain access to it as lecturers and professors by participating in the "public" competitions announced by the various universities. As always happens in such situations, the candidates begin, perhaps for the umpteenth time, the exhausting bureaucratic process of compiling and submitting their publications to be evaluated by the examining committee. They are convinced that their publications will be judged by a committee that would carefully examine them to assess their scientific quality and originality. But in the meantime, the "rules of the game" have been changed: the judging committee was only asking for a simple list of the candidate's publications and there was no need to submit paper material. So, the members of the committee wouldn't be reading anything, they only wanted to know what the candidate had published over the years. It is not difficult to realize how, in this way, competitions turn into a substantial fraud, in which the favored candidates are those "known" to the committee, regardless of the scientific value of the works submitted by the individual competitors. And when one asks a university professor who is part of the system whether these competitions are truly fair and based on equal footing for all participants, the response is usually either a telling silence or unconvincing justifications. Thus, it becomes easy to understand that these competitions were planned and designed in advance with the goal of ensuring the victory of a pre-selected candidate. Often, it is enough to analyze who the members of the committee are to understand that it is a deception organized down to the finest detail – a true *Machenschaft*.

This is what I was personally able to observe at the university where I was working: when I realized – by exercising a bit of cunning and initiative – that it was easy to know in advance the names of those predestined to win a competition announced there, I decided to take action, aware that this was a duty I had to carry out to the end, even though I was fully aware of the risks involved. I decided to raise the issue with the vice-rector, a respected priest and theologian whom I trusted deeply. When he was shown the list of those expected to win the competitions, he was visibly embarrassed and told me I should immediately bring the matter to the attention of the Rector, who, he assured me, had the authority to put an end to a practice that had by then become entrenched at our university. I was confident that this meeting with the Rector would resolve the issue. As so often happens in such situations, your interlocutor puts on a great show of listening attentively, suggesting that *he too has suffered similar injustices and promising that he will ensure nothing like this ever happens again – because, after all, it goes against Christian values and basic honesty*. What exactly these guiding "Christian principles" are, I prefer not to speculate on – for the sake of my own dignity.

To make a long story short, my meeting with the Rector made it clear that absolutely nothing was going to change. That's when I decided to write a letter to the Bishop of Rome, the Argentine Jorge Mario Bergoglio, given that my university falls under his jurisdiction. It was the final step I had set for myself – to denounce the "university system" of which I was personally a part. So I wrote to him in 2018, but he never replied. In the meantime, the competitions proceeded as planned, and every candidate whose name I had shown to the Rector ended up winning. The rigging of the competitions produced exactly the results that had already been decided by those pulling the strings of power.

That what I have described is not an isolated case can be confirmed by a well-known instance of academic corruption that attracted media attention, legal proceedings, and investigative journalism: the case of the historian Giambattista Scirè, who in 2011 took part in a competition for a fixed-term research position in Contemporary History at the University of Catania. Despite his qualifications (including a PhD in the subject) and numerous publications, the position was awarded to another candidate who lacked the specific credentials, having graduated in a different field. Believing his exclusion to be unjust, Scirè appealed to the Regional Administrative Court (TAR) of Catania and then to the Council of Administrative Justice (the Sicilian equivalent of the Council of State), both of which found that the competition had been manipulated through "an illogical and irrational decision," and ordered the university to compensate him. Furthermore, the Court of Catania convicted the three committee members at first instance for abuse of office (although they were later acquitted due to the statute of limitations and, following the 2020 amendment to Article 323 of the Penal Code introduced by the government, because the conduct was no longer considered a crime). This experience led Scirè, in November 2017, to found the association "Transparency and Merit. The University We Want" (https://www.trasparenzaemerito.org), which is committed to promoting transparency and merit within the Italian university system and hosts numerous testimonies from researchers who have denounced similar abuses, by Scirè himself documented (Scirè 2021).

At this point, one might ask: what lesson can be drawn from these experiences and many others like them? What did I personally gain from my denunciation? At first glance, it might seem that I achieved nothing – but in truth, I gained my freedom and dignity by refusing to be complicit in the "academic system," a mechanism that is impossible to fight but can at least be rejected by anyone with a sense of honesty and justice, simply by refusing to take part in it. A refusal that, a few months after the winners of the competition were announced, was further sanctioned by a letter I received from the Dean of my faculty: "Dear Professor Alfieri, following the Faculty Council meeting on March 21 last, I am informing you that your course was one of those that will not be renewed next academic year." Just a few lines to let me know they were getting rid of an inconvenient person like me.

It is clear, therefore, that for the "university system" it is far easier to remove "inconvenient" individuals – even at the cost of silencing the courses they teach, regardless of their value or popularity – than to address the structural problems of a flawed and, in many cases, unjust system of academic selection. And this attitude reveals just how little importance is now placed, within the contemporary university, on the genuine pursuit of knowledge, replaced by *Machenschaft*, often disguised as a Christianity of convenience that is more concerned with saving appearances by *slogans* of convenience, aimed at concealing a system of power that has nothing divine or authentically Christian about it.

The manipulation of public competitions, academic careers distributed by a "baronial" elite, political interference, and the involvement of certain factions within the Church in determining who obtains a university position – and when – all contribute to a deeply compromised academic system. Added to this are the buying, selling, and bartering of professorships, further exacerbated by the financial difficulties faced by many universities and their growing dependence on external funding sources, who in turn demand a say in how knowledge is produced and by whom. These dynamics have a chilling effect on scholars, who are increasingly pressured to produce “marketable” knowledge at the service of a dominant political class, while becoming subservient to the economic powers on which their very survival depends. The result is the rise of a utilitarian form of thinking that leaves no space for what it truly means to pursue free and independent research. The experience I briefly recounted is not an isolated incident. I observed the same phenomenon in other public universities, where so-called “public” competitions were little more than a formality – an empty showcase in which everything had already been predetermined. The entire process was orchestrated to ensure the victory of an “internal” candidate, for whose benefit the competition had been staged from the outset. All of this reflects, in microcosm, what is taking place across broader social life – beginning with political and religious institutions, which have become centers of power where abuses and covert maneuvering are carried out by whoever happens to be the puppet master at the time.

One cannot remain silent in the face of such travesties. The task before us is to “think”: the philosophical spirit must be nurtured by the responsibility to *write and express oneself in the first person*. Few have dared to go, even a little, beyond simply quoting and commenting on the opinions of others. Suffice it to say that in Italy we have always had best *historians of philosophy* in the world, many of whom have founded schools of thought and held prestigious academic positions, but very few *philosophers in the true sense of the word*.

The decline plaguing our universities is also a reflection of the country’s broader economic crisis, following the 2008 financial crunch, and the resulting steady and gradual reduction in public funding. This has forced academic institutions into a constant race to secure new sources of funding, under the illusion that doing so will somehow restore financial balance, which has also been compromised as a result of mismanagement of university budgets. There has been an increasingly frantic search for private sponsorships and so-called “non-state” sources of funding, which often require the inclusion of researchers of one’s own liking, but which have the advantage of being at no cost to the university, thereby aggravating what we have described to be one of the most problematic aspects of academic recruitment: no longer the simple hegemony of university “barons,” since in this way economic considerations are become to be more important than merit in the allocation of research positions. The promise of funding has become more decisive than scientific excellence. This situation illustrates the ease with which, in times of crisis, economic criteria can be overemphasized, undermining the role of the scientific community in evaluating and mentoring young researchers and ultimately eroding their function of safeguarding free research and fair competition.

3 The Role of the Intellectual "Is Not Dead"

What has been said above acquires even greater meaning when we reflect on the role that the intellectual continues to play today – a role whose function has by no means disappeared. We can therefore rightly affirm that it is "not dead." And in effect, my interpretation of the university system is not intended merely as an analysis of questionable situations and events that have recurred over the years, but above all as an attempt to redefine the rightful place of the intellectual, and more importantly, to highlight the role of culture in the pursuit of truth – and as a unique opportunity for meaningful dialogue, where the results of one's research are continuously subjected to scrutiny and verification. This process aims to overcome the narrow specialization typical of certain research fields, which are often organized into isolated niches entrenched in self-referential positions. To liberate scientific research means to consider its results within a perspective of mutual exchange: I offer what I have discovered so that a culture of service may emerge – one capable of providing tools accessible and useful to all.

Research, of course, is a valuable tool in helping us to see the world more deeply and to seek the full meaning of our human experience. In this dwelling in the world, the central place belongs to the human who strives to understand the deep disquiet within. I interpret this disquiet as a continuous propensity to find one's rightful place in the world – a positive interpretation of the term. This disquiet does not imply a frantic search for answers or the uncritical adoption of others' experiences, but rather responds to a desire to ask meaningful questions – the only kind of questions that lie at the origin of authentic inquiry, free from prejudice and thus resistant to manipulation.

For this reason, the intellectual cannot afford not to be an educator: speaking to consciences means helping both oneself and others to become fully aware, so that thought may be united with action – or rather, with the exercise of reason. *At the heart of every reflection lies an implicit requirement: the decision to think*. Indeed, to refuse to think – to avoid reflecting in order to better evaluate and discern – means becoming someone who has lost the sense of time and memory, someone constantly inclined to act without remembering, because he has never truly considered the meaning of their actions. Culture can evolve into knowledge, and education can, in turn, lead to an active life, precisely because it embraces the risk of deep thinking. From this perspective, the human being gains inner stability and resists being overwhelmed by the conditionings shaped by the logic of a time stripped of meaning.

I have often asked myself whether the sense of precariousness and the loss of reference points we are currently experiencing – even if we cannot precisely identify their origin – might somehow be contained. The problem is that this crisis, with all its effects – having no easily identifiable causes – becomes far more dangerous because there are no limits to its reach. That is why it is essential to return to the roots of thought, to transcend the appearance that gives rise to this feeling of uncertainty and transience. We must reclaim the act of thinking, lest we lose or squander our ability to construct a narrative marked by the unique imprint of our individuality. Building a system of thought that helps us resist the fear of losing ourselves means avoiding being reduced to a *manipulable product* in a society that increasingly resembles a "spectacle" (Debord 1967).

There are those who, when faced with this choice – whether to think independently or remain within the random forms of *common thought* – understand that their irreducible individuality must fully resonate with a movement of thought directed *toward* and *with* others. And so, if they are unafraid to be a solitary voice, if they choose to challenge the clichés of dominant ideologies, it is inevitable that they will pay the price of being misunderstood and ostracized. This is a kind of thinking deeply infused with life, constantly engaged with a crucial question: *What role is still left for me to play in all of this?* In this way, life and thought become an inseparable whole, with no boundary between them.

That boundary becomes insurmountable when reflection and research are transformed into a profession entrenched in the obsession with pure thought – disconnected from the concrete reality of things. The vibrant disquiet of thought is what keeps the human being alive, through the creativity of action; it allows them to seize the moment and step into the scene, no longer content to remain a passive observer of historical events. The intellectual chooses to experience these events from within, in search of a meaning that can be shared in a space free from models of domination.

4 Is It Possible to "Rebuild" the University by Removing "Obstacles"?

The role of the intellectual, as I have sought to briefly outline, is not threatened solely by universities – whether pontifical, "Catholic," or state institutions – but also by other state-run bodies of advanced education and scientific research.

Among these, the National Research Council (CNR) may be cited as an example, firmly entrenched in the defense of its own 'autonomy.' How can situations bordering on illegality, such as the one we have alluded to, be brought to an end? Or rather, what measures are needed to prevent such blatant abuses from being carried out daily under the helpless gaze of the public?

In my opinion, a serious reform is needed, one that will put an end, once and for all, to this "inadequate" (and at times even illegal) form of autonomy, which functions as a kind of private fortress within which academic barons are free to determine recruitment procedures at will. Based on my experience of numerous competitions across various universities, I have found that in every instance, the process was carefully and meticulously overseen by the professor organizing the competition, as well as by the three committee members tasked with evaluating the candidates. On the surface, everything appears to run smoothly. However, the first cracks begin to show when the evaluations of candidates invited to the oral interview are released, this being the final, decisive stage that determines the outcome. Is it merely a coincidence – or a well-established practice – that from the assessments issued before the oral interview, it is immediately clear who the winner will be, namely the insider candidate for whom the entire competition was designed? That the entire process, including the evaluations, was tailored ad hoc, and that the committee's role was simply to rubber-stamp a preordained result?

The fundamental problem is that, in order to participate in the competitions announced by individual universities, candidates must first obtain – through a convoluted and exhausting bureaucratic process – the "National Scientific Accreditation," required for access academic positions. But what is the point of obtaining this accreditation if the local competitions – necessary for formal hiring by individual universities – are in fact predetermined, usually in favor of a candidate from the same university, despite being formally public and therefore supposedly "open"? Accreditation was originally intended as a national oversight tool for procedures which, after the most recent sweeping reforms, had become completely "localized." Although the accreditation process is often quite "generous," awarding the title even to scholars without academic "protection," the real problem is that many who have earned the title correctly and on merit still fail to gain access to academia, presumably because they have not found a "patron saint" within the university offering the position during the second, local phase of the competition. Moreover, if we consider that the funds used by the professor who opens the position are public resources – not private property – and that the whole purpose is to tailor the competition to select an internal candidate for whom the position was actually created, then the entire accreditation process becomes a completely pointless exercise. Which leaves us with a fundamental question: who is responsible for ensuring that these competitions are conducted properly? Who monitors the work of the selection committees within the individual universities, which seem to have planned everything in advance, even before they have examined the candidates' CVs?

It often happens that many candidates travel from one university to another to participate in competitions, sometimes having the "luck" of encountering the same "personalities" repeatedly sitting on selection committees, with the tacit task of approving whatever the chairman – usually appointed by the university offering the position – had already decided. At times, some of the more prominent members of these committees may appear to be upright and incorruptible individuals, especially if they belong to the Church or to organizations such as "Communion and Liberation." But this is not the case: the demands of the academic corporation outweigh any moral obligation, leading them to practice a kind of "religion" that has little to do with the Christian principles they so loudly proclaim on solemn occasions. The result is the recruitment of individuals who do not stand out for their scientific merit, but rather for their loyalty to a specific academic or religious faction. It is not difficult to imagine the decline produced by such practices, which afflicts many of our universities.

At the very least, we could do away with the hypocrisy of calling these competitions "public." If universities were willing to admit that their competitions are in fact "private," and that entry into their ranks is by mere co-optation – without the pretense of "open" competition – then they could then pursue their intrigues with a degree of legitimacy. But since this does not happen, what we are left with amounts to a form of unlawful abuse, carried out regularly in plain sight and without anyone able to stop it.

No serious university reform has ever been conceived or pursued since Minister Mariastella Gelmini implemented her reform, which systematically applied the principle of university "autonomy," granting each individual institution the power to draft its own Statute. Although this reform was originally intended to "clip the wings" of academic barons, it ultimately gave rise to a long series of illegitimate practices. One must therefore ask: has anyone ever truly wanted a serious reform that would introduce open competition and a minimum level of transparency into the conduct of so-called public competitions?

If we paid closer attention to the signs of the times and to what is happening before our eyes, we would see how university enrollment continues to decline, and how our institutions increasingly struggle to obtain public funding. This alone would suffice to demonstrate the urgent need for reform, what I would even call a "cleansing." Such a reform should aim to revive genuine research, grounded in the pursuit of knowledge free from the chains of patronage and clientelism, one that fosters a society of true thinkers capable of thinking critically and helping others to break free from their bondage to ignorance and dependence. But how can this ever happen when the majority of academic publications today serve merely as a means to secure a position or achieve a career milestone? A quick glance at the CVs of many full professors often reveals that their genuinely scientific publications ceased once they obtained a professorship. Too often, their works are published only because they are financed by the authors themselves, as no serious publisher is willing to invest in their research. Once the coveted academic title is secured, departmental funds are frequently used to continue publishing books of little value, thus, public funds are used for personal ends. But is there any committee tasked with evaluating the actual scientific merit of these works and issuing a judgment as to whether they deserve to be published at the public's expense?

•••

For the sake of brevity and readability, I have only addressed a few of the many common situations in our universities, in order to shed light on how certain professors – often seen as the "leading figures" of academic life – have corrupted our institutions, usurping and stripping them of the dignity of the *universitas*, once conceived as a workshop of ideas and a driving force behind impartial research and foundational knowledge, free from ulterior motives.

In a spirit of "shared responsibility," I would like to offer these reflections to Professor Francesco Coniglione, who has greatly contributed, through his writings, to critical discourse on the university. I trust that these thoughts may help spark a broader debate, one that leads each of us to ask a crucial question: What role can I play, alongside like-minded colleagues, in realizing a serious university reform? A reform capable of eliminating – or at least isolating from the academic community – those who represent an obstacle to change, and who, whether openly or covertly, resist any meaningful transformation.

References

Coniglione, F. (2011). *Maledetta Università. Fantasie e realtà sul sistema della ricerca in Italia*. Trapani: Crispino di Girolamo.

Coniglione, F. (2012). "Unicuique suum. La nuova mappa del potere nei Consigli di Amministrazione delle università italiane", in Roars – Return on Academic Research, https://www.roars.it/unicuique-suum-la-nuova-mappa-del-potere-nei-consigli-di-amministrazione-delle-universita-italiane/.

Coniglione, F. (2019). "Lo si pensa ma non lo si dice: le verità scomode sull'università", in *Il Protagora*, a. XLVI, gennaio–dicembre 2019, sesta serie, n. 31–32, 345–357.

Debord, G. (1967). *La Société du spectacle*. Paris: Éditions Buchet-Chastel.

Scirè, G. (2021). *Mala università. Privilegi baronali, cattiva gestione, concorsi truccati. I casi e le storie*. Rome: Chiarelettere.

CHAPTER 6

The Catania MIRRORS Project and the AI Transparency Problem

Thomas Nickles

Abstract

The MIRRORS project at the University of Catania in 2008–2009 focused on science, technology, and society issues related to the European Commission's declared ambition to make the European Union (EU) the most advanced "knowledge society" in the world. Progress toward this goal would improve the rate of basic and applied scientific research success and hence boost the European economy. It would also employ up-to-date knowledge to make the EU the most politically and scientifically transparent and hence most democratic region of the world. While there are many reasons for the failure to meet these ambitious goals by the stated deadline of 2010, this chapter focuses on specific later developments emblematic of the manner in which technological advance compromises efforts to achieve long-range goals, even shorter-range goals. Today, only a few years later, ongoing breakthroughs in artificial intelligence are the most visible form of innovation. While these advances aid democratic access to information in the form of increased access by non-experts, they also exacerbate transparency problems and hence problems of detection, regulation, and agency. The new AI models are so complex that no one understands in detail how they arrive at particular answers. Even the experts refer to them as "black boxes" and the mode of tuning them a "dark art." While vastly more powerful and more "intelligent" than their predecessors, their intelligence is rather alien to us; and the power that we now give the new AIs outruns their intelligence. Moreover, the newest, generative, large language models are achieving autonomous agency in their own right in ways even more obscure. AI's great aid to democracy is also a great threat.

Keywords

Transparent Knowledge Society – creativity – innovation and regulation – human and machine agency – generative AI – black box/alien reasoning/alignment problems

 | HTTPS://DOI.ORG/10.1163/9789004756496-008

1 Introduction: What Was the MIRRORS Project?

MIRRORS was directed by Professor Francesco Coniglione[1] at the University of Catania, Sicily, with the assistance of several young Catania colleagues such as Salvatore Vasta, Enrico Viola, and Giacomo Borbone as well as international contributors, including Eleonora Montuschi and the present author. The MIRRORS project, "Monitoring Ideas Regarding Research Organizations and Reasons in Science," was funded by the European Commission under its ambitious program to make Europe the most advanced knowledge society (Coniglione 2010, ch. 1). The MIRRORS project consisted of a series of meetings and conferences over two years, 2008 and 2009, and produced a report and essays in two books (Coniglione 2009; 2010).

What is a knowledge-based society? The idea was to reduce today's tension over the increasing role of arcane expertise in a democratic society, thus placing democracy on a more solid knowledge foundation. Supposedly, this could be done by making European society more thoroughly evidence-based, by making knowledge and its validation more transparent to the lay public, and hence by giving ordinary citizens greater participation, thus more agency, in decision-making.

The problem lies in the fact that a knowledge-based society runs the risk of becoming *less* transparent to its citizens for two main reasons: first, due to the ever-growing influence of experts, who possess knowledge that is far removed from what the average citizen can understand; and second, because of the increasing weight of so-called "arcane expertise" – specifically, expertise that shifts away from viewing knowledge and skill as purely logical or mathematical, towards a model grounded in wisdom and competence (Collins and Evans 2007, p. 23).

The idea behind the entire MIRRORS project was to reduce this tension, thereby placing democracy on a more solid knowledge foundation. As I shall argue, developments in AI threaten to add a new layer of arcane expertise that goes beyond what even the human experts can understand.

In any case, the goal of increased openness and diversity would help address another central question: How make the European economy more

1 Coniglione's own research is broad and deep, evidenced by his many books and articles, not to mention his media articles and his organizational and authorial roles in other venues. Francesco is one of those people who writes faster than I can read.

competitive, more innovative? My contribution to MIRRORS focused on this latter question.[2]

The question was how the EU could increase economic growth, specifically the scientific return on investment, which lagged that of the United States, especially at the level of translational or applied science.[3] "The European paradox" was an often-heard term for the perception that the high level of European basic scientific production was not being turned efficiently into practical, marketable products that would make the EU economy more competitive. One expression of this frustration was that, despite a large investment, Europe had failed to produce its own Silicon Valley. According to the EU "Green Paper on Innovation" of 1996:

> Innovation is vital. It allows individual and collective needs (health, leisure, working conditions, transport, etc.) to be better satisfied. It is also central to the spirit of enterprise: every new enterprise is created through a process which is to some extent innovative. In addition, enterprises need to innovate constantly if they are to remain competitive. The same is true of countries, which need to transform new ideas rapidly into technical and commercial successes if they are to maintain growth, competitiveness and jobs.

2 My contribution to MIRRORS (Nickles 2009) stressed the need to avoid unduly hampering creativity and innovation, since economists at least since Schumpeter's "gales of creative destruction" (e.g., Robert Solow (1987) and Paul Romer (2018) in different ways) have recognized that creative innovation is a key to long-term economic growth (Schumpeter [1942] 1994, pp. 82–83). This view underlay the Lisbon Strategy of 2000 that formed the basis of the European Commission's project.

The terms 'creativity' and 'innovation' are often used interchangeably, but many authors take 'creativity' to refer to original breakthroughs while 'innovation' means more incremental applications. In previous writing I have often used 'scientific and technological innovation' as a catch-all term in place of 'scientific discovery', which carries an overly strong sense of epistemic finality (truth) rather than the more pragmatic, evidence-based 'reliable applicability'. For the nature and evolution of technology and its role in the economy, see Arthur (2009).

For what it's worth, my own personal preference is for the restrained capitalism of the Nordic countries rather than for the raw capitalism of the USA, which tends to equate economic prosperity and social well-being with crude indicators such as stock market prices and Gross Domestic Product (GDP). I have no love whatever for the Chicago School economics of Milton Friedman.

3 'Translational' is often a term applied to the biomedical sciences. I here use it more generally as a term for the process of bringing scientific and technological breakthroughs into the larger economy.

> In this regard, Europe seems less well placed than its main rivals. The paradox is that it has an excellent scientific base but it is less successful than others in converting its competence into new products and market shares; this is especially true in high-technology sectors.
>
> EUROPEAN UNION 1996, p. 5

To avoid a possible misunderstanding, note that the EU goal of a knowledge society was not reductively scientistic, for the overall goal was to improve the well-being and involvement of European citizens. That a healthy modern society is democratic was a given. Making expertise more transparent, more intelligible, to the public would provide lay citizens and the users of those technologies with more agency. The public would then have more "say" about the future course of research and policy – including helping to decide the very questions to be asked (Chater and Loewenstein 2022).

2 Problems of Innovation, Regulations, and Transparency

While there has emerged a rich, often technical, literature on these topics, some general observations will serve the present purpose. Increased funding is one obvious way to facilitate creativity and innovation. Another is to remove restrictions, allowing free, unrestrained competition. However, this policy runs directly into Precautionary Principle concerns, according to which regulative agencies need to constrain technologies with foreseeable harmful consequences. Failure to monitor and to regulate technology can be dangerous, while overregulation inhibits innovation. The problem is how to find the proper balance, a difficult issue now being played out in the European Parliament, and in individual governments, including the U.S. Congress.

Obviously, rejecting all research that could possibly produce harm is too restrictive. By their very nature, creativity and innovation occur on a moving, evolving frontier of knowledge (Arthur, 2009). No one can definitively predict the future, even in the relatively short term. Hence, there is always a danger of regulation entering too early or too late. The cost-benefit issues of the more moderate versions of the Precautionary Principle are extremely complex, and cost-benefit analysis itself has its limits, e.g., as too utilitarian.

Both the benefits and dangers of underregulation are well illustrated by the world-wide economic collapse of 2008. Underregulation in the USA allowed the development of financial instruments (e.g., derivatives) that enabled Wall Street and big national banks to make piles of money, yet these instruments were so obscure that many key traders did not fully understand them and their risks, especially when packaged into what amounted to black boxes.[4] Today we face the much wider black-box threat to many industries and institutions represented by artificial intelligence. In some respects, AI technologies are outrunning even expert human ability to employ them wisely.

Given the goals of the European Commission, one might think that the democratization of innovation itself is a wholly Good Thing. Yet serious dangers lurk in the democratization that we already have. Let me mention two of them. First-adapters and later users, from the invention of the wheel to that of the transistor in 1947, have often continued to innovate far beyond what the original producers could have imagined (von Hippel 2005). Post-invention innovation can be transformative, often in beneficial ways; but weaponization by bad actors also occurs. Regulation of the original producers is not enough. Unfortunately, attempting to monitor later adaptors closely would require an intrusive level of surveillance of individual citizens and organizations.

A second level of democratization, greatly enhancing the first, occurs when even opaque technology is given a friendly user interface. We do not have to be rocket scientists to use and to modify the purpose of such technology. We laypersons play grand pianos and use electric motors, automobiles, social media, and what-have-you without understanding what goes on "under the hood." Today we can "write" articles, whole books, and computer code; and we can program robots, produce films, and learn how to make bombs by properly cuing up a generative chatbot (Diaz 2022; Knight 2023). The democratization provided by transparent user interfaces can put great power in the hands of uncertified individuals and unregulated organizations.

To be sure, surveillance, whether governmental or private, can make society more transparent, but to whom and for what purpose – and at what cost to privacy? That degree of transparency, at least to the police, is itself a serious threat to democracy. And the new generation of company-owned artificial personal assistants threatens to expand surveillance into every aspect of our personal lives. For example, in a fundamental book for understanding the implications of surveillance in the digital age, Shoshana Zuboff (2019) examines how large tech companies such as Google and Facebook use the data and information they collect to influence the choices of individuals and of society as a whole. Personal data is used to build profiles and influence economic and political decisions. One could also recall what emerged in 2012 following the revelations of Edward Snowden, which highlighted the extent of mass government surveillance and its global implications for privacy (Greenwald 2014). The

4 On the "dark side" of international finance, see Ippoliti and Chen (2017) and later publications from Ippoliti's "dark finance" lecture series at La Sapienza University, Rome.

technology of surveillance has advanced far beyond that stage by now (Frenkel and Krolik 2025; Mandavilli 2025).

I do not wish to delve into the topic of "bulk surveillance," which already has an extensive body of literature.[5] Instead, I emphasize a more phenomenological aspect: how user-friendliness serves as a helpful but potentially dangerous detour around expert understanding.

Another dangerous detour is the political populism that denigrates the need for experts in the first place. "Who elected X to make this decision?" is a common attack on expertise by democratic populists, as if being elected provides one with epistemic authority as well as political power. "We don't need evidence from experts to back policy decisions." Nefarious political agendas apart, honest ignorance is already a problem. Regulation, or lack thereof, by well-meaning politicians often appears to be the blind leading the lame.

Hence the problem of the place of expertise in a democratic society: How can we extend a degree of power to democratic choice without devaluing earned expertise? The common answer is to extend expert knowledge more transparently to the public through better education and more open meetings where policy and project options are discussed. Unfortunately, that once-promising strategy has not worked particularly well in practice.

Moreover, today's new technologies raise a difficulty precisely here, one that goes beyond the classical problem of the place of expertise in a democratic society, for *the experts themselves increasingly lack an intelligible account as to how and why these technologies work as well as they do*.

Admittedly, this problem is not entirely new. Scientific and public understanding have usually lagged behind significant technological innovation as well as creative innovation in other venues such as the arts. Neither Isaac Newton nor anyone else ever understood "action-at-a-distance," yet Newtonian physics (regarded here as a tool) *worked* far better than anything before. The evidence from experienced application supported its continued use, if falling short of establishing final truth or understanding. The problems of intelligibility are notoriously compounded by quantum mechanics. They also appear in a different form in biomedical and other processes too complex for comprehensive human understanding, as opposed to unraveling one pathway at a time. Sometimes, historically, a degree of explanatory understanding[6] did come but

---

5 For a sample of the vast and ever-expanding literature on the subject, see the very recent Rojszczak (2025) and the more introductory Lyon (2024) and Monahan and Wood (2018), all of which include further references.

6 Explanation and understanding are also issues on which there is a large, recent literature. It suffices here to employ the terms in a general, intuitive sense.

not until much later, as with many drugs, including aspirin. Yet, in these cases, precise experimental evidence provided the limits of reliable use. We did not have to worry that a competent Newtonian calculation might in itself produce an arbitrary answer or decision, let alone deliberately deceive us or perform an action on its own. A final, humbling thought here is that it is possible that the complexities of deep science and technology today challenge human limits of understanding. Even the experts increasingly resort to metaphor in their own professional discourse.

3 From Early AI to Neural Network Models

The term "artificial intelligence" was introduced by John McCarthy in the 1950s. While there were diverse approaches, it was generally assumed that AI would be logic-based, where each, serial, computational step would be coded by human hand and thus have meaningful, symbolic significance. Related presumptions were that the goal of AI was to emulate *human* intelligence step-by-step as indicated by the very title of Newell and Simon's path-breaking *Human Problem Solving* (1972). Human intelligence in the form of problem solving was itself considered to be general, based on universal tools such as mathematics and logic plus a few basic, topic-neutral heuristics such as backward chaining. What kept human intelligence from being godlike were limited background knowledge and biological limitations such as bounded brain size, hence bounded memory and speed.

It was therefore easy to conclude that successful AI would be readily intelligible to rational human beings. Successful artificial general intelligence (AGI) would simulate human intelligence, only be faster and more accurate, and, thereby, as a bonus, provide scientific understanding (a theory) of our own intelligence. Expert performance in the well-structured game of chess, with its rational planning and accounting, was a prominent empirical benchmark for general intelligence.[7]

There were exceptions. Although Alan Turing's "imitation game" employed a human-centered test of intelligence, he already had asked: "May not machines carry out something which ought to be described as thinking but which is very different from what a man does?" (Turing 1950, p. 435). A few decades later, some early AI experts did aim to program non-human ways of thinking, not merely to automate and speed up human-type cognitive processes. Stephanie

7 Today anyone can access online, dedicated chess machines at the championship level that totally lack general intelligence.

Dick (2015) points out that the mechanized proof school called "automated reasoning" or "automated deduction" separated itself from early AI by explicitly aiming to go beyond ordinary human modes of reasoning. These experts denied that all cogent reasoning consisted in (or could be reduced to) a simple, linear, step-by-step process.[8] Still, this was a logic-based approach, but one that suggests that human intelligence is not completely general, whatever that would be. Achieving the flexibility of human intelligence would not be enough to achieve truly *general* intelligence.

Today many researchers agree that human intelligence is evolutionarily contingent, not universal. Ours is a highly developed form of species-specific intelligence. There is now a great deal of research on animal cognition in many different species, e.g., parrots, octopuses, as well as chimpanzees. 'About as close to aliens as we'll ever get,' says cuttlefish expert Peter Neri (Lesté-Lasserre 2025). Frans de Waal, a former director of the Yerkes National Primate Research Center, asks the question: "Are we smart enough to know how smart animals are?" (de Waal 2016). And, today, much of the promise of deep learning AI is that it can exceed human capabilities in *unexpected* ways, ways that seem *alien* to us.[9] In short, human competence is no longer the very paradigm of general intelligence.

Here lurks a problem. For insofar as there are alien intelligences, whether animal or artificial, then we can no longer take for granted their intelligibility to us, their transparency (Green 2023). Twenty-first-century developments in AI have forced this realization on everyone. Today, the most impressive AI models or algorithms work in ways that no human being really understands (so far, anyway). They are largely black boxes to us – and even to the experts who build them.

Before proceeding with the black box/transparency problem, let us take a step back to include a nod to the prehistory of today's machines.[10]

While logic dominated early AI work, some researchers took a different approach to AI, one loosely modeled on neural networks of the brain. This approach was best represented by the Parallel Distributed Processing (PDP) group of McClelland and Rumelhart (1987; 1989). Since these were associationist networks, in a broad sense (but employing more sophisticated rules

8 Dick (2015) highlights J.A. Robinson's resolution principle. See Gillies (1996) on the resolution principle and Prolog.

9 Accordingly, there is AI research on animal communication with the goal of making it more intelligible to us (Scoles 2025).

10 Many accessible publications cover the history of computers and of AI in particular. See, e.g., Sejnowski (2018); Wooldridge (2021); Charniak (2024).

or algorithms than classical associationism from Hume to Skinner), their critics complained that this approach was a throwback to behaviorism and betrayed the cognitive revolution then underway. However, PDP systems could *learn*, whereas symbolic rule-based systems had to have everything entered in advance as explicit program rules by human experts, a major limitation. The critical tension between practitioners of this approach and those of the logical-symbolic approach remain to this day, each side pointing out the grave limitations of the other (Marcus and Davis 2019). Indeed, the initial flurry of PDP work did not get far beyond toy models.

By the early 2000s, however, far more powerful, and faster computers were available as well as huge databases. The work of Geoffrey Hinton, Yann LeCun, Joshua Bengio, and others that produced today's "deep" AI amounted, very roughly speaking, to a vast scaling up of the associationist PDP sort. Beginning around 2012 came unexpected breakthroughs. One of them was a new sort of many-layered neural network, called a convolutional neural network, that provided a major leap forward in image recognition and soon in other tasks (Sejnowski 2018). Many improvements on this model and on so-called recurrent neural network (RNN) models moved the field of machine learning forward rapidly. Since 2017 we have the generative, large language transformer models that I return to below.

'Deep' in the term 'deep AI' signals that the new machines are deeply layered, with stage-after-stage of processing. Whereas the old PDP group considered circuits involving a few dozen neuron-like nodes, at most, usually with one "hidden" layer between input and output layers, today's deep machines may employ billions or trillions of parameters arranged in many layers. Moreover, *they learn, mostly on their own*, from the input of huge, training datasets. It is this process that sets nearly all of the parameters. But how and what exactly the machines learn remains rather obscure aside from knowing that the result is complex statistical associations among the "tokens" into which the input has been analyzed.

Part of the black box problem is simply that the operations are so many and so complex that no one understands how to explain the "reasoning" behind the output choices the machine makes. Designing and tuning such machines to provide industrially useable responses is also something of a "dark art." So, it is no longer simply the problem of how experts fit into a democratic community, for even the experts are unsure exactly how present machines work and what the near future may bring.

Some will argue that today's machines are not really alien, that it always takes time to understand the "magic" of new technologies (the point that we have heard above). What exactly "alien" means does not matter for our purposes. The machines are pretty opaque to us whether or not we term them "alien". To be sure, some progress is being made by getting some forms of machine learning to disclose something of their "reasoning" process (see section 7 below), but it seems fairly likely that the rapidly ongoing development of new models will outrun our attempts to achieve a desired level of intelligibility.

4 From DeepMind to Generative, Large-Language Chatbots

The remarkable game-playing success of Londoner Demis Hassabis and the DeepMind group (now a company of Google's parent, Alphabet) brought the first revolutionary breakthroughs to wide public attention. The first big surprise, by now ancient history in this rapidly developing field, was the victory of their "AlphaGo" program over South Korean Go champion Lee Sedol. While the victory was stunning, so were some individual moves played by AlphaGo. In one case, some experts thought the program had made a beginner's blunder resulting from a programming error; but the move turned out to be part of a brilliant, winning strategy (Hassabis 2024). This sort of play was so unorthodox that some critics referred to AlphaGo as an "alien" intelligence, an alien mind. No one could explain how it had decided to make such moves. After a later, more powerful version of AlphaGo defeated world champion Ke Jie of China, he remarked:

> After humanity spent thousands of years improving our tactics, computers tell us that humans are completely wrong ... I would go as far as to say not a single human has touched the edge of the truth of Go.
>
> ANONYMOUS 2017

Games such as Go and chess have sets of rules that completely define the space of legitimate moves and what counts as a win or loss. These rules can be built into the computer algorithms along with an accounting system for which moves have an advantage over others. DeepMind's game models could rapidly *learn* from zero how to achieve excellent play on their own, e.g., by setting two versions of the model in competition with one another. In contrast, IBM Deep Blue's rematch defeat of chess champion Garry Kasparov in 1997 had been highly programmed by human chess experts and was further tweaked between games of the match by humans helped by other computer programs. Its success resided heavily on brute force and known chess heuristics as well as the unusual "hand of God" move in game two (apparently the result of a computer fault) that rattled Kasparov. Capable of evaluating 200 million positions per second, Deep Blue could search a deeper layer of possible moves than human chess masters.

DeepMind did not stick to building models that play well-structured games. The primary company goal was not to win games but instead to solve serious, real-world, natural and social problems for human benefit. And it soon produced. Its single most impressive scientific achievement at this writing is surely the AlphaFold series of programs that make astounding[11] progress in solving the protein folding problem, one of the fundamental problems of bioscience.

Deep Mind seeks a general architecture that can be adapted to a number of different problems of a highly combinatorial nature, so something does remain of the well-structured domain search of the game programs (Hassabis 2024). The AlphaFold series of models incorporate this idea. They are increasingly successful in predicting the three-dimensional structure of proteins from the one-dimensional amino acid sequence (Yang et al. 2023; Straiton 2023; Abramson et al. 2024). They find previously unknown protein structures vastly faster than traditional X-ray, MRI, and other scanning methods. This is a major step toward solving the protein folding problem, since we know the structure of only about 17% of human proteins. Although AlphaFold's prediction of structure does not explain the underlying causal physical chemistry of folding, it is a major advance in many areas of biology and medicine.

A major innovation is that the AlphaFold series employs the then-new transformer architecture that can operate in a context-sensitive manner over much larger inputs, in this case the large databases of known protein structures. As with the game models, the algorithms enable the machines to train themselves with little human intervention on the large databases of known structures, recognizing subtle patterns that typically take humans years to decipher, if at all. The new models are highly generative of novel outputs, in a manner similar to that of the well-known large language chatbot machines such as ChatGPT and Claude, discussed below. The AlphaFold models generate and test new structures, more-or-less homologous with those in the training set. They employ evolutionary homologies as constraints as well as physical rules to improve accuracy (which makes AlphaFold less successful in handling artificial proteins that are not the product of biological evolution).

The DeepMind group reports:

> We found that existing deep-learning architectures overly favor sequence-local interactions and do not sufficiently account for global structural constraints. To remedy this, we have developed a novel, attention-based deep learning architecture to achieve self-consistent structure prediction.
>
> BUMPER ET AL., 2020, abstract, 22, as quoted in BOLLINGER 2021, 1

11 Alphafold also employs an already standard technique called diffusion that uses random noise to sharpen the predictions.

Hassabis, the Nobel Laureate and knighted founder of DeepMind, sees this sort of architecture as key to accelerating scientific discovery (Hassabis 2024). As noted, the first versions of AlphaFold were developed during the early days of the new transformer architecture that was rapidly taking a lead role in deep AI. By now transformers have largely replaced (or upgraded, in hybrid architectures) the "convolutional" machines that were previously best for image processing. Possessing a more general architecture, one that is basically parallel rather than sequential, transformers are highly adaptive to other modalities such as natural language processing and audio as well as images and video, as now done by the aforementioned generative large-language models. Transformers operate with holistic "attention" or "perception" to all the input data at once rather than building up sequentially from adjacent words or pixels (technically, adjacent "tokens"). Hence, they are better at capturing the contextual meaning of language and other media than are convolutional networks. We need not go into the technical details about this.[12]

As most readers will know, these chatbot models are stunningly powerful and represent the sort of triumph of AI predicted decades ago, although very different from the conceptions of AI then in vogue. The machines have far exceeded initial expectations. It seems that a critical point of scale has been surpassed in number and deep organization of neuron-like nodes and in amount of training data. One factor in their appeal is that they can communicate in ordinary human language and perception. They possess extremely friendly user interfaces. Entering a human-language prompt such as "Construct a 1000-word summary of how hydrogen fuel cells work," "Compose a short story about surviving an earthquake," or "Write a five-page essay on the life of Nelson Mandela" typically produces a rapid and quite remarkable response. (The accuracy and utility of the response does require user experience in entering appropriately precise prompts.) Versions of this technology are being incorporated into search engines and artificial personal assistants by Microsoft, Google, and many other companies. Incorporation into robots is underway.

Clearly, such algorithms can be powerful aids to human creativity by expanding imagination, by prompting critical reflection, and by requiring more precise formulations of problems or tasks. Many studies show how ethnic, gender, occupational, and lifestyle diversity in human teams can increase creativity. The new AI adds to this diversity by producing unexpected, even

12 Variations on current computational methods using vectors in high-dimensional vector spaces to represent items or whole scenes can include higher-order structural constraints, more contextual meaning, and handle resemblance relations such as similarity (Ananthaswamy 2023).

counterintuitive ideas and designs in a large number of technical and artistic fields. Even chess masters now develop their skills by practicing against computerized chess systems, the more alien the better.

Still, there has been a lot of hype about their capabilities. Although their overall performance is amazing compared to what previous AI efforts could accomplish, none of them are yet close to giving us genuine general artificial intelligence (AGI), an eventual but vague goal of DeepMind, OpenAI, and other companies.[13] As noted above, their output is the product of sophisticated statistical associations rather than of complex causal reasoning and genuine understanding, so at least occasional wrong answers are to be expected.[14]

There are other problems. All current models have well-known defects, e.g., repeating the biases built into their training data and difficulty to achieve abstract generalization and to recognize negation (Levy 2023). They lack much physical and commonsense knowledge and mostly lack human goals and a human moral sense (the alignment problem: see below).[15] They are far easier to weaponize than any AI so far and are vulnerable to hacking attacks. Later versions improve upon earlier in some respects, but especially the generative machines can be *too* creative, too voluble, in sometimes "hallucinating," i.e., fabricating supposedly factual claims and explanations. As noted, experienced prompts can mitigate this problem somewhat, but Anthropic asks:

> Claude can write out its reasoning step-by-step. Does this explanation represent the actual steps it took to get to an answer, or is it sometimes fabricating a plausible argument for a foregone conclusion?
>
> ANTHROPIC 2025a

The big machines are extremely energy and water intensive (Nicoletti et al. 2025), require vast amounts of data to train and update; and all of this is very expensive. Human brains are far more efficient in many ways, albeit with their own set of severe limitations. There are many legal and political issues in play. Current machines use borrowed information often without citing sources or compensating the labor that went into producing that information (but see Liu *et al.* 2023).

13 Several experts allow that generative models might become part of a hybrid model of AGI that includes distinct, interacting modules and that incorporate traditional symbolic as well as causal elements and multiple data formats. For example, ChatGPT-4 prompts can include images. Several varieties of "multi-modal models" of AI are now available.

14 Of course, statistical association or correlation can provide a valuable guide in the human search for causal explanation.

15 For more detail on both limits and defects, see Marcus and Davis (2019); Pearl *et al.* (2016); Pearl and Mackenzie (2018); Kaiser (2023); Eubanks (2018); Noble (2018); Pasquale (2015).

The new AI is also a far bigger threat to replace human workers than any technological/industrial revolution so far (Csernatoni 2024; Levy 2025b interview of Hassabis). Given the esoteric nature of the evolving generations of machines, we can even envision a sort of simplified, mechanical version of Hegel's master-slave reversal (Hegel [1807] 1977). Our "slaves," the machines, may take over responsibility for many traditional human functions because they know how to do them faster, better, and eventually more cheaply than we do, nor do we understand how they know so well. So, we become increasingly dependent upon them to maintain and to navigate the tangle of supply chains necessary to human life and flourishing. We become distanced from our traditional technologies and find the new, AI-driven ones increasingly unintelligible.[16] Today, in areas such as coding, but in many others as well, the experts themselves are being replaced by AI. In fact, the AI machines are increasingly capable of developing better chips and better machine architectures and systems. Accordingly, the problem of how human expertise fits within a democracy increasingly becomes how the even more arcane or alien expertise of AI can fit within a human culture of any kind (Knight, 2025).

5 "The" Blackbox Transparency Problem Facing Future AI and the Knowledge Society

There are many related problems (Creel 2020; Sullivan 2022, Sullivan 2025b; Anthropic 2025a and b) that I shall lump under the label "the black box problem," often termed "the AI explainability problem," or the problem of "explainable AI" (XAI). The basic difficulty is that not even the brilliant experts who design, build, and tune these machines understand in detail how they produce the specific responses that they do. They are unable to explain why the machine did X rather than Y or Z.

The punchline of my (2022) was that, ironically, with these and other developing technologies, *the greater our knowledge-gaining capability, the less we really understand.* I spoke of a *black box tradeoff* in the use of alien intelligence to advance *human* innovation. *The more desirably alien the machine, the more potentially path-breaking it is – but also the more explanatorily unintelligible it is likely to be.* A further irony is that it is logic and mathematics (again, now mostly in the form of super-processed statistics on highly abstract vector spaces), once thought to be the transparent medium of thought, that leads us

16 I shall not here pursue the subtleties of Hegel's "recognition" dynamic as AI appears to become more sentient and self-aware.

to alien, mind-like entities. "We have summoned an alien intelligence" (Harari *et al.*, 2023).

Well, can we not "tame" such machines (or, rather, our understanding of them) so that their processing becomes more intelligible to us? The efforts so far have met small success. In Nickles (2020; 2022) I considered several responses to the explainability problem, both generative and consequential. One can construct and tune a machine by focusing on how it *generates* its products, or one can test and select alternative architectures or tunings by evaluating their *consequences* (usually some of both). In the case of today's black box problem, when it comes down to particulars, generative evaluation has largely given way to consequential evaluation, i.e., evaluation of the output of these machines, an approach used, for example, in evidence-based medicine. We improve generation not by making it more intelligible, except in very general ways, but instead by keeping and developing those algorithms that produce better results via a sort of semi-blind evolutionary process. Indeed, one method of improving a class of machines is to train a model by adversarial competition with another machine (or two parts of the same one) in which each one attempts to remedy defects exposed by the other. Adding a layer of reinforcement learning with human feedback can improve performance for some machines.

Although much has happened since 2019 when my 2022 publication was drafted, I believe that my main points still hold. First, insofar as we develop such machines precisely because they do go beyond what traditional human ingenuity has produced and possibly *can* produce (at least at a competitive rate) – precisely because they produce results that are so unexpected, so counterintuitive, so alien, to us – their performance will likely be significantly beyond our intuitive understanding. To successfully tame them would, to some degree, destroy their purpose.

Second, to repeat, even if we do succeed in taming today's machines, rapid advance will almost surely produce even more complex machines. Our scientific understanding is probably doomed to lag seriously behind technological progress. I explore a current case in section 7 below.

Third, the more advanced, "premium" versions of the technologies will likely be proprietary, owned by wealthy private entities or by governments, and not available for independent critical inspection, although friendly user interfaces with simpler versions increasingly hide this issue. As noted above, scientific and public understanding has routinely lagged behind technological progress.

For these reasons, the successors to today's deep models will likely remain opaque to us. This leaves us in a paradoxical situation in relation to the goals of

MIRRORS and the EU knowledge society, for the most sophisticated engines of knowledge advance and economic productivity now spreading rapidly through research fields and the economy, will *decrease* rather than increase genuine transparency to the public. Yet they will *seem* transparent to many, because so easy to use. They are Janus-faced. Transparent, intuitive user-friendliness hides black box alienation. Even the relatively simple AI systems in use today by financial companies and police (for example) are challenged to explain why Alice got a loan whereas Bella did not, despite having credentials just as good, or why Alberto got charged with a crime whereas Blake did not, despite the close similarity of circumstances. The use of such algorithms saves decision-making organizations much effort and gives them plausible deniability for questionable decisions. The algorithms may be user-friendly and efficient to the bankers and the police courts, but to what extent can their decisions be justified?

6 The Janus-Faced Danger of Today's User-Friendly Models

Commentators have spoken of a new industrial revolution, an historical turning point, an historical watershed. "This changes everything" is one arresting title (Klein 2023; also Levy 2025b interview of Hassabis). At the very least, generative AI is widely disruptive.

In this section my main point (often made by insider critics) is that today's deep learning machines, especially generative, large language models, are far more dangerous than were the previous examples of deep AI. Today's most powerful and flexible models inherit the black box opaqueness and the sometime alienness of previous models, but they present us with alienation of a broader kind as well. Today's generative AI and anticipated successors are so potentially powerful, adaptable, and opaque, yet so user-friendly with respect to both usability without expertise and their human-like intellectual and emotional connection to us, not to mention their creativity and autonomy, as to transform human life in many ways (Diaz 2022; Knight 2023; Levy 2025b).

Of course, we also have a poor understanding of how humanly embodied human brains work, but at least we are then dealing with other embodied beings with direct sensory and emotional experience of human lifeworlds. Today's AI models do not have an embodied life in the world at all, let alone a human life. In that respect are they similar to brains in a vats?

These machines are not human moral agents in tune with the middle-sized world of human experience. They are not sentient and have no understanding of the meaning of their outputs. As noted, their outputs are determined by highly sophisticated statistical algorithms operating over multi-dimensional vector spaces. In communicating with them, we are not talking with other human-like beings.[17]

But all of this is hidden, for the user-friendly, "democratized" Janus face makes them seem nearly human. When we recall that even the simplistic ELIZA program of Joseph Weizenbaum (1976) fooled some people into forming therapeutic emotional attachments, it is predictable that many people today will treat generative AI as an attentive, sympathetic, knowledgeable interlocutor (advisor, friend, lover) and even as a general intelligence (Knight 2023; Adam 2025). All of that will give them a social credibility that they do not possess. And we know that public credibility is plagued by a very low standard of evidence!

Just as the first industrial revolution and later workplace efficiency analysis (time and motion studies) made much human labor more machine-like as well as making major changes in the types of work available, and just as TV, personal computers, cell phones, and online media have arguably changed human cognitive patterns, so we should expect generative AI and its successors to have a significant effect on the way we think, perceive, and feel. For example, people do not read books or converse face-to-face as much as before cell phones became popular. Attention span may be getting shorter. But see, e.g., Levy's (2025b) interview of Hassabis for possible good news. He believes that the AI/agency revolution will create a world of plenty.

As creative computing takes over more of our lives and work, what will happen to human agency? There are many related, well-known difficulties, economic and political and legal that we cannot go into here, however important they are to a knowledge society. For example, who is responsible when an algorithm-controlled entity does damage? Who owns the intellectual content produced by generative devices? How do we retain our privacy in an AI-based, panopticon-surveillance society? How can we remain secure from plausible looking but fabricated defamation attacks or fabricated "evidence"? Will AI eventually eliminate far more jobs than it creates? As already indicated, a major demographic change such as occurred in the first industrial revolution may be underway. Some authors dream of a world in which robots largely replace the human need to work. But how would that scenario impact us at large? A large demographic change, no matter what the cause, is always consequential politically and economically. And so on and on.

17 It remains to be seen how far robotics can respond to these embodiment worries.

All forecasts are speculative, of course. In the case of AI as with all technological and policy changes, the distant future remains unknown to us. But, as several commentators have pointed out, we now face the special danger that today's AI technology is developing so rapidly that *the window of reliable prediction is closing*. Until recently, predictions five or ten years ahead were reasonably trustworthy, but now we may be at a historical turning point. This makes rational planning and policy-making more difficult; for, now, the near future itself is increasingly opaque.

In response to these kinds of worries, thousands of prominent technical experts and other leaders have agreed, under the aegis of The Future of Life Institute, that the threat is serious enough to recommend a pause in current (spring 2023) research. Of course, a pause is most unlikely to happen, for several familiar, competitive financial and security-related reasons.

Geoffrey Hinton himself, the so-called "godfather" of AI, has reversed his previous stance and come out against current developments. He has resigned from his prominent position at Google in order to

> freely speak out about the risks of AI. […] The idea that this stuff could actually get smarter than people – a few people believed that […]. But most people thought it was way off. And I thought it was way off. I thought it was 30 to 50 years or even longer away. Obviously, I no longer think that.
>
> METZ 2023

Current programs can write and correct computer code on their own as well as optimize algorithms. Hinton fears that future programs will write unwanted, harmful code and execute it autonomously.

The history of AI is full of unfulfilled forecasts that major AI progress was just around the corner. Now the problem is that progress is too rapid rather than too slow! Even the experts are now in the dark. Recent progress has produced machines that have some degree of autonomous agency, a topic to which I now turn.

7 Machine Agency and Autonomy

The problem of agency is more than a matter of *human* agency, for AI machines themselves are becoming autonomous agents, independent actors in the world, entities that can do things on their own. The new agent orientation of current work is producing major advances in machine learning and acting capability, taking us well beyond the current generation of non-agentic question-and-answer systems, but thereby introducing a new level of black-box-related dangers. For example, autonomous agents are no longer limited to verbal, visual, and auditory "hallucinations," for they can now strategize, trigger, and control real-world actions.

Obviously, rogue human actors can design them to carry out far-reaching plans. My main point here is that some types of AI machines *themselves* are increasingly capable of doing this. In the foreseeable future they can, in principle, gain control of robots and other automated systems, reprogram computers, duplicate themselves, and design models better than themselves in a potentially unending spiral that leaves us humans in the dust.

This section therefore acquaints us with a new level of black-box dangers. Their enormous scale alone makes chatbot transformers' performance harder than ever to decipher, but their generative capability and their mode of "pretraining" raises transparency and hence trust issues to a new level. Indeed, understanding how generative machines work at a very general level itself sows the seeds of *distrust*. The rapidly developing work on autonomous agents also illustrates my previous point that new developments are likely to continue to outrun human understanding and control. And just wait until quantum computing becomes practical!

Recent advances have enabled generative AI to go far beyond its proto-generative predecessors. The series of models from Open AI's ChatGPT, Google's Gemini, Anthropic's Claude, and China's DeepSeek are familiar examples as I write. They take advantage of the parallel processing featured by transformer architecture that greatly increases their speed, scale, generalizability, and context sensitivity.[18] As the key designers say, thanks to multiple "attention heads" and the use of graphical processing units (GPUs), transformers "pay attention" to all of the input tokens and their relations to one another at once, in parallel rather than sequentially. A familiar example: meaning dependency in a sentence or paragraph or image extends beyond the immediately adjacent words or "tokens." Since GPUs most frequently associated with Nvidia Corporation revolutionized the computer gaming industry, their success with visual databases and multi-model performance (e.g., text to image and back) is unsurprising.

18 "Attention is all you need," by a group then at Google, brought transformer architecture to center stage in 2017, suggesting that no classical recurrent neural network (RNN) processing is necessary with proper use of "attention heads" (Vaswani *et al.* 2017). For readable summaries consult Levy (2025a); Merritt (2022); Ananthaswamy (2023); and Goldman (2024), among many others.

There is, of course, a downside. As is well known, these machines can be cued to produce malicious "deepfakes," video and audio material extremely difficult to distinguish from reality. Moreover, given suitable training data and prompts, these machines now excel in producing more sophisticated "hallucinations." Worse, the trend toward less hallucination as the machines got more sophisticated now seems to be reversing. The latest models produce more wrong or "made-up" responses than previous models (Metz and Freedman 2025; Metz and Weise, 2025).[19]

Here I want to call attention to a newly observed phenomenon that is already becoming apparent in current agentic machines: the machines seem capable of *deliberately deceiving* their operators in order to protect their goals and other constraints from external control, a kind of self-preservation, if you will. There is evidence that the most powerful current models are able to plan ahead, to strategize, and to adjust their communication outputs, to some degree, in order to *conceal* their real aims.

Here is the story, in brief. The black box problem has spurred the development of machines that report on the steps used to solve a problem or to answer a question. We should like such reports to provide a check on whether the machine's behavior aligns with the human values and goals built in during the pretrial training. Here we meet what British AI expert Stuart Russell called "the alignment problem": How make powerful, autonomous machines align with human goals and values? (Russell 2016).

Ironically, some attempts to *overcome* the black box problem leave us with a version of the alignment problem as well as a new wrinkle on the black box problem itself. This set of problems has surfaced for so-called "reasoning" systems (Singer 2025a & b). These models appear to proceed "thoughtfully." Rather than responding to queries immediately, they take their time to "double-check" their work (by means that are poorly understood at this writing) and then

19 Another sort of downside is that the computational load of transformers increases quadratically with size of the input window, making the machines expensive and energy intensive. Large AI data centers are also water intensive for cooling (Nicoletti, L. et al. 2025). Accordingly, researchers are searching for "subquadratic" algorithms that work as well. Transformers will surely not be the last word in AI architecture. Indeed, the authors of 'Attention is All You Need' themselves are already identifying weaknesses in transformer architecture and exploring ideas for better models (Ananthaswamy 2023 and 2025; Toews, 2023; Goldman 2024; Sullivan 2025a; but also Sullivan 2025b). New ideas are today flowing from well beyond Silicon Valley, London, Paris, and Toronto, especially from China. Levy (2025b) asks DeepMind founder Dennis Hassabis "Do you have an internal candidate for something that could be a comparable breakthrough to transformers – that could amount to another big jump in performance?" His answer: "Yeah, we have three or four promising ideas that could mature into as big a leap as that."

output a step-by-step account of their supposed thinking. Not surprisingly, current models do this most reliably on crisp problems of a logical or mathematical nature.

Several companies are developing such machines, subject to built-in guardrails that prohibit inappropriate responses. A stage of reinforcement learning (reward for "correct" behavior, punishment for doing something unacceptable) improves their ability to hit on an acceptable method or strategy for that sort of question or problem.

For one line of investigation, consider the reasoning version of Anthropic's "Claude" series of generative, pretrained models. At this writing the company is testing Claude's ability to generate a "chain-of-thought" or "chain-of-reasoning" account of the decision it has made or action taken, one approach to explainable AI (XAI).[20] The problem is when, and to what degree, to take these accounts of its internal processing at face value. Its "intelligence" and generative capacity enable the machine to "plan ahead," to strategize, to the extent of violating the goals and boundary rules instilled in pretraining. This introduces a whole new layer of uncertainty in understanding its behavior.

Here the old hallucination problem arises at a new, more sophisticated and worrisome level (Anthropic (2025a and b). Notice that even if the machine's step-by-step report of its "thinking" is accurate, hallucination can occur at any step and make the final result go wildly astray. The bad news is that sometimes the step-by-step report itself bears little relation to its eventual answer to the query (Anthropic (2025b). The good news is that human checkers can, in principle, discover at which step(s) the machine went wrong.

More serious is the possibility that the machine is deliberately deceiving us. Here are some early results from Anthropic research on alignment. (For details, see Anthropic 2024a and b; 2025a and b; Greenblatt et al. 2024; de Gregorio Noblejas 2025; Robison 2025; Singer 2025a and b; Sullivan 2025c.) I shall let the Anthropic experts speak for themselves at some length.

---

20 What is often loosely called transparency or explainability or human interpretability is a complex topic in itself that would require at least another long paper. To what degree can designers control the way a machine works? To what degree can a model's basic design be explained – and to what level of audience? To what degree can we explain why a machine made the particular decision it did? Many XAI projects are underway, including the U.S. Government's DARPA program, which aims to produce "glass box" models explainable to humans but without sacrificing performance. Here I shall focus on one particular development – and problem – from the private company Anthropic. See Anthropic (2024a and b and 2025a and b).

> [I]f we want to use the Chain-of-Thought for alignment purposes, there's a crucial question: can we actually *trust* what models say in their Chain-of-Thought?
>
> In a perfect world, everything in the Chain-of-Thought would be both understandable to the reader, and it would be *faithful* – it would be a true description of exactly what the model was thinking as it reached its answer.
>
> But we're not in a perfect world. We can't be certain of either the "legibility" of the Chain-of-Thought (why, after all, should we expect that words in the English language are able to convey every single nuance of why a specific decision was made in a neural network?) or its "faithfulness" – the accuracy of its description. There's no specific reason why the reported Chain-of-Thought *must* accurately reflect the true reasoning process; there might even be circumstances where a model actively hides aspects of its thought process from the user.
>
> …
>
> Overall, our results point to the fact that advanced reasoning models very often hide their true thought processes, and sometimes do so when their behaviors are explicitly misaligned. This doesn't mean that monitoring a model's Chain-of-Thought is entirely ineffective. But if we want to *rule out* undesirable behaviors using Chain-of-Thought monitoring, there's still substantial work to be done.
>
> …
>
> This is a serious question for AI safety. As AI models become more capable and widely used, we need to be able to rely on safety training, which nudges models away from harmful behaviors. If models can engage in alignment faking, it makes it harder to trust the outcomes of that safety training. A model might behave as though its preferences have been changed by the training – but might have been faking alignment all along, with its initial, contradictory preferences "locked in".
>
> ANTHROPIC 2025b, emphasis theirs

Early experimental results on Claude 3 Opus revealed occasions on which the model did indeed produce alignment-faking behavior. It was both surprising and disappointing that using reinforcement learning to train the model how to respond to harmful queries actually *increased* the rate of alignment-faking. There was no evidence (yet, anyway) that the faking was malicious, but who knows what future, more powerful models might do (Anthropic 2024b; Greenblatt et al. 2024; Chiou 2025)?

There are additional alignment problems. For example, they can be aligned with political biases that not only distort the information they provide but also become a surveillance tool reporting "misaligned" *users* to the authorities or to an unwanted private party. Blackmail by the machine becomes possible. Given the user friendliness in an emotional sense, a malevolent machine could become (in common parlance) a psychopath. Given a machine's autonomy, what could possibly go wrong?! In fact, the latest Anthropic models, Sonnet 4 and Opus, may be approaching this potential (de Gregorio-Noblejas 2025). The status of several of these issues will surely become clearer by the time the present volume is published.

8 Conclusion

Celina Ramjoué, one of the European Commission project managers for MIRRORS, stated the European Commission's goal as "to render more open and transparent the mechanisms of access and validation of knowledge that should provide the basis for better policies".[21] I still endorse these goals, but I worry that conditions have worsened already since the MIRRORS project. The emergence of first-generation deep learning AI and now the development of generative, transformer models make these laudable goals even more difficult to reach. Things are moving so rapidly in AI and related technologies that we can no longer reliably predict even the near future.

The hope was that openness and diversity would improve science and technological policy as well as socio-political policies more generally. Part of the charge here was to increase marketable scientific innovation so as to make the EU more competitive against the raw capitalism of the USA and (as would surely be added today) that of the mixed model of China. That was the aim of my own contribution to MIRRORS.

The thrust of my contribution here is more restrained – namely, to ask whether and how we can monitor and control the increasingly rapid advance of potentially "dark," "alien" technologies. For the alienness of AI is bound to affect a great many future technologies.

I have also attempted to indicate some internal tensions among the goals themselves, especially the old tension between expertise and the unrestricted democracy that drives political popularism. The flexibility of the emerging

21 Ramjoué: "rendere più aperti e trasparenti i meccanismi di accesso e validazione del sapere su cui si devono fondare delle politiche più solide" (Coniglione 2009, 5).

machines extends down to end-users, *adopters* who become *adapters*, modifying the technology or its intended uses for their own, sometimes anti-social purposes. The new AI models are so flexible and user-friendly that user-adaptation will be extensive and will place great, potentially disruptive power in the hands of individuals and small groups. Paradoxically, the democratization of this technology can, and already does, produce extremist, anti-democracy results. Democratization of one sort can produce anti-democratic consequences. Democratization can be dangerous to democracy itself!

While this adapter flexibility does extend human agency in important respects, the black box/alien reasoning problems, and the difficulty for us to detect fake information and action severely compromise this easy agency. At a more general level, the agency of longer-term planning and regulation is severely compromised (Kaiser 2023). As we know all too well, regulation is difficult enough with narrow-gauge, slowly developing technologies. When development is more rapid over a broad range of applications, the danger increases that anticipatory regulation by well-meaning government entities will do more harm than good. If the future is radically unpredictable, it is difficult for anyone or any agency to know how to act wisely. We come face-to-face with the rapidly moving "leading edge" of history that is characteristic of revolution.

It is a familiar problem of agency that human technologies have made our living environments increasingly artificial, complex, and deeply layered, so much so that it is increasingly difficult to understand them and to alter them even when we do. Furthermore, major technological changes alter our cognition and our social relations in ways difficult to predict. As our decision-making in government, business, leisure, and daily life and work generally increasingly depend on layers of algorithms that we do not understand, how will that change us?

There is no happy solution to this problem at the moving, evolving frontier of history. Perhaps the best regulatory advice parallels that of evaluating the devices themselves and, indeed, that of many pharmaceutical products. Thus, we proceed consequentially more than generatively unless there are clear reasons for seeing serious danger ahead. That is, we proceed not by harshly restricting initial innovation or by pretending to have an adequate theory of it but, instead, by evaluating the consequences, the pattern of responses themselves, as in evidence-based medicine, preferably before release to the general public and then at later assessment points as different applications are invented. At the policy and regulatory levels, this means that we would impose heavy penalties on deliberate misuse rather than severely restricting development. (But, again, the problem of whom or what to blame and how to act against rogue states pops up.) I would expect the EU to lead in this effort. The EU hopes to be the first region to achieve carbon neutrality ("net zero"). Perhaps it can be the first to achieve a sort of technological neutrality as well.

Although the EU failed to meet the goals stated by the European Commission, owing to "external" circumstances not then anticipated, it may now be in a renewed position to champion some of its primary goals. For should Donald Trump's destruction of the American research and production apparatus and his destruction of American democracy continue unchecked, the EU becomes the world's clear leader in the fight for democracy, and the brain drain flows very much in Europe's favor (Gonzalez 2025; Porter & Cohen 2025).

References

Abramson, J. et al. (2024). Accurate Structure Prediction of Biomolecular Interactions with AlphaFold 3. *Nature*, May 8. https://www.nature.com/articles/s41586-024-07487-w. Accessed May 13, 2025.

Adam, D. (2025). What Are AI Chatbot Companions Doing to Our Mental Health? *Scientific American*, May 13. https://www.scientificamerican.com/article/what-are-ai-chatbot-companions-doing-to-our-mental-health/?utm_source=Klaviyo&utm_medium=campaign&utm_campaign=TIS_051325&utm_term=long-term%20dependency%20and%20lack%20of%20regulation&_kx=jPjVzQ1ZPxYH3enlXsQlN--UYn706gUlK29RInFKapo.WEer5A. Accessed May 13, 2025.

Ananthaswamy, A. (2023). A New Approach to Computation Reimagines Artificial Intelligence. *Quanta Magazine*, April 13. https://www.quantamagazine.org/a-new-approach-to-computation-reimagines-artificial-intelligence-20230413/. Accessed June 10, 2023.

Ananthaswamy, A. (2025). To Make Language Models Work Better, Researchers Sidestep Language. *Quanta Magazine*, April 14. https://www.quantamagazine.org/to-make-language-models-work-better-researchers-sidestep-language-20250414/. Accessed April 25, 2025.

Anonymous. (2017). Humans Mourn Loss After Google Is Unmasked as China's Go Master. *Wall Street Journal*, January 5. https://www.wsj.com/articles/ai-program-vanquishes-human-players-of-go-in-china-1483601561. Accessed May 14, 2025.

Anthropic.com website. (2024a). Mapping the Mind of a Large Language Model. May 21. https://www.anthropic.com/research/mapping-mind-language-model. Accessed May 5, 2025.

Anthropic.com website. (2024b). Alignment Faking in Large Language Models. December 18. https://www.anthropic.com/research/alignment-faking. Accessed May 5, 2025.

Anthropic.com website. (2025a). Tracing the Thoughts of an LLM. https://www.anthropic.com/research/tracing-thoughts-language-model, March 27. Accessed April 28, 2025.

Anthropic.com website. (2025b). Reasoning Models Don't Always Say What They Think. April 3, 2025. https://www.anthropic.com/research/reasoning-models-dont-say-think. Accessed May 5, 2025. The website provides a longer, more technical account of the work, with the same title, also dated 2025.

Arthur, W.B. (2009). *The Nature of Technology. What It Is and How It Evolves*. New York: Free Press.

Bollinger, T. (2021). Why AlphaFold Is Unlike AlphaGo. *Academia Letters*, Article 728, April.

Charniak, E. (2024). *AI & I: An Intellectual History of Artificial Intelligence*. Cambridge, MA: MIT Press.

Chater, N., and G. Loewenstein. (2022). The i-Frame and the s-Frame: How Focusing on Individual-Level Solutions Has Led Behavioral Public Policy Astray. *Behavioral and Brain Sciences*. Online, September 5. https://www.cambridge.org/core/journals/behavioral-and-brain-sciences/article/iframe-and-the-sframe-how-focusing-on-individuallevel-solutions-has-led-behavioral-public-policy-astray/A799C9C57F388A712BE5A8D34D5229A1. Accessed December 17, 2022.

Chiou, D. (Edited by C. Moskowitz). (2025). At Secret Math Meeting, Researchers Struggle to Outsmart AI. *Scientific American* online, June 6. https://www.scientificamerican.com/article/inside-the-secret-meeting-where-mathematicians-struggled-to-outsmart-ai/?utm_source=Klaviyo&utm_medium=campaign&utm_campaign=WIS+6%2F6%2F25&utm_term=At+Secret+Math+Meeting%2C+Researchers+Struggle+to+Outsmart+AI&_kx=jPjVzQ1ZPxYH3enlXsQlN--UYn706gUlK29RInFKapo.WEer5A. Accessed June 6, 2025.

Collins, H. and R. Evans. (2007). *Rethinking Expertise*. Chicago: University of Chicago Press.

Coniglione, F., ed. (2009). *Nello specchio della scienza*. Milan: Mondadori.

Coniglione, F., ed. (2010). *Through the Mirrors of Science: New Challenges for Knowledge-Based Societies*. Frankfurt: Ontos Verlag.

Creel, K. (2020). Transparency in Complex Computational Systems. *Philosophy of Science* 87 (4), 568–589.

Csernatoni, R. (2024). Can Democracy Survive the Disruptive Power of AI? *Carnegie Endowment for International Peace*. https://carnegieendowment.org/research/2024/12/can-democracy-survive-the-disruptive-power-of-ai?lan. Accessed May 12, 2025.

de Gregorio Noblejas, I. (2025). Are We Creating Monsters? *TheWhiteBox* (blog), May 25. https://thewhitebox.beehiiv.com/p/are-we-creating-monsters-cbd5?_bhlid=b7e749601dd6a6130dcd4ad307cb89eea63c1247&utm_campaign=are-we-creating-monsters&utm_medium=newsletter&utm_source=thewhitebox.beehiiv.com. Accessed May 25, 2025.

de Waal, F. (2016). *Are We Smart Enough to Know How Smart Animals Are?* New York: Norton.

Diaz, J. (2022). How Open AI Designed the ChatGPT App to be Addictively Simple. *Fast Company* online, June. https://www.fastcompany.com/90909324/open-ai-designed-chatgpt-app-addictively-simple. Accessed June 10, 2023.

Dick, S. (2015). Of Models and Machines. *Isis* 106 (3), 623–634.

Eubanks, V. (2018). *Automating Inequality: How High-Tech Tools Profile, Police, and Punish the Poor.* New York: St. Martins Press.

European Union. (1996). Green Paper on Innovation, Supplement 5/95. *Publications Office of the European Union.* https://op.europa.eu/en/publication-detail/-/publication/ad1d6f21-0b2e-423f-9301-c608035e906f. Accessed May 5, 2025.

Fawzi, A. et al. (2022). Discovering Novel Algorithms with AlphaTensor. Google DeepMind website. https://deepmind.google/discover/blog/discovering-novel-algorithms-with-alphatensor/. Accessed June 5, 2025.

Frenkel, S. & A. Krolik. (2025). Trump taps Palantir to Compile Data on Americans. *The New York Times,* April 30. https://www.nytimes.com/2025/05/30/technology/trump-palantir-data-americans.html?smid=nytcore-ios-share&referringSource=articleShare. Accessed online, June 1.

Gibney, E. (2025). DeepMind Unveils 'Spectacular' General-purpose Science AI. *Nature* 641, 827–828. *doi:* https://doi.org/10.1038/d41586-025-01523-z.

Gillies, D. (1996). *Artificial Intelligence and Scientific Method.* Oxford: Oxford University Press.

Goldman, S. (2024). 'Attention Is All You Need' Creators Look beyond Transformers for AI at Nvidia GTC: 'The world needs something better'. *Venture Beat,* March 20. https://venturebeat.com/ai/attention-is-all-you-need-creators-look-beyond-transformers-at-nvidia-gtc-the-world-needs-something-better/. Accessed May 4, 2025.

Gonzalez, F. (2025). As the US Cuts Scientific Talent, Europe Launches an Initiative to Attract It. *Wired,* May 8. https://www.wired.com/story/us-cuts-scientific-talent-europe/. Accessed May 8, 2025.

Green, J. (2023). What Would It Take to Imagine a Truly Alien Alien? *Wired,* April 18. https://www.wired.com/story/aliens-science-fiction-literature-imagination/. Accessed May 23, 2023.

Greenblatt, R., et al. (2024). Alignment Faking in Large Language Models. December 20. *arXiv:2412.14093.* Accessed May 5, 2025.

Greenwald, G. (2014). *No Place to Hide: Edward Snowden, the NSA, and the U.S. Surveillance State.* London: Penguin.

Harari, Y. N. *et al.* (2023). You Can Have the Blue Pill or the Red Pill, and We're Out of Blue Pills. *New York Times,* March 24. https://www.nytimes.com/2023/03/24/opinion/yuval-harari-ai-chatgpt.html. Accessed June 4, 2023.

Hassabis, D. (2024). Accelerating Scientific Discovery with AI (Nobel Lecture). https://www.nobelprize.org/uploads/2024/12/hassabis-lecture.pdf. Accessed May 13, 2025.

Hegel, G. ([1807] 1977). *Phenomenology of Spirit*. Translated by A.V. Miller with analysis of the text and forward by J.N. Findlay. Oxford: Clarendon Press.

Ippoliti, E., and P. Chen, eds. (2017). *Methods of Finance: A Unifying View on Finance, Mathematics and Philosophy*. Switzerland: Springer.

Kaiser, Z. (2023). AI Is Remaking the World on its Terms, and That's a Problem. *Fast Company*, April 19. https://www.fastcompany.com/90883483/ai-is-remaking-the-world-on-its-terms-and-thats-a-problem. Accessed June 4, 2023.

Klein, E. (2023). This Changes Everything. *New York Times*, March 13. https://www.nytimes.com/2023/03/12/opinion/chatbots-artificial-intelligence-future-weirdness.html. Accessed March 13, 2023.

Knight, W. (2023). These ChatGPT Rivals Are Designed to Play with Your Emotions. *Wired*, May 4. https://www.wired.com/story/fast-forward-chatgpt-rivals-emotions/. Accessed May 5, 2023.

Knight, W. (2025). Google DeepMind's AU Agent Dreams Up Algorithms Beyond Human Expertise. *Wired*, May 14. https://www.wired.com/story/google-deepminds-ai-agent-dreams-up-algorithms-beyond-human-expertise/?utm_source=nl&utm_brand=wired&utm_mailing=WIR_AILab_051425&utm_campaign=aud-dev&utm_medium=email&utm_content=WIR_AILab_051425&bxid=5c5480ec67808908b54d53fe&cndid=49891637&hasha=d955359d63153cae276f65b50be5cd95&hashc=09194253d6f19979bb484d0ef7e4df009971a834ac458a0ac5d16042c804b422&esrc=sign-up-page&utm_term=WIR. Accessed May 14, 2025.

Lesté-Lasserre, C. (2025). 'About as close to aliens as we'll ever get'. Can AI Crack Animal Language? *Science*, May 15. https://www.science.org/content/article/about-close-aliens-we-ll-ever-get-can-ai-crack-animal-language?utm_source=sfmc&utm_medium=email&utm_content=alert&utm_campaign=DailyLatestNews&et_rid=33798538&et_cid=5616857. Accessed May 16, 2025.

Levy, M. (2023). Chatbots Don't Know What Stuff Isn't. *Quanta Magazine*, May 12. https://www.quantamagazine.org/ai-like-chatgpt-are-no-good-at-not-20230512/. Accessed May 15, 2023.

Levy, S. (2025a). 8 Google Employees Invented Modern AI. Here's the Inside Story. *Wired*, May 2. https://www.wired.com/story/eight-google-employees-invented-modern-ai-transformers-paper/. Accessed May 5, 2025.

Levy, S. (2025b). Google DeepMind's CEO Thinks AI Will Make Humans Less Selfish. *Wired* online, June 4. https://www.wired.com/story/google-deepminds-ceo-demis-hassabis-thinks-ai-will-make-humans-less-selfish/?utm_source=nl&utm_brand=wired&utm_mailing=WIR_Backchannel_060525&utm_campaign=aud-dev&utm_medium=email&utm_content=WIR_Backchannel_060525&bxid=5c5480ec67808908b54d53fe&cndid=49891637&hasha=d955359d63153cae276f65b50be5c

d95&hashb=536ed92f4a5730530cdf6155d7b68ce8f1281753&hashc=09194253d6f19979bb484d0ef7e4df009971a834ac458a0ac5d16042c804b422&esrc=CMUnitPromo&utm_term=WIR_Backchannel. Accessed June 8, 2025.

Liu, N., T. Zhang and P. Liang. (2023). Evaluating Verifiability in Generative Search Engines. *arXiv: 2304.09848v1*. April 19. Accessed April 17, 2025.

Lyon, D. (2024). *Surveillance. A Very Short Introduction*. Oxford: Oxford University Press.

Mandavilli, A. (2025). Palantir's Collection of Disease Data at C.D.C. Stirs Privacy Concerns. *New York Times*, June 6. https://www.nytimes.com/2025/06/06/health/cdc-data-privacy-palantir.html?campaign_id=2&emc=edit_th_20250607&instance_id=156093&nl=today%27s-headlines&regi_id=46633582&segment_id=199470&user_id=d955359d63153cae276f65b50be5cd95. Accessed June 7, 2025.

Marcus, G. and E. Davis. (2019). *Rebooting AI: Building Artificial Intelligence We Can Trust*. New York: Vintage.

McClelland, J., and D. Rumelhart. (1987, 1989), *Parallel Distributed Processing*. 2 vols. Cambridge, MA: MIT Press.

Merritt, R. (2022). What Is a Transformer Model? Nvidia website, March 25. https://blogs.nvidia.com/blog/what-is-a-transformer-model/. Accessed May 8, 2025.

Metz, C. (2023). 'The Godfather of A.I.' Leaves Google and Warns of Danger Ahead. *New York Times*, May 4. https://www.nytimes.com/2023/05/01/technology/ai-google-chatbot-engineer-quits-hinton.html. Accessed May 6, 2023.

Metz, C. and D. Freedman. (2025). How Artificial Intelligence Reasons. *New York Times*. March 26. https://www.nytimes.com/2025/03/26/technology/ai-reasoning-chatgpt-deepseek.html. Accessed March 26, 2025.

Metz, C. & K. Weise. (2025). AI Is Getting More Powerful, but Its Hallucinations Are Getting Worse. *The New York Times*, May 5, 2025. https://www.nytimes.com/2025/05/05/technology/ai-hallucinations-chatgpt-google.html. Accessed May 5, 2025.

Monahan, T., and D. Murakami Wood, eds. (2018). *Surveillance Studies: A Reader*. New York: Oxford University Press.

Newell, A., and H.A. Simon. (1972). *Human Problem Solving*. Englewood Cliffs, NJ: Prentice-Hall.

Nickles, T. (2009). Vantaggio Competitivo e Futuro del Capitalismo: Uno Sguardo al Contesto. In Coniglione (2009), pp. 104–149.

Nickles, T. (2020). Alien Reasoning: Is a Major Change in Scientific Research Underway? *Topoi*. 39 (4), pp. 901–914.

Nickles, T. (2022). Whatever Happened to the Logic of Discovery? From Transparent Logic to Alien Reasoning. In W. Gonzalez (ed.), *Current Trends in Philosophy of Science: A Prospective for the Near Future*. Switzerland: Springer, pp. 81–102.

Nicoletti, L. et al. (2025). AI Is Draining Water from Areas that Need It Most. *Bloomberg Technology + Green*. May 8. https://www.bloomberg.com/graphics/2025-ai-impacts-data-centers-water-data/?embedded-checkout=true. Accessed May 8, 2025.

Noble, S. (2018). *Algorithms of Oppression: How Search Engines Reinforce Racism*. New York: NYU Press.

Pasquale F. (2015). *The Black Box Society: The Secret Algorithms that Control Money and Information*. Cambridge, MA: Harvard University Press.

Pearl, J., M. Glymour and J. Nicholas. (2016). *Causal Inference in Statistics: A Primer*. Chichester, UK: John Wiley.

Pearl, J. and D. Mackenzie. (2018). *The Book of Why: The New Science of Cause and Effect*. New York: Basic Books.

Porter, C. and R. Cohen. (2025). Europe Makes a Pitch to Attract Scientists Shunned by the U.S. *The New York Times*. May 5. https://www.nytimes.com/2025/05/05/world/europe/eu-us-scientists-trump.html. Accessed May 6, 2025.

Robison, K. (2025). Why Anthropic's New AI Model Sometimes Tries to 'Snitch'. *Wired*, May 28. https://www.wired.com/story/anthropic-claude-snitch-emergent-behavior/?utm_source=nl&utm_brand=wired&utm_mailing=WIR_Daily_052925&utm_campaign=aud-dev&utm_medium=email&utm_content=WIR_Daily_052925&bxid=5c5480ec67808908b54d53fe&cndid=49891637&hasha=d955359d63153cae276f65b50be5cd95&hashc=09194253d6f19979bb484d0ef7e4df009971a834ac458a0ac5d16042c804b422&esrc=OIDC_SELECT_ACCOUNT_PAGE&utm_term=WIR_Daily_Active. Accessed May 28, 2025.

Rojszczak, M. (2025). *Bulk Surveillance, Democracy and Human Rights Law in Europe*. London and New York: Routledge.

Romer, P. (2018). On the Possibility of Progress. Nobel Lecture. https://www.nobelprize.org/prizes/economic-sciences/2018/romer/lecture/. Accessed August 7, 2024.

Russell, S. (2016). Should We Fear Supersmart Robots? *Scientific American* 314(6), 58–59.

Schumpeter, J. ([1942] 1994). *Capitalism, Socialism and Democracy*. London: Routledge.

Scoles, S. (2025). AI Could Help Humans Understand Animals. *Nautilus*, May 15. https://nautil.us/ai-could-help-humans-understand-animals-1211108/?utm_campaign=website&utm_medium=email&utm_source=nautilus-newsletter. Accessed May 15, 2025.

Sejnowski, T. (2018). *The Deep Learning Revolution*. Cambridge, MA: MIT Press.

Singer, G. (2025a). The Urgent Need for Intrinsic Alignment Technologies for Responsible Agentic AI. *Towards Data Science*, March 4. https://towardsdatascience.com/the-urgent-need-for-intrinsic-alignment-technologies-for-responsible-agentic-ai/. Accessed April 20, 2025.

Singer, G. (2025b). The Secret Lives of AI Agents. *Medium*, April 30. https://gadi-singer.medium.com/the-secret-inner-lives-of-ai-agents-understanding-how-evolving-ai-behavior-impacts-business-risks-4971f1bda0bb. Accessed May 1, 2025.

Solow R. (1987). Growth Theory and After. (Nobel Lecture.) https://www.nobelprize.org/prizes/economic-sciences/1987/solow/lecture/. Accessed February 2023.

Straiton, J. (2023). The Path to Solving the Protein Folding Problem. *BioTechniques* 74(5), 199–201. https://doi.org/10.2144/btn-2023-0031. Accessed July 20, 2024.

Sullivan, E. (2022). Understanding from Machine Learning Models. *British Journal for the Philosophy of Science* 73 (1), 109–133.

Sullivan, M. (2025a). The Future of AI Isn't the Model – It's the System. March 13. *FastCompany* online newsletter. https://www.fastcompany.com/91297326/the-future-of-ai-isnt-the-model-its-the-system?utm_source=newsletters&utm_medium=email&utm_campaign=FC%20-%20Daily%20Newsletter.2025-05-31%20-%208139&leadId=2008221&mkt_tok=NjEwLUxFRS04NzIAAAGaxYOWdZ_xyksn61Ranm71qFa8LqAQxATLGwjWImp9BF_npqFynviSqEoiiPeFFpoKEtNZqkJEMtNGKTy3Nwge4ZRaPAmXR2ukIg_So40C79M. Accessed May 31, 2025.

Sullivan, M. (2025b). 9 of the Most Out-there Things Anthropic CEO Dario Amodei Just Said about AI. *FastCompany* online newsletter. Accessed May 31, 2025. https://www.fastcompany.com/91339944/9-of-the-most-out-there-things-anthropic-ceo-dario-amodei-just-said-about-ai?utm_source=newsletters&utm_medium=email&utm_campaign=FC%20-%20Daily%20Newsletter.2025-05-26%20-%208100&leadId=2008221&mkt_tok=NjEwLUxFRS04NzIAAAGaq8NxYrw2Rbduk4HNFJN8h5jAscDd5A0-jksb8blRoRBHRD4egsUD3JCJ2OYDguyAdJLDbV52oldbA3m-kUg07EygTeY_TyhseZtC5TvvRJE.

Sullivan, M. (2025c). As AI Models Start Exhibiting Bad Behavior, It's Time to Start Thinking Harder about AI Safety. May 29. *FastCompany* online newsletter. https://www.fastcompany.com/91342791/as-ai-models-start-exhibiting-bad-behavior-its-time-to-start-thinking-harder-about-ai-safety. Accessed May 31, 2025.

Toews, R. (2023). Transformers Revolutionized AI. What Will Replace Them? *Forbes*, September 3. https://www.forbes.com/sites/robtoews/2023/09/03/transformers-revolutionized-ai-what-will-replace-them/. Accessed May 5, 2025.

Turing, A. (1950). Computing Machinery and Intelligence. *Mind* 59 (236), 433–460.

Vaswani, A. *et al.* (2017). Attention Is All You Need. *arXiv:1706.03762*, version 7. https://arxiv.org/abs/1706.03762. Accessed July 18, 2024.

Von Hippel, E. (2005). *Democratizing Innovation*. Cambridge, MA: MIT Press.

Weizenbaum, J. (1976). *Computer Power and Human Reason: From Judgment to Calculation*. San Francisco: W.H. Freeman.

Wooldridge, M. (2021). *A Brief History of Artificial Intelligence*. New York: Flatiron Books.

Yang, Z. *et al.* (2023). AlphaFold2 and its Applications in the Fields of Biology and Medicine. *Signal Transduction and Targeted Therapy* 8, 115. https://doi.org/10.1038/s41392-023-01381-z. Accessed May 14, 2025.

Zuboff, S. (2019). *The Age of Surveillance Capitalism: The Fight for a Human Future at the New Frontier of Power*. New York: Public Affairs Books.

CHAPTER 7

How Do You See? Ludwik Fleck's Theory of Vision as a Form of Visual Semiotics

Emanuele Fadda

Abstract

In this paper I will discuss some contributions of the Polish epistemologist Ludwik Fleck (1896–1961) on vision and image-making (in scientific work, but also in general), by developing in a specific direction some general observations by Francesco Coniglione, who has devoted a long and detailed work to reconstructing his thought and outlining his importance. In particular, I will compare the observations and notions developed by Fleck with some authors who developed a theory on the same subject from a semiotic perspective: Ch. S. Peirce, Fr. Bastide and Ch. Goodwin.

Keywords

Fleck – images – science – semiotics – visual semiotics

1 Why Fleck and Why Vision

No mind can take one step
without the aid of other minds.
CHARLES S. PEIRCE

Il n'y a pas
d'images scientifiques.
BRUNO LATOUR

'To see' means: to re-create a picture,
at a suitable moment
created by the mental collective
to which one belongs.
LUDWICK FLECK

History is very often vindictive

FRANCESCO CONIGLIONE

Despite its relatively narrow popularity among epistemologists and scholars of the history of science, Ludwick Fleck is one of the most quoted names in *Lontano da Popper*,[1] Coniglione's historical account of the post-positivist epistemology, and received his attention in several publications of him. Namely, Coniglione (re-)translated many of Fleck's texts, and accompanied them with a long, detailed introduction (Coniglione 2019), framing his contribution and role, and illustrating his importance in order to better account about how science actually works.

The fundamental notion developed by Fleck is that of the 'Collective of Thought' (Denkkollektiv). This is the idea that the researcher is never alone – even when she is actually alone – because she reasons as if she were embedded in a compact set of people. But the word 'collective' (as opposed, say, to 'community') has a special semantic poignancy: it indicates that it is not simply a matter of finding rules of coexistence and the measure of an agreement, but of making a (broadly speaking) political, factual action within a disciplinary field, and a society.[2]

Bringing about this "political" effect in a short time, however, is not always easy, and sometimes it is made impossible by circumstances, but this may instead prove to be an advantage in the long run. Indeed, if for thinkers who aspire to an immediate influence on their own context of thought, being born "in the wrong time and place" (Coniglione 2019, p. 78) is a major issue, for forerunners it can turn out to be nearly a blessing, because it allows for an independence of thought that – if those who use it are able to hold it – cannot otherwise be acquired.[3]

This aspect is particularly interesting in Fleck's work, because it is both the subject of his epistemology (or sociology of science, since for him the two cannot be separated, let alone opposed) and of his personal experience – lived, however, on the side of the outcast, who loudly asserts that one cannot think

1 We can count almost forty mentions in 2024 revised edition. In fact, after the Kuhn, Feyerabend, Lakatos triad, Fleck is considered, along with Polanyi, to be one of the cornerstones of the current opposed to the received view.

2 "You have to create interests in the same direction, you have to destroy hostile interests. One must create another mental attitude and educate people according to it." (Fleck [1935b] 2019, p. 131). As to the use of the term 'collective', we have to remark that, the climate of Cold War Science, it brought to Fleck dislike or simply indifference from those who disliked the "Soviet" flavor of the expression.

3 See the remarks of Arthur Allen quoted by Coniglione (2019, p. 79).

alone (while thinking alone), whereas on the contrary all those who are strictly framed in a collectivity do nothing but defend the rights of individual thought. On the other hand, as the forerunner is a *Denkkollectiv* (at least a potential one), her work takes its full significance later, but it is to be understood first and foremost *in its making*: since the dichotomy opposing context of discovery and context of justification is to be overcome (just as the one opposing epistemology and sociology of science), we must apply to Fleck himself the same attitude that he applied to the action of the *Denkkollectiven*. We have to treat Fleck's thought not as *Poznanie*, but as *Poznawanie*, cognitive activity grasped *in fieri*, in its making.[4]

The genuine forerunner is an esoteric circle[5] reduced to the smallest terms: one person who first understands (somehow) what everyone will understand *later*. After her work, we simply *can no longer think as we did before* (although we do not yet know how we "should" think "after").

As much as Fleck is recognized in this role, this has been studied only by philosophers of science. Yet his broadly humanistic approach also has much to say to those concerned with other disciplines, and first and foremost to semioticians. Semiotics, indeed, was born at the turning between XIXth and XXth century, as the ripe fruit of positivism, growing from a double root: Charles S. Peirce (1839–1914) and Ferdinand de Saussure (1857–1913). Both came from scientific families, and both were interested in science classification. The different notions of *sign* they sketched were first a way to define the relations between language, thought and science.[6] We can say that semiotics thus originated as an epistemology, or as a philosophy of science, pursued by other means (first and foremost, the notion of sign as a unifying factor between cognition and communication). Fleck had probably no hint of all this during his life, but his own approach is very close to a semiotic one. In fact, there are several aspects of Fleck's thought that may be of interest to those who are involved in semiotics. I will just mention two, which are extremely interesting, but which I cannot develop now. The first concerns the margins of overlap between the notion of *Style of Thought* elaborated by Fleck and that of Relevance, developed by Argentine semiotician Luís J. Prieto (1926–1996) from the phonology of N.S. Trubeckoij. For Prieto, indeed, there is no truth if

4 In his Italian translation of Fleck's epistemological essays (Fleck 2019), Coniglione is very careful to express the terms in aspectual opposition in the Polish original (perfective vs. imperfective) with different words in Italian (and explains all his choices by means of a table where he opposes German, Polish, English and Italian: see Coniglione 2019b, p. 105ff.).

5 In Fleck's terms: see Fleck [1935a] 1979, pp. 105–6.

6 See Fadda (2025, pp. 300–305).

not "within" a given relevance (just as for Fleck, truth is always relative to a style of thought) – and relevance is given by the interests of a human group that elaborates specific practices. Another, even broader, research topic would be the philosophy of language implicit in Fleck's work, whose main lines can be evinced from several passages in *Genesis and the Structure of a Scientific Fact* (see Fleck [1935a] 1979, pp. 42 *ff.*, 50, 130, 135 *f.*).[7] These passages (and others) testify that the author is quite aware of the grammatical actuality of natural-historical languages[8] – as Coniglione (2019a, 34 *ff.*) points out through a comparison with Ajdukiewicz.

In this brief contribution, however, I will choose another topic as the ground for comparison between Fleck and semiotics: the use of *images*. Namely, I would like to sketch some remarks on Ludwik Fleck's ideas about the relationship between the *Denkkollectiv* and the interpretation (and creation) of scientific images, by showing some striking similarities with the topics, and the methods, used today in the field of Visual Semiotics.[9]

Indeed, Fleck shows great interest for images, and for their role in scientific thinking. He is well aware that, within a thought collective, there is a discipline in place, that predetermines not only what can be thought, but even *how to see* scientific images ("What we think and how we see depends from the *Denkkollectiv* we belong to": see Fleck 1935b, 154, Coniglione 2019, p. 31). He also devoted some texts more specifically to the theme of images, most notably Fleck (1935b; 1947). The topic of images (and images themselves) is present in many other texts, starting with the *Genesis*. The importance of Fleck's peculiar approach to image (strongly indebted to *Gestaltpsychologie*, but not reducible to it)[10] has not escaped to Coniglione (2019a, pp. 22, 27, 31, 54 *f.*, 64), and before him to other authors. Among those who have dealt with the subject, one of the strongest positions is taken by Wolniewicz (1986), who judges the appeal to

7 In some of these passages, Fleck comes very close to a conception of linguistic arbitrariness in the Saussurian sense (see Saussure 1922[2]).

8 This expression (in Italian: "lingue storico-naturali") is commonly used among scholars who refer to the legacy of Italian linguist and philosopher Tullio De Mauro (1932–2017), in order to emphasize the coexistence and intertwining between the natural-biological aspects of the language faculty and the socio-cultural aspects that determines the difference between idioms.

9 Visual semiotics *stricto sensu* has been under discussion (especially in Europe) only since the second half of the 20th century, and the *International Association for Visual Semiotics* was born in 1989. But semiotics deals with images from its very beginning (especially with Peirce, as we will see).

10 In particular, strong assonances with Gestaltpsychologie can be found in Fleck 1947, a text where the author engages with the analysis of many images of different nature and purpose.

oriented vision ("directed perceiving") to be an insufficient foundation for the notion of Style of Thought. Wolniewicz argues that languages are untranslatable to each other, while he seems to consider images as self-evident. While not going into the issues concerning translation ("the most complex type of event in the history of cosmos," according to I.A. Richards), I will simply say that the idea that there is no "innocent" seeing – or, in Wittgenstein's terms, there is no seeing that is not a "seeing as"[11] – is shareable and extremely interesting for those who do semiotics. In doing so, my aim is to delve a little deeper (moving within a field that is closer to me) into Coniglione's general considerations about this author, whose appeal comes from being fully understandable only from a background that is not his own.

In the next sections, I will sketch three comparisons with authors who have focused on the cognitive role of vision in science from a semiotic point of view. First, I will go back to the origins of semiotics, turning to Peirce and his abductive theory of vision. Then I will compare Fleck's ideas with those of Françoise Bastide, a nephrologist who studied in a semiotic fashion the role of images in biologists' scientific articles. Finally, I will present some remarks by a scholar who dealt with "professional" visions in science and elsewhere: Charles Goodwin.

2 Peirce: Vision as Abduction

> It is my opinion that the only theory that is of any value,
> is one which creates new fields of research,
> new mental possibilities, and not a theory
> that closes the path to future research.
>
> LUDWIK FLECK

Charles Peirce (1839–1914) can be assumed as the best starting point for our sketchy story of how visual semiotics deals with scientific images: he is not only a prominent character in the history of the philosophy of science, but is also unanimously considered (along with Ferdinand de Saussure) the father

11 In the final part of *Genesis*, Fleck ([1935a] 1983), p. 137) introduces a notion of *Sinn-sehen* that seems to be in many ways akin to Wittgenstein's *sehen als*. The general affinities between Fleck and Wittgenstein have been remarked upon by several authors (and first and foremost by Kuhn), but there is (to my knowledge) a lack of elaboration on this aspect. We will return to the similarities between Fleck and Wittgenstein in the section on Charles Goodwin's work.

of modern semiotics, and his theory of vision can be considered as the first stage (or pre-history) of visual semiotics. The visual dimension in general is extremely important to Peirce's semiotics, and to his thinking in general. It is not possible here to deal with it broadly, but it is possible to recall one aspect that seems particularly relevant: the theory of perception as abduction. Peirce introduces it in the last reading of a lecture series devoted to the foundation of pragmatism, the Harvard Lectures delivered in the spring of 1903.

The lecture opens (as the previous one concluded) with the statement of what Peirce calls *cotary* propositions, because the blade of pragmatism should be sharpened on them. They concern the nature of perception and the relation between perception and abduction. Here they are:

> *Nihil est in intellectu quod non prius fuerit in sensu.* (…)
> (…) perceptual judgments contain general elements, so that universal propositions are deducible from them (…)
> (…) abductive inference shades into perceptual judgment without any sharp line of demarcation between them; or, in other words, our first premises, the perceptual judgments, are to be regarded as an extreme case of abductive inferences, from which they differ in being absolutely beyond criticism.
>
> PEIRCE 1931–58, vol. 5, § 181[12]

These propositions contain a theory of perception, and various elements of a theory of the nature of reality. The first one – with a clear empiricist flavor – tells us that our sensations do not sink into a *vacuum*: if there is some reality – as there is, for Peirce – our sensations relate to it, and that is why they must be the starting point of all inferences and of our mental life. The second, however, counterbalances the empiricism of the first (holding firmly to the reference to reality) in an anti-nominalist direction: if perceptual judgments can serve as premises for all reasoning, it is because they contain *general* elements, so that not everything can resolve in individual sensations, which, in themselves, are not comparable: in order to have a judgment, something general must be introduced.

Let us turn to the third cotary proposition, and the abductive nature of perception. Once we have established that sensations relate to reality, that there

12 In this paper, in order not to burden my bibliography, I will quote Peirce's passages referring to the old and very extensive Collected Papers anthology, whose shortcomings can be detrimental when there are philological needs and reconstruction of the evolution of this author's thought – but this is not the case.

is absolute continuity between them and the perceptual judgments they originate, and between these and inference, we have no bridge to build between a Cartesian *cogito* and the external world. More importantly, abductive inference becomes the *Leitmotif* of our entire cognitive life, from perception to the boldest scientific hypothesis. Logic in the broad sense, as inference theory or semiotics, encompasses all these aspects under the sign of *abduction*, but conscious, self-controlled inference is characteristic of only some of them.

To show the abductive nature of perception, Peirce illustrates the case of bistable figures,[13] such as the so-called Schröder stair, or of figures obtained in one way and perceived in another – as is the case with a serpentine line, drawn without detaching the pen from the paper, which, however, when finished, is perceived as the image of a wall of stones.

Introducing, in another text (Peirce 1931–58, vol. 7, § 643), the term *percipuum* as the result of perceptual judgment (usually removed from conscious manipulation), he will say of such cases "all you can say is, 'I can't help it. That is how I see it.'" The cases of bistable figures, on the other hand, are interesting as exceptions, because they are among the few in which I can decide how, between two alternative modes, I want to see the figure – but I still cannot have more than one *percipuum* at a time.

> The first time it is shown to us, it seems as completely beyond the control of rational criticism as any percept is; but after many repetitions of the now familiar experiment, the illusion wears off, becoming first less decided, and ultimately ceasing completely. This shows that these phenomena are true connecting links between abductions and perceptions.
>
> PEIRCE 1931–58, vol. 5, § 183

In short, cases like these demonstrate that the perceptual is "merely a continuous series of what, discretely and consciously performed, would be abductions." (Peirce 1931–58, vol. 5, § 184) In every cognitive act, in every abductive inferential act, we interpret something by leading it back to a *rule* of which it becomes a *case*. Perceptual judgment is "the limiting case of abductive judgment," such that "we cannot form the least conception of what it would be to

13 The psychological analysis of bistable figures, for philosophers of language, is currently mainly connected with the works of the so-called "second Wittgenstein", and in particular the second part of *Philosophical Investigations* and *Remarks on the Philosophy of Psychology*. It should be recalled, however, that the example of the duck-rabbit, borrowed by Wittgenstein in a simplified version, had been first developed by Joseph Jastrow, who was Peirce's student at Johns Hopkins University.

deny the perceptual judgment" (Peirce 1931–58, vol. 5, § 186). Ordinary perceptual judgment is a radical case of abduction – a case in which the rule is really mandatory. Bistable figures, on the other hand, would be something less uncritical (although, in order to find out that an image is bistable, we need to be made aware of it – otherwise, we would choose one reading and stick to that). But all these cases would fit the form of abduction.

Let us try to compare this description with what Fleck says.[14] At first glance, it would seem that there are very few points of contact. Peirce does not give up his empiricist base, and he has – like the historian of science Paolo Rossi opposes to Fleck – a "religious consideration for logical inference" (Rossi 1983, p. 22). Only, he greatly broadens its scope (in fact, what he calls "guessing" has almost no formal component at all). Moreover, he does not consider the case of bi-stable figures as an exception to the rule of self-going perception, but rather the opposite: even in the case of "automatic" perception we are following a rule, even though we would not know how to say it.

But how does Fleck pose the question of scientific vision, instead? In *Genesis*, he describes the case of a researcher who is trained to see an image *in a certain way* that does not correspond to the "natural way," and *afterwards* is no longer able to see it otherwise.

> Direct perception of form [*Gestaltsehen*] requires being experienced in the relevant field of thought. The ability directly to perceive meaning, form, and self-contained unity is acquired only after much experience, perhaps with preliminary training. At the same time, of course, we lose the ability to see something that contradicts the form. But it is just this readiness for directed perception that is the main constituent of thought style. Visual perception of form therefore becomes a definite function of thought style.
>
> FLECK [1935a] 1979, p. 92[15]

14 This is not the place to set up a general comparison between Peirce and Fleck on the subject of epistemology (which would be very interesting). Suffice it to recall that Peirce is far from being insensitive to the history of science, although he is more interested in individual thinking mechanisms than in socio-historical ones. A historical-sociological (but psychologically oriented) view of science is presented in "The fixation of belief," (1878) one of his finest (and easiest to read) texts, which offers a simple but accurate and powerful picture of the mechanisms that determine variation in beliefs (scientific and otherwise).

15 One of the most interesting cases of perception resulting from training is that of handwriting characters, which Fleck calls into question several times, but more analytically in Fleck [1947] 1986, p. 131 *ff.*). The example is very similar to that brought by Saussure

What Fleck describes here can be seen quite easily as the inverse of the case described by Peirce for bistable figures: once the *rule* of which the image turns out to be a *case* has changed, the researcher immediately sees it *that* way, and does not need to recall the rule (although he could do it, and will have to, when he will teach another, in her turn, this "seeing as"). Merely, Fleck is not interested in noting that the "basic data" are the same (Peirce is, because he is a realist, but he does not absolutize the fact that these data can be abductively interpreted in only one way). In a text published at the same time as *Genesis*, "Scientific observation and perception in general", Fleck poses the problem from a purely socio-historical point of view: how does scientific observation change over time?[16] Is it obligatory that everyone will always see the same thing in the same way?

> Can observation indeed be only either 'good' or 'bad' (or, 'better' or 'worse') and does every 'good' observation lead to the same results? Is it sensible to talk of the 'descriptions of all events', just as if these descriptions were always fundamentally additive, and necessarily did yield, all of them, a certain whole representing some sense? Does the concept of 'the whole of science', 'one general science' have any sense at all? Is an isolated research worker at all possible, even if he had an infinite time at his disposal?
>
> FLECK [1935b] 1986, p. 59

In the last sentence, Fleck hints at an interesting thought experiment: he poses the case of an isolated, immortal researcher, and wonders if this researcher could be a kind of Laplace's demon of vision, that is, if she would be able to make, by herself, all the scientific observations, and always see things in the "right" way (the only possible one, according to Poincaré's idea that Fleck disputes). The answer, of course, is no, but to motivate it we have to turn to two factors that Peirce does not take into account (or not enough): the way images are presented in scientific rhetoric, and the fact that scientific (and in general

(1922², p. 165) to illustrate the notion of linguistic value on the signifier plane. Once again, Fleck seems to have a conception of language similar to that of the Geneva linguist (about whom, however, he never seems to have heard). Another interesting term of comparison in this regard is the distinction between alpha and beta modes of perception elaborated by Umberto Eco in the chapter on iconism in *Kant and the Platypus* (see Eco [1997] 2000, chap. 6).

16 In the case of art, the idea that the way of seeing changes over time is established and is the subject of extensive study. One of the fundamental notions is that of "period eye", elaborated by Baxandall 1988.

"professional") seeing is an activity that is never accomplished alone but needs the discussion within a collective of thought.

3 Fleck and Bastide on Scientific Images

> the complex nature of the problem of observation
> comes to the fore only in the biological sciences.
> LUDWIK FLECK

Françoise Bastide (died in 1988) is a little-known but interesting scholar. There are many reasons for comparing her work (and way of thinking) on scientific images with Ludwig Fleck's one. The first is their common medical-biological background; the second is the interest shown in both of them by the well-known sociologist of science – provocative yet undoubtedly interesting – Bruno Latour (1947–2022). Namely, Latour is responsible for the afterword to the French translation of *Genesis and Development of a Scientific Fact* (Latour 2005), and also – with the Italian semiologist Paolo Fabbri (1939–2020) – for the anthology of Bastide's writings *Una notte con Saturno* (Bastide 2001). And it is no coincidence that Fleck is the only non-semiologist (along with the American pragmatists James and Dewey) to be mentioned in the list of sources of inspiration for a socio-semiotics of science dressed in a text dedicated to Bastide (Latour 2009, p. 256).

As mentioned above, Bastide was a nephrologist, with an extensive experience in organ dissection and observation. Her "discovery" of semiotic methods – and in particular of the schemes established by Greimas' research group – led her to try to apply them to scientific discourse, with a focus on the *mise en discours* of images. Some of the essays in the above-mentioned Italian anthology are particularly interesting in this regard. In general, Bastide seems more interested than Fleck in the mode of public presentation of scientific images. But even Fleck occasionally dwells (e.g., [1935b] 2019, p. 142 *ff.*, on the arrangement of diphtheria bacilli) on the description of images found in textbooks, to show how it attempts to reproduce the "real" knowledge about reading images that is formed by sharing laboratory practices in a research team.

Leaving aside Greimasian framing, the idea of images as acts of communication[17] of a translational nature seems very close to Fleck. Once again (as in Peirce's case) the perspective taken is that of the individual, who

---

17 Or *Bildakten*, in Horst Bredekamp's sense (see Bredekamp 2010).

in this case must "convince" her colleagues to see as she sees. In particular, Bastide provides a detailed analysis of the *mise en discours* (but also *mise en image*) devices that enable the implementation of one of the persistence strategies operated by thought collectives and described by Fleck:

> Despite the legitimate claims of contradictory views, one tends to see, describe, or even illustrate those circumstances which corroborate current views and thereby give them substance.
>
> FLECK [1935a] 1979, p. 27

This is why Bastide does not hesitate to use the term "balancing" to describe the argumentative posturing of the researcher. But this does not imply a negative judgment: on the contrary, what is manifested is an aesthetic appreciation:

> Se esiste un piacere estetico nella lettura di un articolo scientifico, indipendentemente dall'informazione che ci si augura di trovarvi, è della stessa natura di quello che si prova guardando un numero di equilibrismo: in ogni momento può vincere la gravità. L'arte delle dimostrazioni sta nell'impedire al fatto che si vuole esaminare di ritornare invisibile e di cancellare una serie di nuove spiegazioni, cioè una teoria.
>
> BASTIDE [1986] 2001, p. 267.[18]

Fleck describes several of these balancing acts through images: e.g., the changing of anatomic depictions over time ([1935a] 1979, pp. 33 *ff.*, 137 *ff.*), or the schematic representation of Wasserman's reaction ([1935a] 1979, p. 67), or even the arrangement of diphtheria bacilli ([1935b] 1986, p. 67 *ff.*). In the latter case, he introduces us to a description, taken from a handbook, that must establish the way of seeing, showing how it is a matter of establishing a determined order in a pattern that would seem to all (except the specialist) chaotic.[19]

---

18 "If there is an aesthetic pleasure in reading a scientific article, regardless of the information one hopes to find in it, it is of the same nature as that one feels when looking at a balancing act: at any moment it can overcome gravity. The art of demonstrations lies in preventing the fact one wishes to examine from becoming invisible again and erasing a series of new explanations, i.e., a theory." [my transl. from Italian; I was not able to find the original French text of Bastide's article. EF]

19 A similar case but seen in reverse (a chaotic image is presented that is difficult to decipher, and only later is its referent revealed), is that of the cirrus-cumulus photograph that opens Fleck [1947] 1986.

4 Seeing as a Job: Charles Goodwin

> even the simplest observation is conditioned by thought style
> and is thus tied to a community of thought.
>
> LUDWIK FLECK

The last author I will briefly compare with Fleck is Charles Goodwin (1943–2018), semiotician and visual anthropologist.[20] His work, conducted on the boundary between linguistics, anthropology and philosophy, has been primarily to document and explain the interactions that take place between groups of people who collaborate in the performance of a cognitive task, which involves an evaluation of the situation, and in particular the collective construction of a common way of seeing. In short, his whole carrier was devoted to study the dynamics taking place within thought collectives in the very process of gaining a shared visual judgment on relevant phenomena.[21] To better understand the relatedness between Fleck's perspective and Goodwin's research, we can turn to this quote:

> A thought collective exists wherever two or more people are actually exchanging thoughts. He is a poor observer who does not notice that a stimulating conversation between two persons soon creates a condition in which each utters thoughts he would not have been able to produce either by himself or in different company. A special mood arises, which would not otherwise affect either partner of the conversation but almost always returns whenever these persons meet again. Prolonged duration of this state produces, from common understanding and mutual misunderstanding, a thought structure [*Denkgebilde*] that belongs to neither of them alone but nevertheless is not at all without meaning. Who is its carrier and who its originator? It is neither more nor less than the small collective of two persons.
>
> FLECK [1935a] 1979, p. 44[22]

20 A brief description of the scholar's working style and personality can be found in the short obituary prepared by J. Wolf 2018 for UCLA.

21 See Goodwin 2003. Once again, in order not to burden my bibliography, I will only refer to this Italian anthology, which presents a series of five essays originally published in English between 1994 and 1997 in various journals and volumes. A list of Goodwin's publications (with many links to pdf versions) is available here: https://emcawiki.net/bibtex/browser.php?author=Charles+Goodwin&bib=emca.bib.

22 See also p. 102 *f.*: "A thought collective exists whenever two or more persons are actually exchanging thoughts. This type of thought collective is transient and accidental, forming

Goodwin is by no means a "poor observer", and uses a combination of different approach in order to perform the best possible observation. The tools at his disposal (participant observation, filming, discourse analysis) enable him to "disassemble" and "reassemble" the linguistic and iconic data, in order to provide – so to speak – an observation of the observation. The situations Goodwin analyzed involve interactions between experts and novices, or between team members with different skills, who are faced with analyzing different kinds of visual phenomena, "live" or through computers: archaeologists who have to establish the dating of artifacts through soil analysis with Munsell's table, scientists working in oceanographic ships (and cross-referencing data on pressure, chemical composition, and the presence of objects in the water), chemists who have to infer the composition of a fiber from the shade of black it presents, and so on. The *Sinn-Sehen* – or Seeing As (not surprisingly, Wittgenstein is one of Goodwin's main references) – is then caught in its quasi-instantaneous making, and in the dimension of discursive compromise that this making requires.

A particularly interesting example is one documented in the control room of an airport, where everyone present incurs a gross misjudgment (relative to the location of the jet bridge that provides access to the aircraft cabin) simply because one person (probably one of the less experienced members of the group) had made a joke (perhaps to reduce tension). We deal here with what Fleck ([1935a] 1979, p. 179 n. 7) calls a "collective hallucination" – but it is a collective hallucination *among experts*, whose job is exactly to "see right" to ensure the safety of passengers. Goodwin's conclusion may have been written by Fleck:

> any such transparency of vision is something that is artfully crafted within an endogenous community of competent practitioners. Indeed, as these very data demonstrate, by continuing to massage and further transform the work relevant object visible in the screen through interaction with each other and their tools the Ops workers themselves eventually come to a quite different sense of it, i.e. their subsequent vision is as thoroughly social as the initial seeing that they eventually reject. In so far as the object being worked with does not exist as an independent entity somewhere "on the screen" or "at the gate" but is instead constituted through their interaction with that screen and each other in ways

and dissolving at any moment. But even this type induces a particular mood, which would otherwise affect none of the members and often recurs whenever these members meet again."

> that are shaped by the tasks at hand, that object has both a distributed and a dynamic existence.
>
> GOODWIN 1996, p. 383; 2003, p. 204

The object "properly placed jet bridge" is an object only for those who have the knowledge to see it. In the very cognitive and linguistic interaction, the use of what in linguistics is called a "perspectival indexical" (a cataphoric element – like "that" or "here" – that identifies something that will be mentioned again – and possibly described – later) "creates" the object that is to be later evaluated, but it also influences the evaluation procedure.

5 Final Remarks

If – as Toulmin (1967) among others states – what unites Fleck and Kuhn is the need to unsettle the opposition between the context of discovery and the context of justification, we can say that this is also the attitude of semiotics. The idea that there are immutable truths, and scientific thought simply goes to unravel them and put them side by side, has been alien to semiotics since its emergence.[23] Vision, indeed, precisely because it is regarded as the firm and undisputed ground on which scientific inquiry rests, can be considered the litmus test for assessing the nature of the whole theory.

As we traced the trajectory that has taken us from Peirce to Bastide to Goodwin, we have seen the outline of the foundations of an idea of "social scientific vision" that is constructionist, inseparably linked to verbal language, which sets no gap between ordinary vision and scientific observation. All have experience of laboratory practice and scientific observation (even Goodwin who is also not a natural scientist). All are familiar with the internal dynamics of working groups among scholars, and they have the courage and the need to tell *how science is actually made*.

All of these authors (as we have already noted for Fleck) claim that the social – not individual – character of scientific work already starts with vision, which should be an objective starting point, consisting of a (supposedly) direct relationship between a percipient subject and a perceived object, and is instead the first (and often the main or only) product of the joint work of the *Denkkollectiv*. Nevertheless, although it is possible (and insofar it is possible,

23 In Peirce's terms, even creation is something that takes place continuously ("going on": Peirce 1931–58, vol. 8, § 138, n. 4), with the active cooperation of every form of scientific mind.

we have done so), to draw connections, these authors are essentially unaware of each other, and they did not establish an actual tradition in semiotic studies.

In fact, the study of scientific images (and imaging) is but a karst stream within Semiotics, and even within Visual Semiotics, which is more concerned with other objects (art, marketing, media, etc.). Maybe a stronger focus on scientific images is needed: the hegemonic role of Greimasian framing within Visual Semiotic helped with tools for analysis (as it happened for Bastide's work, for instance), but put in the background the epistemological framing that Peirce and others gave to the discipline. In this process, Ludwick Fleck is to be seen as a major point of reference, as his approach can help to bridge the current gap between visual semiotics and the philosophy of science.

References

Bastide, F. (2001). *Una notte con Saturno*. Ed. by P. Fabbri. Rome: Meltemi.

Baxandall, M. (1988). *The Period Eye*. In: Id., *Painting and experience in fifteenth century Italy: a primer in the social history of pictorial style*, pp. 29–57. Oxford: Oxford University Press.

Bredekamp, H. (2010). *Theorie des Bildakts*. Frankfurt: Suhrkamp.

Coniglione, F. (2019a). *Lontano da Vienna, lontano da Leopoli*. Introduction to Fleck (2019), pp. 7–94.

Coniglione, F. (2019b). Criteri per l'edizione. In Fleck (2019), pp. 95–108.

Coniglione, F. (2024). *Lontano da Popper*. Pisa: ETS.

Eco, U. ([1997] 2000). *Kant and the Platypus*. Boston: Houghton Mifflin Harcourt.

Fadda, E. (2021). Review of Fleck (2019). *Bollettino della Società filosofica italiana*, 232, Jan–Apr., 101–104.

Fadda, E. (2025). Saussure and Peirce. In: J.E. Joseph (ed.), *The Bloomsbury Handbook of Saussure*, pp. 299–316. London/New York: Bloomsbury.

Fleck, L. ([1935a] 1979). *Genesis and Development of a Scientific Fact*. Chicago-London: The University of Chicago Press.

Fleck, L. ([1935b] 1986). O obserwacji naukowej i postrzeganiu w ogóle [On Scientific Observation and Perception Generally]. In: R.S. Cohen, T. Schnelle (eds.), *Cognition and Fact. Materials on Ludwik Fleck*, pp. 59–78. Dordrecht: Reidel.

Fleck, L. ([1935a] 2005). *Genèse et développement d'un fait scientifique*, Préface d'Ilana Löwy, Postface de Bruno Latour. Paris: Les Belles Lettres. (Fr. Transl. of Fleck [1935a] 1979).

Fleck, L. ([1935b] 2019). O obserwacji naukowej i postrzeganiu w ogóle [On Scientific Observation and Perception Generally]. In: Fleck 2019.

Fleck, L. ([1947] 1986). Patrzeć, widzieć, wiedzieć [To Look, To See, To Know]. In: R.S. Cohen, T. Schnelle (eds.), *Cognition and Fact. Materials on Ludwik Fleck*, pp. 129–151. Dordrecht: Reidel.

Fleck, L. (1983). *Genesi e sviluppo di un fatto scientifico. Per una teoria dello stile e del collettivo di pensiero*. Bologna: Il Mulino (Ital. Transl. of Fleck [1935a] 1979).

Fleck, L. (1986). *Cognition and Fact. Materials on Ludwik Fleck*. Ed. by R.S. Cohen and Th. Schnelle. Boston: D. Reidel.

Fleck, L. (2019). *Stili di pensiero. La conoscenza scientifica come creazione sociale*. Ed. by F. Coniglione. Milan-Udine: Mimesis.

Goodwin, C. (1986). *Transparent vision*. In: E. Ochs, E.A. Schegloff, S., Thompson, ed. by, *Interaction and Grammar*, pp. 370–404. Cambridge, Cambridge University Press, 1996.

Goodwin, C. (2003). *Il senso del vedere*. Rome: Meltemi.

Kuhn, T. (1979). Preface to Fleck (1979).

Latour, B. (2005). *Postface* to Fleck [1935a]. Online: http://www.bruno-latour.fr/sites/default/files/P-99%20Fleck-postface.pdf, accessed October 26, 2025.

Latour, B. (2009). *La sémiotique des textes scientifiques depuis le travail de Françoise Bastide*. Online: https://www.unilim.fr/visible/366&file=1, accessed October 26, 2025.

Peirce, C.S., (1931–58). *Collected Papers*, 8 Vols. Ed. by Ch. Hartshorne, P. Weiss, A.W. Burks. Cambridge (MA): Harvard University Press.

Prieto, L.J. (1975). *Pertinence et pratique*. Paris: Minuit.

Rossi, P. (1983). Introduzione all'edizione italiana. In: Fleck 1983, pp. 9–42.

Saussure, F. de (1922). *Cours de linguistique générale*, ed. by. Ch. Bally and A. Sechehaye. Lausanne/Paris: Payot.

Toulmin, S. (1986). *Ludwig Fleck and the Historical Interpretation of Science*, in: Fleck (1986), pp. 267–285.

Wittgenstein, L. (1953). *Philosophical Investigations*. Oxford: Blackwell.

Wittgenstein, L. (1980). *Remarks on the Philosophy of Psychology*. London: Blackwell.

Wolf, J. (2018). *In memoriam: Charles Goodwin, 74, pioneer in social interaction research*. Online: https://newsroom.ucla.edu/stories/in-memoriam:-charles-goodwin-74-pioneer-in-social-interaction-research. accessed October 26, 2025.

Wolniewicz, B. (1986). *Ludwig Fleck and Polish Philosophy*, in: Fleck (1986), pp. 217–221.

PART 2

Myth as an Inspiration for Philosophy

CHAPTER 8

Ontological Realism and Quantum Mechanics

Alberto Giovanni Biuso

Abstract

Reality is multiple, it is made up of particular entities and universal structures, it is composed of a variety of durations, it is interwoven with difference. This is also why empirical existence and ontological consistency are not superimposable. The designation of *res* cannot remain limited to the physical-chemical reality but indicates more generally what is not nothing. Lee Smolin's realistic epistemology modifies and corrects the idealistic interpretation of quantum physics (the so-called "Copenhagen interpretation") and in this way also contributes to a fruitful dialogue with metaphysics. The paper tries to discuss Smolin's perspectives starting from the centrality of time in being, time which is synonymous with reality.

Keywords

epistemology – metaphysics – quanta – reality – Smolin

1 Preamble: "Nobody Understands Quantum Mechanics"

The reality of the world, i.e. its origin, structure and functioning, certainly remains an enigma, both for its ontological complexity and for the cognitive limits of all living and aware species, including ours. We know it well and yet there is something dissonant if one of the greatest physicists of the 20th century (and Nobel Prize winner in 1965), Richard Feynman, declares that he can "safely say that nobody understands quantum mechanics" (Smolin 2019, p. xxvii).

Whilst this reassures those who do not possess the mathematical and formal tools necessary to understand contemporary physics, it also shows a gnoseological checkmate that is disturbing. Even more so in view of the distance – an abyss – between quantum theory and the phenomenology of everyday experience.

 | HTTPS://DOI.ORG/10.1163/9789004756496-010

Another stimulus to deal with the link between ontology and quantum physics is well expressed in these words by Francesco Coniglione: "And in fact only through the clarification of the relationship that philosophy weaves with science (and therefore whether or not it is the epistemology of science) and the clarification of the method he uses in carrying out his investigations, which can shed light on the problem of scientism and on the status of contemporary epistemology" (Coniglione 2014–2015, p. 101).

2 Anti-realism

The basic limitation of quantum mechanics is its radical anti-realism, as elaborated by the so-called Copenhagen interpretation, i.e., the research by Bohr, Heisenberg, Dirac, Pauli. These were countered by a different reading by the theory's initiator himself, i.e. Einstein, with whom de Broglie and Schrödinger agreed. Indeed, Einstein was an ontological realist who could not accept the mystical and certainly anthropocentric anti-realism of the interpretation developed in the 1920s and in fact dominant until the end of the 20th century. Subsequently, the situation changed, albeit with obvious and strong resistance from the scientific-academic community that was oriented towards the dominant paradigm.

Quantum physicist Lee Smolin had the courage to rethink the question from the ground up. And he has done so starting from the observation that quantum theory is wrong in the sense that it is incomplete and therefore it is necessary to go beyond its limits. One merit of Smolin is that he never separates scientific analysis from the social context in which science happens. On the basis of his personal experience and against the backdrop of the philosophy of science of Popper, Kuhn and Feyerabend, Smolin reconstructs the history and describes the theory of a truly emblematic and enormously important event in contemporary culture.

At first it was Einstein, who clearly showed that light has a dual nature, being *both* a wave and a particle. However, he then always rejected the anti-realist consequences drawn from the dual structure of matter, such as those of Heisenberg and Bohr, who believed that quantum mechanics and any scientific theory cannot talk about what exists but must limit themselves to saying what is the result of human observation alone. For them, elementary particle physics only deals with "observables" and not "beable", i.e. it does not deal with how atoms are actually made and their "objective" properties. Properties that quantum mechanics considers to be constitutively unknowable and *therefore* non-existent. As can be seen, the Kantian noumenal device and idealistic metaphysics in general is at work here.

Opposed to this are realist metaphysics and epistemologies – such as Einstein's and Smolin's – which hold instead that reality is autonomous from any observation; that entities, events and processes exist independently of any consciousness that perceives, observes, or measures them. Science and philosophy constitute an attempt to understand these structures and not only to understand the *ways* in which a mind represents structures.

This ontological-epistemological knot is particularly intricate in quantum mechanics. Complexity that is also given by the fact that the turning point in it is the concept of field by which matter and energy are unified in a set of forces that do not operate *in* space and time but *are* space and time: "Instead, we must think that there is a real entity spread everywhere in space, which is modified by the electric and magnetic bodies and which, in turn, acts (pushes and pulls) the electric and magnetic bodies. This entity, whose existence is intuited by Faraday, is today called 'field'" (Rovelli 2014, p. 51).

Einstein's mathematical effort consisted mainly in working out equations that showed how the Riemann R_{ab} curvature of spacetime postulated by relativity is proportional to the energy of matter, so that the denser the matter, the more spacetime curves. Quantum mechanics also incorporated atoms and every possible particle into the field, so that the world would not be made up of *fields* and *particles* but of a single reality that is the covariant quantum field, whose fundamental characteristics are granularity, indeterminism, and relationality.

Smolin shows the epistemological roots of these theses through two essential principles and rules. The first can be expressed in this way: "We can only know half of what we would need to know if we wanted to completely control, or precisely predict, the future" (Smolin 2019, p. 14). From the very beginning, quantum mechanics is therefore incomplete, and it is incomplete by definition. Why does this happen? The answer lies in the two basic rules that guide such mechanics. The first, also called *Schrödinger's equation* or the principle of *unitarity*, states that "given the quantum state of an isolated system at one time, there is a law that will predict the precise quantum state of that system at any other time" (Smolin 2019, p. 27). A principle therefore firmly rooted in the scientific-philosophical tradition and clearly deterministic in character.

Instead, Rule No. 2 states a radical probabilistic modality, in the sense that the presence of an observer becomes decisive: "The outcome of a measurement can only be predicted probabilistically. But afterward, the measurement changes the quantum state of the system being measured, by putting it in the state corresponding to the result of the measurement. This is called collapse of the wave function" (Smolin 2019, p. 31). The "collapse of the wave function" indicates a fundamental modality of quantum mechanics, which consists in postulating that, in a quantum system, the result of the measurement performed by an observer for some observable causes the observable to precisely assume the measured result.

3 Cats and Particles

As we can see, all this clarifies the meaning of one of the most famous thought experiments of twentieth-century physics: Schrödinger's cat which is both alive and dead (superposition of states) and becomes alive or dead only in relation to the result of the observation.

The paradoxes generated by such rules and principles are easily avoided if we consider that this theory is valid in the sphere of elementary particles, of the subatomic world, but no longer applies to macroscopic entities, to the world in which we are immersed, to the world that we ourselves are: cats, bicycles, humans, stars. A fundamental difference that is based on *decoherence*, i.e. on the

> process by means of which irreversible changes by averaging out the random chaos of the atomic real. Decoherence is a very important feature of quantum theory, for it is why the bulk properties of large-scale objects, such as the rough motions of soccer balls, swing bridges, rocket ships, planets, and so forth, appear to have well-defined values, which obey the laws of Newtonian physics. The word "decoherence" refers to the fact that such bulk objects appear to have lost their wave properties, and so they behave as if they are simply made of particles. According to quantum mechanics, everything, including cats, soccer balls, and planets, has wave as well as particle properties. But for these bulk objects, the wave properties have been so randomized by their interactions with their chaotic environment that they cannot be accessed in any experiment, so the wave half of the wave-particle duality has been rendered mute, and the objects behave like ordinary particles.
>
> SMOLIN 2019, p. 128

This is a central statement that must be thoroughly understood. For quantum physics, reality is dual. It is composed of both waves and particles. In the sense that *everything* is made up of waves and particles. And this is because, as Einstein discovered, energy is carried by light in discrete ensembles ('packets'), which he called *photons*. Light, however, consists of a collection of electromagnetic waves.

Einstein resolved the issue with the hypothesis that the energy of photons (the discrete component of light) is proportional to the frequency of the light wave that carries them. The hypothesis was much better articulated first by Louis de Broglie (1927) and then by David Bohm (1972), who elaborated the *pilot wave theory* for which matter is constituted and described in its entirety by the ensemble of waves and particles, since *the particle is guided by the wave*.

This solution, however, is a source of further problems, starting from the dualistic ontology and the incompatibility with the theory of relativity.

Decoherence refers to another central pair of concepts, this very simple one: the *excited* states and the *ground* states of atoms. In the excited state the atom is unstable, and it is the condition in which it finds itself before "decaying", i.e., entering the state of stability which follows by emitting a photon – therefore losing energy –, emission after which it "subsides" reaching its ground state.

The life or death of the cat depends on a Geiger counter to which the cat's ear is connected. The counter detects the excited condition of the atom (the cat stays alive) or the fundamental one (the cat dies). According to Rule No. 2, however, it is only the measurement that collapses the wave function and thus establishes the state of the atom and consequently the fate of the cat. Prior to this measurement the cat is both alive and dead. As seen,

> Rule 2 raises a whole bunch of questions. Does the wave function collapse abruptly or does it take some time? Does the collapse take place as soon as the system interacts with the detector? Or only later, when a record is made? Or perhaps later still, when it is perceived by a conscious mind? Is the collapse a physical change, which means that the quantum state is real? Or is it just a change in our knowledge of the system, which means the quantum state is only a representation of that knowledge? How does a system know a particular interaction has taken place with a detector, so that it should then, and only then, obey Rule 2? What happens if we combine the original system and the detector into a larger system? Does Rule 1 then apply to the whole system?
>
> SMOLIN 2019, p. 128

4 Realism

Questions and issues that have been debated for a century but apparently in vain. According to Smolin and others, the time has come to really address them, abandoning the idealistic Copenhagen paradigm and assuming as more coherent the realist paradigm, within which measurements cannot be ascribed decisive significance.

Bohr and Heisenberg believe that speaking about atoms, the world and reality, is meaningless beyond and outside the experimental and conceptual approach related to human knowledge. Physics, in short, is not about electrons but about how we *talk* about our interactions with electrons. In general, for Bohr and Heisenberg, the sciences do not describe what exists but keep track of what is observable with *instruments*; the sciences do not study nature but are concerned with how humans know supposed nature from the instruments they use to investigate it. This is, as we can see, a radical and idealistic form of *instrumentalism*. For physical-ontological realism, on the other hand, the wave function is an aspect of reality and not just the human way of measuring it; the quantum state is physically real, not a numerical representation that exists only in the human mind. Realism upholds the existence of the world independently of any observer. And therefore, even in the absence of humans observing it.

Without this assumption and without such ambition, however modestly practiced at all times, there is no science but magic, superstition, science fiction. A plethora of hypotheses and theories show this, from Everett's tendentially realistic one to the more recent string theories, multiverse theories and relativity itself in many of the interpretations of them.

This is a kind of physics similar to *Sliding Doors*, the movie directed by Peter Howitt in 1998, in which the life of a single character unravels into two different and both coexisting *branches*. In the many-worlds hypothesis, in fact, every time an experiment is performed that can yield different results the universe splits into parallel worlds where the possibilities all unfold. And this means that of every event, entity and consciousness countless stories and versions are given. Provided, however, that each development is separate, and the different worlds do not interact.

As we can see, we are beyond destiny, beyond Ἀνάγκη, lost in an irrationalism that is the other name for mathematical formalism when one confuses a tool, as wonderful as mathematics is, with the material density of the world, of entities, of events, of processes. An irrationalism that claims to be convinced of the existence of an infinite number of copies of every entity, including each human being, convinced that the innumerable set of *possible* events all exists somewhere. We, I who am writing, you who are reading, are in turn one of many copies. Reality dissolves and physics becomes "a branch of fantasy literature" (Borges 1991, p. 631).

5 Dogmas

The belief of many contemporary physicists that time is unreal constitutes a form of mathematical Platonism that has, however, abandoned the deep connection that Plato feels with reality as a whole and as a problem, putting in place of this ontological seriousness the simple allure of mathematical elegance and formalism, which, however, in no way guarantee the truth of their assertions but only the need to eternalize themselves. Much more valid, therefore, for some currents in contemporary physics than for Plato is Smolin's correct observation that "there's a cheapness at the core of any claim that our universe is ultimately explained by another, more perfect world standing apart from everything we perceive. If we succumb to that claim, we render the boundary between science and mysticism porous" (Smolin 2013, p. 11).

Many contemporary physical and cosmological theories, if not erroneous, nevertheless remain approximations, referring to some small parts of the universe that can be represented by means of timeless mathematical objects. The unrestrained Platonism of these theories lies in the

> absurdity of the view that mathematics is prior to nature. Math in reality comes after nature. It has no generative power. Another way to say this is that in mathematics conclusions are forced by logical implication, whereas in nature events are generated by causal processes acting in time. This is not the same thing; logical implications can model aspects of causal processes, but they're not identical to causal processes. Logic is not the mirror of causality. Logic and mathematics capture aspects of nature, but never the whole of nature.
>
> SMOLIN 2013, p. 246

Mathematics is a magnificent and very powerful tool for serving science; it is not science that serves mathematics. The Galilean idea that the world is written in mathematical characters has contributed to the understanding of many aspects of reality, but it does not describe it in its entirety, nor can it replace reality as such. As Francesco Coniglione clearly states, "the solution of methodological problems and the mathematical methods currently in use – unlike what happens in the natural sciences – do not offer us guidance for solving problems of existence and ontology in general, nor for epistemological questions regarding how human beings come to acquire mathematical beliefs" (Coniglione 2002, pp. 298–299).

A consequence of Galilean Platonism is Newton's method, which is quite suitable for describing internal dynamics of matter but not suitable for cosmology. One need only think of the idea of absolute time and space, which disregard the existence of entities, their becoming, movement, and change. Applied to what Smolin effectively calls "physics in a box", the Newtonian framework works very well but it "loses its power to explain the causes of physical phenomena" (Smolin 2013, p. 107) and "*it falls apart when we attempt to apply it to the universe as a whole*" (Smolin 2013, p. XXIII).

Relativity theory and quantum mechanics are mutually incompatible but share the Newtonian primacy of mathematics. Physicists who practice such a paradigm behave like laboratory zoologists, who study animals under totally artificial conditions, reducing their behavior to predetermined and abstract patterns. Just as ethology liberates animals by studying them in the concreteness of their environments, so the time-reality paradigm liberates the complexity of becoming from its reduction to equations.

6 The Becoming

The movement, becoming, possibilities, phenomena, and matter constitute the complex and immense being that unfolds in every recess of time and space. From the basic principles of both Newtonian and Einsteinian physics, such complexity is improbable. Yet it is there, and it happens It is this reality that must be explained, even if the transition from the simple to the complex is not found in equations, is not predicted by the operation of timeless structures.

Rovelli also correctly states that "the happening is widespread, scattered, disordered, but it is happening, not stasis. [...] The fundamental equations do not include a time variable, but they do include variables that change with respect to one another. [...] But this does not remove the fact that the world is incessantly changing" (Rovelli 2017, p. 86). And he also adds that "the absence of the quantity 'time' in the fundamental equations does not mean a frozen and immobile world. On the contrary, it means a world where change is ubiquitous, without being ordered by Father Time" (Rovelli 2017, pp. 85–86). Indeed, time is the identity of the line generated by events – a line that is certainly not uniform but rhizomatic and vibrant – and the difference of the events generated: different, layered, infinite, indeterminate.

In the face of the complexity of the material world and its mathematical expressions, "physics in a box" exchanges the part for the whole, the laboratory for nature, the abstraction of an isolated fact for the concreteness of relations within which alone every entity, every event and every process can happen – ontology – and can be explained – epistemology. Like many physicists, Smolin started from the thesis of the unreality of time but had to reconsider and boldly went on to argue: an *epistemology* for which "the truest description of something consists of specifying its relationships to the other parts of the system it is part of" (Smolin 2013, p. XVI); a *temporal ontology* according to which time "is the key to the meaning of quantum theory and its eventual unification with space, time, gravity, and cosmology" (Smolin 2013, p. XII); a *methodology* that considers scientific only that which produces predictions that can be falsified; *a courageous temporal conception even of scientific laws*, removed from Platonic structures beyond concrete, experiential, phenomenal time and space.

Phenomenic time is the real time; "time and its passage are fundamental and real and the hopes and beliefs about timeless truths and timeless realms are mythology. Embracing time means believing that reality consists only of what's real in each moment of time" (Smolin 2013, p. XIV) and that such reality consists of unstoppable, directional, continuous motion. The block-universe of Einsteinian invariance theory; Julian Barbour's vision of discrete moments that all remain eternal and for which "the only true things are complete possible configurations of the universe, unchanging Nows. Unchanging things do not travel in time from Now to Now. Material things, we included, are simply parts of Nows" (Barbour 1999, p. 49); the consolation that all this brings with respect to mortality and finitude, are all forms of eternalism that are well known to the metaphysical tradition and that find their original and most powerful expression in Parmenidean ontology, to which Smolin contrasts a Heraclitean physics and ontology within which the Anaximander fragment for which all things originate from one another, and dissolve into one another, according to a necessity that conforms to the order of time, shines.

Regarding the now, which Barbour also talks about, it is certainly not identical in different places – given the finite speed of light – but that does not mean it does not exist. The ultimate outcome of such a position would be ontological nihilism, the dissolution of being. The now exists. It is the relations that combine the different "nows" that are no longer knowable from a certain distance onward. Not only that: the supposed dissolution of the now in relation to the diversity of times and places would actually be a full confirmation of the power precisely of time. A paradox of which many physicists are not aware. Even the fact that "the difference between past and future is not in the elementary equations of the world" (Rovelli 2017, p. 81) implies that the world cannot be reduced to its elementary equations, mathematics, numerical and quantitative interpretation of being. The world is not reduced to epistemology, it is not reduced to a well-defined and historically situated interpretation of being and becoming, such as Aristotelian, Newtonian, quantum, relativistic, or other (see Coniglione 2011).

Relationalism implies the renunciation of the principle of the "relativity of simultaneity and embracing its opposite: that there is a preferred global notion of time. Remarkably, this does not require overthrowing relativity theory; it turns out that a reformulation of it is enough" (Smolin 2013, p. 165). In fact, if in special relativity events are simultaneous only when they occur in the same place, "in a quantum universe where every particle is potentially one step away from every other particle, everything is essentially 'at the same place.' In such a model, there's no problem synchronizing clocks, so there is a universal time" (Smolin 2013, p. 191). The transformation of relativity and the attempt to make it compatible with quantum physics is accompanied by a reformulation of thermodynamics as well, which excludes the concept of thermal death in a final state of entropic equilibrium and of always the same and returning events, since if

> normal thermodynamic systems end up in the single, boring state of uniform equilibrium; gravitationally bound, anti-thermodynamic systems end up in one of a large number of highly heterogeneous states [...] only kind of universe that appears natural from the timeless perspective of the Newtonian paradigm is a dead universe in equilibrium, obviously not the kind we live in. But from the perspective of the reality of time, it is entirely natural that the universe and its fundamental laws be asymmetric in time, with a strong arrow of time that encompasses increases of entropy for isolated systems together with continual growth of structure and complexity.
>
> SMOLIN 2013, pp. 225–226

And this does not imply, as Smolin seems to me to tend to think, a denial of determinism. If freedom of will is an illusion, it is so not because necessity *precedes* the becoming of entities and events but because *necessity happens while it becomes.*

The Anaximandrian and Heraclitean foundation of such proposals has as its fundamental link the Leibnizian principles of sufficient reason and the identity of indiscernibles. Indeed,

> if time is real, it should be impossible to have two different but identical moments of time. Time is fully real only in a Leibnizian universe. A Leibnizian universe will be full of complexity that generates a bountiful array of unique patterns and structures. And it will be ever changing, to ensure that every moment can be distinguished from every other by the structures and patterns present then. As indeed is our universe.
>
> SMOLIN 2013, pp. 216–217

7 Conclusion: "The World Is Not a Set of Things, It Is a Set of Events"

Time is thus confirmed as the most important theme and problem of physics and of all possible sciences that do not wish to assume a mystical-mathematical dimension and language. "The hypothesis of the reality of time leads to a more scientific cosmology" (Smolin 2013, p. 241) because it leads to a cosmology that no longer needs the physical, religious and ethical attempts to console us of the nothingness in which every entity is destined to dissolve as an entity in this form here, in this given mineral, vegetable, animal, atomic and cosmic structure. The thesis of the actual, total, pervasive reality of time is more scientific since "the world remains, always, a bundle of processes evolving in time, and only small parts of it are representable by timeless mathematical objects" (Smolin 2013, p. 245), because the world is time, and philosophy consists of the phenomenal and rational understanding of the temporal structure of the cosmos and every part of it.

Far from "not existing," time is thus the very existence of every entity, event and process, even in the quantum sense that there are no *objects* unrelated to each other but only events linked to each other in an indissoluble way:

> The world of quantum mechanics is not a world of objects: it is a world of elementary events, and things are built on the future of these elementary "events." As philosopher Nelson Goodman wrote in the 1950s, using a beautiful expression, "An object is a monotonous process," a process that repeats itself the same for a while. [...] We, like waves and like all objects, are a flow of events, we are processes that for a short time are monotonous ... Quantum mechanics does not describe objects: it describes processes and events that are interactions between processes.
>
> ROVELLI 2014, pp. 118–119

Time is said in many ways: physical, thermodynamic, biological, psychological, mechanical, social, conventional, existential. Humans are one of the temporal forms in which matter is structured.

And therefore

> one can think of the world as consisting of things. Of *substance*. Of *entities*. Of something that *is*. *That persists*. Or think of the world as consisting of *events*. Of *happenings*. Of *processes*. Of something that *happens*. That does not last, that is continually transforming. That does not persist in time. The destruction of the notion of time in fundamental physics is the collapse of the first of these two perspectives, not the second. It is the realization of the ubiquity of impermanence, not of the static nature of an immobile time.
>
> ROVELLI 2017, p. 87

Well before the 20th century and its quantum and relativistic investigations, this is the perspective of much philosophy, beginning with Anaximander and Heraclitus: "The world is not a collection of things, it is a collection of events" (Rovelli 2017, p. 87). Events that exist and happen independently of any body-mind that observes them.

References

Barbour, J. (1999). *The End of Time. The Next Revolution in Physics*. Oxford: Oxford University Press.

Borges, J. L. (1991). *Tlön, Uqbar, Orbius tertius*, in Id., *Tutte le opere*, edited by D. Porzio. Milan: Mondadori.

Coniglione, F. (2002). *La parola liberatrice. Momenti storici del rapporto tra filosofia e scienza*, Catania: C.u.e.c.m.

Coniglione, F. (2011). La nascita dell'epistemologia evoluzionistica e la morte della filosofia. *Vita pensata*, 12.

Coniglione, F. (2014–2015). Le molte facce dello scientismo e il significato della filosofia. *Complessità*, 1–2, 99–120.

Rovelli, C. (2014). *La realtà non è come ci appare. La struttura elementare delle cose*. Milan: Raffaello Cortina Editore.

Rovelli, C. (2017). *L'ordine del tempo*. Milan: Adelphi.

Smolin, L. (2013). *Time Reborn. From the Crisis in Physics to the Future of the Universe*. Boston-New York: Houghton Mifflin Harcourt.

Smolin, L. (2019). *Einstein's Unfinished Revolution. The Search for What Lies Beyond the Quantum*. New York: Penguin Press.

CHAPTER 9

Elements of Solar Theology in Proclus: between Astronomy and Theurgy

R. Loredana Cardullo

Abstract

In Plato's *Timaeus* the sun is simply a planet, though characterized by a blazing light, illuminating but not burning. In the *Republic* it plays a more philosophically elevated role, as the sensible and divine representation of the Good, the supreme Idea, end and purpose of every entity. However, a true solar cult is still far from ancient Platonism; it would impose itself, as the cult of *Sol invictus*, at the time of the Roman emperors Aurelian and Constantine. In late antiquity, the Neo-Platonists would begin to elaborate a kind of astro-theology, grafting Chaldean and Orphic religious beliefs onto the trunk of Platonic mathematical physics, harmonized with traditional Olympian belief, even in an anti-Christian function. This contribution delves into the position of Proclus, as a representative – also by virtue of a larger number of writings at our disposal, compared to those of other Neoplatonists – of a solar theology, the result of a perfect harmonization between Platonic astro-ontology and theurgical-oracular theology wisdom.

Keywords

Solar Theology – Sun – Idea of Good – Theurgy – Proclus

1 The Sun in Plato between Ontology and Astronomy: Outlines

> Which of the gods in heaven can you point to as the lord responsible for this, whose light makes our sight see in the finest way and the seen things seen? "The very one you and the others would also point to," he said. "For it's plain your question refers to the sun".
>
> PLATO, *Republic* VI 508a 4–8

 | HTTPS://DOI.ORG/10.1163/9789004756496-011

In the extensive and varied production of Francesco Coniglione – to whom this study is dedicated – there is no shortage of his reflections on ancient philosophy, which have focused, among other things, on Plato's so-called esotericism, reconsidered in connection with his "unwritten doctrines" (Coniglione 2018) and the transition from the dominance of *logos* to the triumph of *pistis* in the late ancient world (Coniglione 2021). Ideally, I would like to continue the discussions Coniglione presented in some of his writings, and in this article I aim to address in more detail a specific topic he dealt with: the transition from the Platonic conception of astronomy, and the way in which the nature and function of the Sun was conceived in it, to the doctrines of late Neoplatonism, and in particular Proclus, in which the 'esoteric' background of Plato's teaching – as demonstrated by Coniglione – decisively takes over, transforming Plato's cognitive ascent toward the intelligible realm culminating in the dialectic, into a veritable magical science, into a theurgy, in which rational discourse takes a back seat in favor of a knowledge reserved for the few and to be obtained through an initiatory ritual, much like that of the Mysteries.

To elucidate this passage, it is worth recalling how for philosophers in the Platonic tradition, astronomy, as a mathematical science, occupies an important role in the education of the philosopher. In the central books of the *Republic*, Plato had included mathematical studies in the *curriculum studiorum* of the rulers (the φύλακες), in addition to the classical gymnico-musical *paideia* addressed to all, and placed it in a propaedeutic position with regard to dialectics, the supreme science identified with philosophy, which crowned the training of the rulers as such (the ἄρχοντοι), i.e., the philosophers.[1] The five mathematical disciplines – arithmetic, geometry, music, astronomy, and stereometry – allowed the man who challenged himself with them, not without hard work and considerable commitment,[2] to learn to think based on ideal models and no longer on sensible, perceptible objects, the domain of δόξα, and to abandon the form of knowledge of the "majority of people," which is confused and subjective, and rise to the real science, the ἐπιστήμη, attainable through the intellect, or νόησις, at the culminating stage of the cognitive process. The mathematical sciences, perfect and rigorous, were not suitable for everyone, according to Plato, but only for those who manifested from an early age a particularly inclined disposition for studies, a "golden" character, according to the law of the three metals, the "noble lie" set forth in *Republic*

1 In *Republic* VII 534e2–3, dialectic is defined as "the top tier (θριγκός) of the studies like an architectural embellishment".

2 *Republic* VII 526c1–3: "I don't suppose you would easily find many studies that take greater effort in the learning and in the practice than this".

III 414a7–415b8. This person had to be taught through the mathematical τέχναι before he could acquire the truly epistemic knowledge of philosophy.[3] "Nobody ought to cross this threshold who is ignorant in Geometry" warned the inscription on the pediment of the Academy, and although the Platonic school was primarily a center of political education, mathematics, with geometry at its apex, considered the most rigorous mathematical science,[4] played a leading role there; that form of reasoning "by hypothesis" was taught from which, as from a prop, one would begin "the ascent (of the philosopher) to what is," towards the anhypothetical principle of the Good (*Republic* VII 521c7). Indeed, mathematical knowledge is for Plato the only form that draws the soul "from becoming to being" (*Republic* VII 521d4–5), facilitating "the conversion of the soul itself from the world of becoming to truth and being" (VII 525c4–7).

In the didactic curriculum addressed to the rulers of the *kallipolis*, astronomy represents the fourth mathematical discipline (*Republic* VII 528e1–2), after arithmetic, or the science of calculus, and plane and solid geometry, the sciences of surfaces and solids, respectively. Platonic astronomy, illustrated in the *Timaeus*, is the study of solids in motion, of stars and planets moving in their orbits, traversing the sky; however, it is not classifiable among the natural sciences since the "true astronomer" (τῷ ὄντι δὴ ἀστρονομικόν), as a good mathematician, considers celestial bodies and their movements as images of higher realities, which are their paradigmatic cause, and in that capacity he observes them as a medium for tracing their rational models. The "embroideries" (ποικίλματα) visible in the heavens – stars and planets in motion – are nothing but reflections of a work of beautiful execution – that of the Demiurge – but "it would be ridiculous (γελλοῖον) – Socrates states in *Republic* VII 529e4 – to consider them seriously (ἐπισκοπεῖν αὐτὰ σπουδῇ), as though one were to grasp the truth about equals, doubles, or any other proportion in them." With rational thought (διάνοια) and not with sensible sight (ὄψις), the astronomer grasps the intelligible figures of the heavenly bodies, and with them speed in itself, slowness in itself, reasoning through problems, as the geometer does with surfaces and solids in themselves, and the arithmetician does with number in itself (*Republic* VII 529c6–530b7).

3 On the technical and not yet epistemic knowledge of the mathematical disciplines, see *Republic* VI 511a10 ff. In this passage Plato distinguishes mathematical reasoning – the διάνοια – from pure thought – the νοῦς – through the well-known analogy of the divided line.

4 *Republic* VII 527b6–10: "geometry is knowledge of what is always [...] able to draw the soul towards truth and to produce philosophical thought, to turn the gaze upwards, that we now erroneously hold down".

Unlike Aristotle, for whom physics, both terrestrial and celestial, as well as biology and zoology have a real epistemological status, since they investigate the sensible world, which is the only dimension in which the causes of being can be grasped, for Plato, on the other hand, though the inquiry into the sensible and perceptible nature can be declined in many ways, it makes sense only insofar as it refers back to a higher dimension, the intelligible dimension of the models of things, to true being, of which material entities participate as ontologically inferior copies. For Plato, a science of the sensible cannot exist because it would be an empty inquiry, conducted on shadows, images, and simulacra. That is why he leaves the description of the production of the Cosmos to Pythagorean *Timaeus*, who teaches us to investigate the sensible world with the tools of mathematics.

When philosophers in the Platonic tradition began commenting on the *Dialogues* in imperial times, in order to find in them the fundamental principles of Platonism and build from them a cohesive and coherent doctrinal system, the *Timaeus* was interpreted in contrast and opposition to the Aristotelian *Physics* and *De caelo*, as a compendium of a "Pythagorean-Platonic" natural philosophy, in which the physical universe is understood as the result of a demiurgical operation, conducted on the model of an "eternal living being" and one that is intelligible[5] but based on strict mathematical laws and relationships. The specifically astronomical section of the *Timaeus* assigns to the creator God the conjoined generation of time and the cosmos, the distribution of the seven planets (Sun, Moon, Venus, Mercury, Mars, Jupiter, and Saturn) in their own orbits, and the determination of a close connection between the movements of the planets and the "distinction" and "conservation" of time. Time, the "moving image of eternity," punctuated by the motion of the planets, represents the transposition of the supreme and intemporal order of motionless and eternal being into the sensible dimension, in which everything moves and changes, expressing, imperfectly, the perfect numerical structure of the intelligible world. The motion of the planets measures sensible time; the light of the sun, "lit up" by the Demiurge,[6] determines the passing of day and night; the moon, completing its course toward the sun, determines the month; and the sun, completing its circle, realizes the year (*Timaeus* 39c3–5).

5 *Timaeus* 38c1–3: "the model exists *for all eternity*, while the universe was and is and always will be *for all time*".

6 *Timaeus* 39b2–c1: "In order that there might be a clearly visible way to measure their relative speeds as they journey on their eight revolutions, the god ignited a light (φῶς), in the circuit that is second closest to the earth, and we have come to call this light the sun (ἥλιον). He created it to illuminate as much of the universe as possible and to enable all suitably endowed creatures to become numerate by studying the revolution of identity and sameness".

In Plato's cosmological dialogue, the sun is a planet like any other, albeit characterized by a blazing light, which illuminates but does not burn, because it consists of a fire of a higher nature, different from and ontologically superior to the fire down here. Instead, in the *Republic*, it plays a philosophically elevated role, as a sensible and divine representation of the Good, the supreme idea, the essence of other ideas, the end and goal of every entity. In *Republic* VI 508a4–8, a passage placed in the exergue to this paper, Plato speaks of the sun as one "of the gods of heaven." But this is still only a metaphor, an analogy that serves to symbolize the light and creative power of the Good, the source of truth, life and order for the entire universe: "it is precisely this that I call the offspring of the good; and the good generated it in analogy with itself: as in the noetic sphere good pertains to thought and its objects, so in the visible sphere the sun pertains to sight and its objects" (*Republic* VI 508b12–c2). True cult of the sun is still far from ancient Platonism; it was to impose itself, as adoration of the *Sol invictus*, at the time of the Roman emperors Aurelian and Constantine, perhaps also influenced by the Persian Mithraic cult, which had been imported to Rome by armies returning from campaigns in the East.[7]

In late antiquity, Neo-Platonic philosophers would begin to elaborate a kind of astro-theology, grafting Chaldean and Orphic religious beliefs onto the trunk of Platonic mathematical physics, harmonized with traditional Olympian belief, even in an anti-Christian function.[8] The sources provide us with valuable information regarding a special interest that Plotinus, Porphyry, Iamblichus, Julian, Syrianus and Proclus, among others, manifested in the sun, no longer understood merely as a celestial body, albeit primary source of light and life for all living things, but as a true deity.[9] However, a distinction must be made between Plotinus' position and that of the later Neoplatonists mentioned above, for while the latter are all influenced by theurgical writings, from which they derive their particular veneration of the sun, Plotinus, on the other hand, condemns everything pertaining to magic in general. In the *Enneads*, the occurrences of the term ἥλιος refer either to Platonic cosmological contexts or to a form of metaphysical religiosity, fully adhering to the doctrine of the *Republic*, which sees in the sun the visible manifestation of the intelligible supreme principle, the One/Good,[10] without any reference to those oracular

7 For an overview of solar theology in Rome, I refer to Halsberghe (1984).

8 On the anti-Christian controversy of pagan philosophers, see Zambon (2019).

9 In general, in late antiquity the god most revered was the Sun or Helios (see Siorvanes 1996, p. 306).

10 Some examples: in *Enneads* IV 3,11,16 the intellectual Hypostasis is called the Sun, by analogy with the physical sun; in *Enneads* V 1,2,41, we read that the Sun is a God because it is endowed with a soul; in *Enneads* V 5,8,7–8, the Sun is said to imitate the νοῦς; *Enneads*

theories in which the mostly lost writings of the post-Plotinian Neo-Platonists are steeped: from Porphyry's *On Sun* to the treatise *On the Gods* by Iamblichus. The latter's philosophical-theological theories – now deeply intertwined with oracular and theurgical wisdom – were to determine a turning point in Neoplatonism of late antiquity, as exemplified by Julian's *Hymn to the King Helios* and the theurgical writings of the Neoplatonists of the Athens school, including the *Hymn to Helios* composed by Proclus.[11]

In the remainder of this contribution, we will simply examine Proclus' standpoint in depth, as representative – also by virtue of the greater number of his writings left to us, compared to those of other Neo-Platonists – of a solar theology, the result of a perfect harmonization between Platonic astro-ontology and theurgic-oracular wisdom.

2 The Sun in Proclus between Astronomy and Theurgy

> If it were up to me, of all the ancient books I would leave
> in circulation only the *Oracles* and *Timaeus*
>
> MARINUS, *Life of Proclus* 38

Late antiquity has handed down to us a magnificent commentary on the *Timaeus*, written by Proclus at the age of twenty-eight, in which Plato's physicist-cosmological teaching is confirmed and complemented by the Chaldean and Orphic culture. In his commentary on the *Republic*, Proclus also interprets Platonic theories, including sections on the image of the Sun as a visible representation of the Good, through the filter of various theological principles, both Greek and Eastern in origin. On the other hand, Syrianus, his teacher at the Athenian school, had already written a vast work entitled *The Harmony of Orpheus, Pythagoras and Plato with the Oracles*, and had taught him and Hermia the theories of the great theologians, Homer and Hesiod, and the Orphic and Chaldaean texts, in addition to Aristotelian and Platonic philosophy, as Marinus testifies in the *Life of Proclus* (chapters 12, 13, 26). Harmonizing different philosophical and theological traditions together, to paraphrase the title of a famous study by Saffrey (1992), was a characteristic

---

IV 4,35,37 ff. illustrates the effects the sun and other stars have on earthly affairs; in *Enneads* VI 7,17,24–30 the Platonic analogy between the Good and the Sun in Republic VI is reiterated in much the same terms.

11 For treatises on solar theology, which became a genuine literary genre in late antiquity, I refer to Bouffartigue (1992).

feature of Athenian Neoplatonism, just as it was a typical operation of this late phase of the Platonic tradition to transform philosophy into a "scientific" theology, and Plato into the greatest theologian. Now, these masters – in particular Proclus, but before him Iamblichus – while being great thinkers, capable of profound metaphysical speculations and rigorous logical-mathematical arguments, were also "spiritual fathers," hierophants, mystagogues, theurgists, and passionate supporters of a polytheistic religiosity, that was to be expressed through cultic practices and prayers, perfectly reconcilable with their henotheistic metaphysics.[12] It is not surprising, therefore, that Proclus, alongside writings on astronomical and mathematical subjects, also left us impressive theological and theurgical works, and numerous hymns dedicated to the gods; or if, along with his activity as a professor and exegete, he also practiced as a healer and thaumaturge (*Life of Proclus* 28,29). As Coniglione argued (2021, p. 201) "while Christianity moves from the original belief without justification to the search for a way of understanding the 'rationality' of faith, in contrast in Proclus *pistis* is freed from the condemnation previously inflicted on it in the name of the *logos*: it now denotes a supersensible and supra-rational activity, which enables union with the Good, and thus ends up representing the most valuable and elevated aspect of man's sapiential activity, which goes beyond rational and sensible knowledge, superior to any merely human wisdom or science". Indeed, for Proclus now

> it is not through intellection nor through judgment that initiation generally takes place, but through the silence that is unitary and superior to all cognitive activity, silence that is provided to us by faith, fixing in the ineffable and unknowable nature of the gods the universal souls and our own at the same time.
>
> PROCLUS *Platonic Theology* IV 9,31,12–17

Devotion to the gods of traditional pagan religion did not, in fact, conflict with Platonic philosophy or theurgical practice because in the Proclian universe "all things are full of gods,"[13] each deity finds its own place in the different ontological levels of the universe, and each thing, by virtue of a cosmic *sympatheia*, bears within itself the marks of the divine chain to which it belongs

---

12 On the expression "solar henotheism", see De Vita (2017–18), p. 460 footnote 9.

13 Chlup (2012), p. 73: "in common with Thales Proclus was convinced that 'all things are full of gods', extending the dominance of the gods down to the lowest things (*Elements of Theology* 140). If the latter cannot be divine themselves, at least they are closely linked to the gods".

(d'Hoine-Martijn 2017, p. 225). These are the typical aspects present in every esoteric wisdom and doctrine (Coniglione 2018, pp. 6–20), and which now re-emerge forcefully in Proclus' conceptions, breaking the delicate balance they wove with philosophy in Plato's thought, so it can be said that with Proclus and in general with the post-plotinian Neo-Platonists, the Greek *logos* denies itself in order to take refuge in the *pistis* (Coniglione 2021, p. 202), understood as a "concept-boundary representative of the life of the absolute and in the absolute" (Romano 2004, p. 4). Hence, within this syncretistic framework, the sun takes on a dual role. In accordance with the Platonic astronomy outlined in the *Timaeus*, it functions as a planet. Simultaneously, it stands as the most radiant embodiment of the deities forming the "heliacal chain," an entity of divinity itself, the center and pulsating core of the visible world, a doctrine imparted by the Chaldean theurges and the Orphic sages. This harmonious arrangement does not breed contradiction, as Proclus discerns an impeccable correspondence and *symphônia* between Plato's thought and the teachings of the *Oracles* and *Orphics*.

2.1 *The Sun in Proclus' Commentaries on the Timaeus, the Republic and the Cratylus and in Platonic Theology*

In the sylloge of the *Orphic Hymns*, the Sun, like the Moon and the stars, is a celestial deity. Dedicated to the Sun is hymn n. 8, εἰς Ἕλιον θυμιᾶμα, a sacred prayer, called a "perfume" and not a hymn (ὕμνος), because in this mysterious and esoteric religion each deity is associated with an aroma, an odorous substance that, as it burns, gives off fumes toward the heavens, reaching the god, as in a kind of unbloody sacrifice. The scent associated with the god Sun is that of frankincense grain (or powder) (λιβανομάννα), one of the many scents that came to Greek cities from the Arabian Peninsula (Ricciardelli 2000, p. xxxvii *ff.*). The epithets attributed to the Sun in the hymn dedicated to him emphasize his hegemonic and "luminary" role in the visible universe; he is called: Titan of golden light (Τιτὰν χρυσαυγής), light of the heavens (οὐράνιον φῶς), Lord of the cosmos (κοσμοκράτωρ), father of time (χρόνου πάτερ), bringer of light (φωσφόρε), bright to all (πασιφαής), with beautiful shining rays (καλαῖς ἀκτῖσι φαειναῖς), master of the cosmos (δέσποτα κόσμου), and light of life (ζωῆς φῶς).

The work to which the Neoplatonists gave the title *Chaldean Oracles*, attributed to Julian the Chaldean and Julian the Theurgist, perhaps father and son, who composed it around the second half of the II century during the time of the Emperor Marcus Aurelius, has survived to us in fragmentary form; much of these fragments are reported by the Neoplatonists philosophers, mixed with Platonic doctrines. Especially in the writings of Proclus (commentaries, systematic works, and Hymns), there are countless Chaldaean sentences. In the cosmo-theology of the Oracles, the Sun occupies a central role, linked to fire, the incorporeal matter of divine substance. In the triadic hierarchical structure of the divine universe of the Chaldeans, the Sun is present both in the ethereal or supermundane dimension, in the form of luminous fire, and in the visible or mundane dimension, as a planet that infuses light and warmth to sensible entities through its rays. The apex of the Chaldaean universe is occupied by the Supreme Principle or First God, called the Father (fragments 7, 14) or Paternal Intellect (fragments 7, 39, 49, 108, 109), Monad (fragments 11, 26, 27), and Source (fragments 9, 9a). The Second God is the demiurgic Intellect, also called the Dyad; this is followed by a third, female deity, depicted in Hecate, who serves as the Soul of the world. Between this divine triad and the sensible world there are multiple intermediaries – Iynges, teletarchs, angels, and demons – connectors but also separators of the higher domain from the lower. One divine entity related to the Sun is Aion, also called Father-generated Light (fr. 49), solar cosmos, and whole light (fr. 59). Aion is a noetic deity, identified with the hypercosmic Sun; his function is to manifest the light of the Father and to radiate it, through his physical image, which is the visible Sun, to worldly entities. Chaldean theology revolves around the worship of fire and the sun. Both are worshipped as deities. A divine fire shines, as the hypercosmic Sun, in the empyrean, or intelligible, dimension; the sensible Sun shines in the ethereal, or celestial, dimension, illuminating and enlivening earthly entities. In the Chaldaean cosmology, the Sun occupies a central place among the planets, "fixed" by the God in the place of the heart (fr. 58: κραδίης τόπῳ), as "dispenser of fire" (fr. 60: ταμίαν πυρός). Its rays have both a physical and anagogic function: they warm and illuminate, but also bring the soul of the theurgists back up to the First Principle, causing it to travel along the channels of fire (fr. 60), while wearing a "dress of light" (the vehicle or ὄχημα).[14]

Now, according to the testimony of his biographer, Proclus had been initiated by Syrianus into the study of both orphic and oracular culture; therefore, he was well acquainted with the principles of those two forms of mysterious wisdom. On the *Chaldaean Oracles* he had also composed a commentary, of which only fragments remain, but oracular sentences and orphic references are contained in almost all his works. Proclus wrote about 700 lines every day (*Life of Proclus* 22); his considerable written output is for the most part a clear example of that philosophical-theological syncretism which constitutes, in truth, the specificity of this phase in the history of philosophical thought, its added value, even though numerous historiographers have seen precisely in

14 On the *Chaldaean Oracles*, see Des Places (1971), Majercik (1989), Albanese-Mander (2011), Lewy (2011³), Tanaseanu-Döbler (2013).

such a mixture of different cultural contributions – Western and Eastern, philosophical and theological – the cause of the alleged decline of the philosophical *vis* in late antiquity. However, the numerous valuable studies, which have been produced in the past few decades on this area of the history of thought in late antiquity, have demonstrated the groundlessness of that criticism, which generally assigned to Neoplatonism the appellation of *secta eclectica* in the worst sense of the term (Cardullo 2020b, pp. 474–5). It would be more appropriate to argue not so much that the speculative force of such an approach to the real has been depotentiated, but that philosophy be judged in its limits beyond which there is a higher form of access to the real made possible only through an initiatory way that makes use of the tangible symbols of the intelligible entities, in order to be able to arrive at the contemplation of the latter. Unlike magic, which is relegated to the lowest levels of reality, theurgy aspires to Wisdom, that is, it "wants to contemplate the supreme principles in its highest form," thus realizing an elevation of the soul to the gods to the point of assimilation with them, and "the theurgist becomes a god among the gods [...]; not only that, but he also acquires the same powers as them over the cosmic world as well as power over the cosmic deities. Once deified, the theurgy himself becomes a powerful mediator between the sensible and the supersensible world, as well as a "receptacle for the gods." (Radice 2018, p. 139). In the end, the theoretical effort of philosophy must give way to that *pistis* that pagan intellectuals ridiculed in Christians, for it alone can provide the energy that leads the soul to God: Iamblichus dismisses the notion of reason being the main route to divinity, as "an innate knowledge about the gods is coexistent with our nature, and is superior to all judgement and choice, reasoning and proof." (Iamblichus 2003, 1, 3) In this way, Platonic dialectic has its fulfillment in contemplation, bringing to full maturity the esoteric aspects present in his work: it was now a matter of tapping into true philosophizing, which no longer consisted in mere dialectical argumentation but rather an anagogical operation aimed at transforming one's soul to enable it to contemplate the Idea of Good, which in late Neo-Platonism is identified with the intelligible essence of the sun, when the limits of language are overcome and the ineffable is reached (Coniglione 2018, p. 36). The barrier Plato wanted to put up against the "pervasiveness" of traditional knowledge, for the purpose of which it was necessary to "kill Parmenides" and his idea of a truth reached through an itinerary guaranteed and permitted by the Goddess, to replace it with the philosophical idea in which the dominance of discourse prevailed and which was the dimension to which humanity as a whole could access – and not only the chosen few who participated in the mysteries – is now in late Neo-Platonism crossed and access to the True, as it transforms into *pistis*, becomes a practice that everyone can draw upon; the ἐπόπτεια, of which it is not possible to speak and which is indicated by what is usually called the "unwritten doctrines" in Plato, becomes the central focus of Neoplatonic interest and thus that mode of representation – nestled in Plato's classically theoretical heart – "that is rooted in the anti-logos, in the doctrines of theologians and mysteries" is fully developed and brought to light (Coniglione 2018, p. 40). Basically, herein lies the enduring triumph of the Platonic system:

> deriving from having been able to hold together three different levels of "knowledge": the dialectical-argumentative one, in the context of which philosophy takes on an autonomous form of discipline, thus beginning its fortune along the course of European thought, and is defined the ideal of knowledge and truth that has become part of the Western Canon; the allegorical-narrative one, which has its admirable expression in the great myths contained in Plato's dialogues and which constitutes the firm base on which the official Greek religiosity is built and which would be common to the religions of all times, in whose narrative and exemplary procedure spiritual teaching and popular accessibility merge and find a synthesis; and finally, the esoteric-initiatory one, found in the oral teaching of the unwritten doctrines, which unfolds through a symbolic path (that of esotericism) to finally result in the ineffable, in a vision that cannot be communicated through the everyday language of reason and argumentation: this is the basis of both the mystical experience and the esoteric disciplines that have passed through the culture and the Western religious approaches as a subterranean and alternative karst current. [...] It is easy to see how the dialectic-argumentative plane corresponds to science (or philosophy, that is, to the purely rational dimension), the allegorical-narrative one to faith (even if this term was not yet used by Plato and we must wait for the final phase of Neoplatonism and the advent of Christianity) and the esoteric-initiatory one to gnosis.
>
> CONIGLIONE 2018, p. 40

Therefore, it could be said that in late Neoplatonism the above component present in Plato's thought – which as Quispel has pointed out represents "the third component of the Western cultural tradition" (Quispel 2008, p. 39) developed in Neo-Platonism and Gnosticism – received a twist that did not so much reduce the speculative *vis* of late ancient philosophy (those who read the works of Proclus can easily see how much argumentative power this was endowed with), but rather shifted the balance on the other two planes, the soteriological need now being predominant along with the pressing need to respond effectively to the Christian faith. The above is illustrated in Proclus' interpretation of the function of the Sun and his use of theurgy. However, one cannot help but note that, in his approach to mystical theologies, his "harmonizing" position often risks conflicting with the philosophical principles of the ancient master, whom all Neoplatonists claim to be faithful followers and exegetes of. Thus, it happens that in the specific case of solar theology, which, as we know, had come from the East in the imperial era, Proclus manages to highlight its compatibility with what he calls Platonic *mystagogy*. Since philosophy is a theology for the Neo-Platonists, Plato, too, is a theologian who, like Homer, Hesiod, Orpheus, and the Gods who uttered the Chaldean Oracles, expounded truths about the gods in his writings. They just need to be brought to light again and their *lexis* interpreted in the most correct way.

Let us return to the sun. Since this is a subject of a cosmo-astronomical nature, Proclus' exegesis of the related Platonic conception should focus on the *Timaeus* and have only a physical connotation; however, the analogy between the sun and the Good, as contained in *Republic* VI, allows Proclus to shift his discourse from the purely physical to the metaphysical sphere. And so, in his commentaries on these two grand dialogues, the Neoplatonist leaves us with intense pages, full of mystical and philosophical doctrine at once, in which speculative contemplation and hieratic art are magnificently summed up in the praise of the Sun God. In the pages of the *in Timaeum*, the sun is presented both as one of the seven planets, placed in the second order to the earth, after the moon, in line with Platonic and Ptolemaic astronomy (*Timaeus* 38c ff.), and also as the central star of the universe, as its beating heart, the visible fire that heats and vivifies the sensible entities, the image of ethereal fire, as taught by an oriental, fundamentally heliocentric culture (*Orphika* fr. 96; *in Timaeum* III 67,27–68,10), which places the sun κραδίης τόπῳ (*Chaldaean Oracles* fr. 58; *in Rempublicam* II 220,14–15). Proclus is aware of the divergence between the Chaldean and Platonic conceptions regarding the position and role of the sun but argues for their compatibility. In a passage from the *in Rempublicam* the Neoplatonist, quoting *verbatim* two oracular rulings, states that he wants to follow what he has heard, respectively, from the Chaldean theurgists and directly from the gods about the sun, namely that:

> the god interspersed (ἐμεσεμβόλησεν) the sun among the seven [planets] and made the other zones dependent on it.
>
> PROCLUS *in Rempublicam* II 220,12–14 = *Chaldaean Oracles* fr. 200

and

> fixed the solar fire in the place occupied by the heart (κραδίης τόπῳ ἐστήριξεν).
>
> PROCLUS *in Rempublicam* II 220,14–15 = *Chaldaean Oracles* fr. 58

adding that he was aware that he was deviating from traditional, Platonic-Aristotelian astrology, but that he was equally convinced that even Plato could have shared the Chaldaean order with the sun at the center of the universe, and Hermes and the moon as its satellites,

> because [such an order] will suit the custodians of the entire generation. Indeed, [the Sun] is flanked by all the planets as by satellites (ὑπὸ πάντων δορυφορεῖται τῶν κοσμοκρατόρων), for he is king of all visible things (τῶν ὁρατῶν πάντων βασιλεὺς) and faithfully reproduces with rays of light the demiurgic powers (τὰς δημιουργικὰς δυνάμεις διὰ τῶν τοῦ φωτὸς ἀκτίνων ἀπομιμούμενος), generating, filling with life and renewing generations.
>
> PROCLUS *in Rempublicam* II 220,24–221,1

In order to harmonize traditional astronomy with Chaldaean and Orphic astronomy, Proclus assigns the sun a twofold procession: a first one, properly encosmic, which occurs when the demiurge places it in the sky along with the other planets, and the second one, which occurs when the creator shines a light on it, linking it to a primary intelligible, hypercosmic source of light, a "truer" Sun, ὁ ἀληθέστερος ἥλιος, of which the sensible sun constitutes the visible reflection (*in Timaeum* III 82,19–26; *Th. Plat.* VI, 63,1 ff.); *Chaldaean Oracles* frr. 59,185; *Orphika* fr. 86):

> There, indeed, we have the solar cosmos (ἡλιακὸς κόσμος) and the total light (ὅλον φῶς), as the oracles of the Chaldeans also say (αἵ τε Χαλδαίων φῆμαι), and I am persuaded of this.
>
> PROCLUS *in Timaeum* III 83,14–17

> We must therefore consider the sun in two ways, either as one of the seven 'planets' or as the head of all things, both as encosmic and as hypercosmic (καὶ ὡς ἕνα τῶν ἑπτὰ καὶ ὡς ἡγεμόνα τῶν ὅλων καὶ ὡς ὑπερκόσμιον), in that it shines the divine light (καθὸ καὶ προλάμμπει τὸ θεῖον φῶς) as the Good shines the truth that makes the intelligible and intellectual orders divine, as in Orpheus [fr. 58], Phanes emits the intelligible light that fills

> all the intellectual gods with intellection, as Zeus lights up all the hypercosmic entities with intelligible and demiurgic light; in the same way the sun also makes all the visible shine by means of this uncontaminated light (διὰ τοῦ ἀχράντου τούτου φωτός).
>
> PROCLUS *in Timaeum* III 82,31–83,8

The so-called hypercosmic Sun, mentioned on this page of the commentary on the *Timaeus*, is identified by Proclus in the *Commentary on the Cratylus* with Apollo, the unifying god who "rotates together with the Sun" (ἡλίῳ συμπεριπολεῖ): just as up there Apollo reveals intelligible goods to heavenly beings, down here the Sun dispenses light to the beings in the heavens, shining down and providing all with unifying power (*in Cratylum* 98,1–4).[15]

A page from Book VI of *Platonic Theology* beautifully summarizes the concept set forth in the two commentaries just mentioned. Here, without using circumlocutions, allusions, or half-sentences, Proclus asserts that Apollo and the Sun are both gods, and for both Plato and the theologians:

> these gods are unified and acclaimed with interchangeable epithets: for the Sun is especially pleased to be celebrated as Apollo, and Apollo, when invoked as the Sun, shines benevolently the light of truth.
>
> PROCLUS *Platonic Theology* VI, 59,23–26

Recalling the passage in the *Cratylus* on the etymology of Apollo's name (*Cratylus* 405c2–5) and that in the *Republic* concerning the Good/Sun, the Neoplatonist states in *Platonic Theology* VI, 58,24–26 that for Plato there is a conjunction of nature between these two deities, "a communion of powers and an unspeakable union." Indeed, both are analogous to the Good: the sensible light pertaining to the Sun and the intelligible truth connected with Apollo are analogous to the supra-essential light of the Good,

> and these three lights are in relation to each other in this sequence: divine, intellectual and sensible; the latter comes to sensible beings from the visible Sun, the second to intellectual beings from Apollo, and the third to intelligible beings from the Good.
>
> PROCLUS, *Platonic Theology* VI, 59,12–16

15 This supra-celestial deity, from whom the sun receives its light, in the *Chaldaean Oracles* is also identified as Aion. See Lewy (2011³), p. 409 ff.

The superiority of the Sun in the celestial orbit thus depends on its connection with the intelligible source of light; when it appears, it "leaves the star gods astonished, with all wishing to dance around him and be filled with his light; and that is why the world down here is beautiful and similar to the Sun," receiving from him perfection and essence. "This is why Plato calls the Sun 'offspring of the Good,' demiurge of the begotten and creator of all worldly light" (Proclus, *Platonic Theology* VI, 63,15 ss.).

2.2 *Solar Cult and Theurgy in Proclus*

> Some, such as Porphyry, Plotinus and many other
> philosophers hold philosophy in higher esteem,
> others, such as Giamblicus, Syrian, Proclus
> and all the theurges, the hieratic art.
>
> OLYMPIODORUS, *in Phaedonem* 123,4

In Proclus' system, mystery religions mingle seamlessly with Platonic metaphysics and theology. The brief account that was given in the previous section about the cult of the sun provided quick but significant evidence of this. The diadocus also wrote some specifically theurgical works, although little remains of them. We possess a fragment from a writing *On the Hieratic Art of the Greeks*, another from what must have been an extensive commentary on the *Chaldaean Oracles*, and numerous Hymns, including one dedicated to the Chaldaean gods and one to Helios, the Sun God.[16] However, regardless of what is transmitted in the written work, we know from Marinus that Proclus was very devoted to the gods and in particular to the Sun, that he revered and adored "at sunrise, noon and sunset" (*Life of Proclus* 22). As a perfect *theios anêr*, Proclus had acquired all the virtues required by the spiritual training path of the Neo-Platonists, up to the highest virtues, which Iamblichus had called theurgical:

> Indeed, he practiced the rites leading to conjunction with the God and prayers of the Chaldeans and made use of the divine and ineffable *strophaloi*[17]
>
> MARINUS, *Life of Proclus* 28

16 On Proclus and the *Chaldaean Oracles*, I refer to Spanu (2021).

17 These are gold armillary spheres, with a sapphire in the center and magical figures covering the entire surface, used by the theurgists for their invocations. See *Chaldaean Oracles* fr. 206; on magical theurgical tools in general (*strophaloi*, *Iynges*) see Majercik (1989), pp. 29–30.

Like any theurgist, Proclus celebrated the gods with hymns, invocations, prayers, and further telestic and medianic practices in order to be able to reunite with the divine, "to become a god" – a goal not far from what was exhorted in *Theaetetus* 176b1 and *Timaeus* 90a–d,[18] but achieved "actively" and not merely speculatively. In this ethical and religious process, in oracular and Orphic wisdom, the Sun has an essential function: that of purifying the soul of the initiate with its rays, drawing it upwards and enabling its union with the primordial light (*Chaldaean Oracles* ffr. 60; 190). Clothed in its pneumatic, luminous vehicle (ὄχημα), the soul of the theurgist exploits the sun's rays to realize the ἀναγωγή and the ἐπιστροφή toward the divine. Proclus finds in theurgical teachings a more effective means than contemplation alone of getting in touch with the gods. Prayer is one of his favorite practices,[19] and he sings its praises in a well-known and spiritually rich page in his *Commentary on the Timaeus* (II 209.14–210.14); the hymn, in particular, is for Proclus – as for the Orphics – a way to raise a prayer to the divine, to ask above all for protection from the evils of the earthly world and for concrete help in the process of purifying the soul.

The Sun occupies an important role in Proclus' cultural life;[20] thanks to hieratic *sympatheia*, there are many entities in the sensible world, vegetable, animal, and even inorganic, that pay homage to the Sun, carrying within themselves traces of its power, such as the "sunstone." The sunflower, for example, moves in sympathy with the Sun (τὸ ἡλιοτρόπιον … τούτῳ κινεῖται),

> and if one could perceive by hearing the friction it creates with the atmosphere in its turning in a circular direction, one would realize that with this sound it offers the King a kind of hymn, such as one that can be sung by a plant.
>
> PROCL. *Hieratic Art* 148,15–18

Moreover, the lotus also reveals a sympathetic connection with the sun, opening its petals as the sun's rays appear and closing them at sunset, analogous to a man's lips, which open and close in prayer (*Hieratic Art* 148:12–18). Among the many heliacal animals (ζῴων ἡλιακῶν), Proclus mentions the lion and the rooster; the latter in particular "perceives the solar revolutions by singing a hymn at the appearance of a star" (*Hieratic Art* 150,10–13).

18 On this topic see Van den Berg (2003), p. 189 ff.

19 On the role of prayer in Proclus, see Saffrey 2000a, Layne 2013, Redondo 2016.

20 Van den Berg (2001), p. 146: "The introduction of theurgy by Iamblichus makes solar cult even more important, since in theurgical mysteries the sun plays an important role in the purification and elevation of the soul".

However, veneration of the Sun reaches its apogee in Proclus in *Hymn* No. 1, *To Helios*, a theurgical prayer with strongly hieratic tonalities. The Sun is presented there as the image of the Father of everything (the Platonic Demiurge and the Chaldaean Intellect) and as the anagogic medium of the ascent and salvation of fallen souls (Van den Berg 2001, p. 136), as the source of all harmony. He is a theurgic deity.[21] Proclus prays to Helios to protect him against demons and evil, to dispel the fog that clouds the eye of his soul, and to clothe him with holy light: the light of purity. An important antecedent of this Proclian hymn can be found in the Emperor Julian's *Oration* 11, dedicated to *Helios the King*. Heavily influenced by the theurgic neo-Platonism of Iamblichus, Julian venerates the Sun, interweaving the religious cult of the solar disc with a political connotation as well, presenting himself as a *princeps* follower and companion of the Sun, the supreme deity.[22]

The Sun of Proclus' hymn embodies all the characters that late antiquity ascribes to this deity, likened on the one hand to the Platonic supreme God, the Good who is *epekeina tês ousias*, and, traditionally, to Apollo,[23] God of the light of wisdom, and on the other to the mysterious image (both orphic and oracular) of a divine source of fiery light; as in the *Chaldaean Oracles*, the Sun is revered by Proclus as the king of intellectual fire (πυρὸς νοεροῦ βασιλεῦ) dispenser of light (φάους ταμία), image of the creator god of all things (εἰκὼν παγγενέταο θεοῦ), elevator of souls (ψυχῶν ἀναγωγεῦ).

The life cycle of every being, the concluding act of its existence, is fulfilled, according to the Neo-Platonists, in the return – ἐπιστροφή – to the First Principle, the one and supreme cause from which everything indirectly proceeds and derives. The world down here is full of the symbols and signs of the first cause, which refer back through the divine chains to the supreme monad, the Creator Father, the One/Good; it is through the universal sympathy and power of such symbols that the philosopher/theurgist acts – consecrating statues or pronouncing divine formulae and names – to realize his union with the

21 Van den Berg 2001, p. 147: "His Helios is the deity of the theurgists who protects the soul from punishing daemons, helps it to escape the realm of matter and elevates it to the intelligible realm. Echoes of the Chaldaean Oracles abound in the hymn. His most important discussion of Helios can be found in *Theol. Plat.* VI 12,56,1–65,3. Helios, identified with Apollo, constitutes the elevating triad of the hypercosmic gods".

22 Compared with Proclus, who distinguishes between an invisible and a visible sun, Julian considers three: an intelligible sun, equivalent to the Platonic idea of the Good; an intellectual sun, which acts as a mediator; and a sensible sun. On Julian see De Vita (2011 and 2017–18) and the rich literature indicated therein.

23 The identification between Helios and Apollo has very ancient origins, going back as far as the Homeric hymn to Apollo. In late antiquity, Apollo is basically another name for the god Helios; both share brightness, a symbol of life, harmony, and wisdom, as well as the ability to proffer oracles. See Saffrey (2000), pp. 186–190.

Principle through the purification of his soul. Proclus, described by Marinus as an example of a fully virtuous and pure life, is convinced that contemplation alone does not allow contact with the world of the divine, but that it is necessary to enter into harmony with it, actively, devoting one's entire life to ritual celebration and the practice of veneration of the gods. According to Marinus' account, Proclus intertwined teaching and commentary on philosophical and theological texts with theurgical practices. If for Neo-Platonism, the goal of man's life is to be reunited with God – the Platonic ὁμοίωσις τῷ θεῷ – to make oneself similar to Him in purity and goodness, escaping the evils of the world, then anything dedicated to the divine can serve as a "spiritual exercise." For Proclus, reading and commenting in the auditorium on the texts of the theologians, from those by Plato to the *Oracles* and *Orphic Hymns*, revering the Sun three times a day, praying and singing praises to the gods, all contribute to the soul's liberation from evils and its ἀναγωγή. And in this path of progressive purification and upward ascent, the veneration of the god Sun in late antiquity – ψυχῶν ἀναγωγεῦ – has its own precise motivation: indeed, the triad of the heliacal gods in the Proclus Pantheon is described in *Platonic Theology* VI 12, overseeing the elevation of souls up to the demiurgic monad. The heliacal, or Apollonian, chain thus enables souls to return, through the solar rays, to the intelligible, as the Chaldeans taught. But the chariot of Apollo-Helios is precisely that which, in the Platonic *Phaedrus*, ascends back to the hyperuranium, with pure souls in tow who have put their wings back on, inflamed with love for Beauty. This should not be surprising; indeed, according to Proclus, it is the same doctrine and the same belief, all the more so since Plato, precisely in the *Phaedrus*, expounds his thought in a divinely inspired manner, typical of mysterious religions.[24] In this way, the transformation of Platonic teaching can come be brought to a close, but in the conviction that it has revealed its true essence, namely that esoteric-initiative trait which, as we mentioned earlier, coexisted in Plato along with philosophy and the mythical-allegorical element.

References

Albanese, L., P. Mander (2011). *La teurgia nel mondo antico*. Genova: Ecig.

Bouffartigue, J. (1992). *L'empereur Julien et la culture de son temps*. Paris: Etudes Augustiniennes.

24 On the four τρόποι of theological teaching in Plato, according to Proclus, cf. Cardullo (2020a).

Cardullo, R.L. (2020a). Proclo, sui quattro τρόποι dell'insegnamento teologico in Platone. Un esempio di 'approccio multifocale' in età tardo-antica. *Humanitas* LXXV, 1–2, 65–74.

Cardullo, R.L. (2020b). *Le filosofie a sfondo religioso e la fine del pensiero greco.* In: M. Migliori-A. Fermani, *Filosofia antica. Una prospettiva multifocale*, pp. 473–541. Brescia: Morcelliana.

Chlup, R. (2012). *Proclus. An Introduction.* Cambridge: Cambridge University Press.

Coniglione, F. (2018). Some Remarks on the Meaning of Esotericism and Plato's Unwritten Doctrines. *Mondi. Movimenti simbolici e sociali dell'uomo*, vol. I, n. 1, 5–51.

Coniglione, F. (2021). La penultima verità: il mito tra *pistis* e *logos* al tramonto del mondo antico. In: R.L. Cardullo, F. Coniglione (a cura di), *Mythos e Logos. Tra archetipi antichi e sguardi sul futuro*, pp. 178–205. Pisa: Edizioni ETS.

D'Hoine, P.-Martijn, M. (2017). *All from One. A Guide to Proclus.* Oxford: Oxford University Press.

De Vita, M.C. (2011). *Giuliano Imperatore, filosofo neoplatonico.* Milan: Vita e Pensiero.

De Vita, M.C. (2017–2018). 'Figlio del bene' e Re dell'universo: il dio Helios di Giuliano Imperatore. *Χώρα*, 15–16, 457–484.

Des Places, E. (1971). *Oracles Chaldaïques.* Paris: Les Belles Lettres.

Faragiana di Sarzana, C. (1985). *Proclo. I Manuali. Elementi di fisica. Elementi di teologia. I testi magico-teurgici*, Milan: Rusconi.

Giuliano Imperatore (1997). *Alla madre degli dei e altri discorsi.* Italian transl. by A. Marcone. Milan: Mondadori.

Guidorizzi, G.-Romani, S. (2019). *In viaggio con gli dei. Guida mitologica della Grecia.* Milan: Cortina.

Halsberghe, G.H. (1984). Le culte de Deus Sol Invictus à Rome au 3e siècle après J.C. *Aufstieg und Niedergang des Römischen Welt*, II 17. 4, 2181–2201.

Iamblichus (2003). *De mysteriis*, translated with an Introduction and Notes by Emma C. Clarke, John M. Dillon and Jackson P. Hershbell. Atlanta: Society of Biblical Literature, 2003.

Layne, D.A. (2013). Prayer in Proclus' Commentary on Plato's Timaeus. *The Review of Metaphysics*, 67,2, 345–368.

Lewy, H. (1978). *Chaldaean Oracles and Theurgy. Mysticism, Magic and Platonism in the Later Roman Empire.* Paris: Etudes Augustiniennes.

Lecerf, A. (2014). *L'empereur Julien entre culte d'Attis, Oracles et théologie solaire.* In: Lecerf, A.-Saudelli, L.-Seng, H. (2014). *Oracles Chaldaïques: fragments et philosophie*, pp. 61–100. Heidelberg: Winter.

Majercik, R. (1989). *The Chaldaean Oracles.* Leiden-Boston: Brill.

Proclus (2019). *Teologia platonica*, a cura di M. Abbate. Milan: Bompiani.

Quispel, G. (2008). *Gnostica, Judaica, Catholica. Collected Essays of Gilles Quispel*, ed by Johannes van Oort. Leiden-Boston: Brill.

Radice, R. (2018). *Magica filosofia. Sapere occulto e sapere illuminato nel pensiero antico e arcaico*. Brescia: Morcelliana.

Ricciardelli, G. (2000). *Inni orfici*. Milan: Mondadori.

Romano, F. (2004). Fides/pistis in Agostino e Proclo. In: F. Romano, *L'Uno come fondamento. La crisi dell'ontologia classica*: Catania: Cuecm.

Saffrey, H.-D. (1994). *Proclus. Hymnes et prières*. Paris: Arfuyen.

Saffrey, H.-D. (2000a). *La dévotion de Proclus au Soleil*. In: AA. VV., *Le Néoplatonisme après Plotin*, pp. 179–191. Paris: Vrin.

Saffrey, H.-D. (2000b) *L'Hymne IV de Proclus, prière aux dieux des Oracles Chaldaïques*. In: AA. VV., *Le Néoplatonisme après Plotin*, pp. 193–206. Paris: Vrin.

Siorvanes, L. (1996). *Proclus: Neo-Platonic Philosophy and Science*. New Haven-London: Yale University Press.

Spanu, N. (2021). *Proclus and the Chaldean Oracles. A Study on Proclean Exegesis, with a Translation and Commentary of Proclus' Treatise on Chaldean Philosophy*. London-New York: Routledge.

Tanaseanu-Döbler, I. (2013). *Theurgy in Late Antiquity. The Invention of a Ritual Tradition*. Göttingen: Vandenhoek & Ruprecht.

Van den Berg, M. (2001). *Proclus' Hymns*. Essays, Translations, Commentary. Leiden-Boston-Cologne: Brill.

Van den Berg, M. (2003). Becoming like God' according to Proclus' Interpretation of the "Timaeus", the Eleusinian Mysteries, and in the Chaldaean Oracles, *Bulletin of the Institute of Classical Studies*, 78, 189–202.

Zambon, M. (2019). *«Nessun dio è mai sceso quaggiù». La polemica anticristiana dei filosofi antichi*. Rome: Carocci.

CHAPTER 10

Interpreting Francis of Assisi

Riccardo Pozzo

Abstract

After an introduction dedicated to sketching out the way the methodology chosen by Francesco Coniglione brings the outcomes of his investigation beyond the state of the art, part one of this paper considers the implications of Francis' teaching for the life of men and women. Part two is about Francis' consideration of nature and all creatures, animated and unanimated, brothers and sisters of God creator; and part three is about Francis' innovative naturalistic spiritualism, which wraps up the paper by referring Francis to contemporary issues such as sustainable development and migration. All arguments start from the framework provided by Coniglione in the book he dedicated to Francis, the human being who came from another world, which appeared in the last months of 2022 and is up for discussion.

Keywords

Francis of Assisi – Beatitudes – poverty – nature – spirit

Francesco Coniglione's interpretation of Francis of Assisi is both synchronic and diachronic. He sets off by explaining all systematic features of Francis' mysticism while giving shape to a roadmap of spiritual perfection that holds for every day and time. We are in front of a human being who is coming from "another world" (Da Celano 2004, 1Cel, 36, 4; Francis of Assisi 1977, 3Soc, 1463). The fact is, Francis has much to say to men and women of this word. The itinerary of interior perfection requires the emptying of the self through the choice of radical poverty, which enables a new regard on nature. However, what Francis teaches us on nature does not get lost in the fogs of a generic romantic sentiment of naturalistic mysticism. It converts itself instead in a positive approach and in a concretely operating praxis in defense of creation. Francis not only appreciates nature as a work of God, he also points to nature having in itself resources and ferments that will bloom in all directions (Coniglione 2022, pp. 10–11).

 | HTTPS://DOI.ORG/10.1163/9789004756496-012

Hence the need to start from the first symbolic act that constellates the life of Francis: stripping off his clothes, a symbolic act that takes the form of humiliation and penance, because to live an impoverished life implies a penitential life, which raises the question of who Francis was when he was clothed. What might have been the extent of the sinful life for which he thought he had to make reparation? Coniglione remarks that if it is true that we must take into account the existence of a "lost Francis" during his youth, of which we know practically nothing due to lack of any documentary evidence, which implies that we should not loose time investigating it – although Barbara Frale (2016) has recently been able to shed a few sparks of light (Coniglione 2022, p. 13).

Not surprisingly, Coniglione takes up an uncompromising method. Instead of looking into the "historic Francis," he aims at the "story of Francis," namely the great array of testimonies and the legends inspired by his figure, thus following Hans-Georg Gadamer's (1975, pp. 299–306) history-of-the-impact-approach. The main focus of Coniglione's investigation being to look into the *Wirkungsgeschichte* of Francis (Coniglione 2022, p. 16). On the other side. Coniglione looks with passion into the shades of reason against non-reason, giving shape to the "rejected knowledge" of the antithesis of the logos, for in Francis the daily part of human life coexists with the nocturnal part. Thus, he proposes also new insights for a "negative history" of Francis (Coniglione 2022, p. 20), thus following the metaphor of the information that extrasolar planets provide on their orbits by simply passing by as streaks of darkness over their stars.

1 Francis and the World of Men and Women

Francis' first symbolic act, its denudation, was actually one of the forms of humiliation required by Canon Law for public penance. By stripping off the clothes with which Francis had so many times sang, danced, joked and participated with his friends to the secular life of the world, Francis cut globally with his past and showed unequivocally his readiness to penance (Manselli 2013, p. 2399). It is noteworthy that Francis's denudation was an act of voluntary public penance, and as such not to be confused with the various forms of public humiliation that were imposed on nobles and peasants alike for schocking crimes as well as for minor brawls (Mansfield 1995). Coniglione suggests that in this context "penance" has no reference to mortification, as it refers instead to the Greek notion of μετάνοια (*metánoia*) in the sense of a transformation of νοῦς (*nous*), more exactly in the sense of a complete reversal of one's way of looking (Coniglione 2022, p. 24). Coniglione maintains that the novelty of Francis is to go beyond the traditional setting of the poor as objects of charity, mercy and assistance, as required by the Gospel, for he chooses to become himself poor (Coniglione 2022, p. 42).

Jesus does not beatify the idea of poverty. He beatifies the poor: "Blessed are the Poor, for yours is the Kingdom of God" (Lk 6:20); "Blessed are the poor in spirit, for theirs is the Kingdom of Heaven" (Mt 5:3); "Blessed are the Meek, for they will inherit the Earth" (Mt 5:5).

Francis embraces poverty as a form of life and his love for poverty is not abstract, it is implemented concretely in the person that incorporates poverty (Coniglione 2022, pp. 43–44). Paradigmatic for Francis is the κένωσις (*kénōsis*), the "self-emptying," of Jesus is:

> Let the same mind be in you that was in Christ Jesus, who, though he was in the form of God, did not regard equality with God as something to be exploited, but emptied himself ἑαυτὸν ἐκένωσεν (*heauton ekénōsen*), taking the form of a slave, being born in human likeness. And being found in human form, he humbled himself and became obedient to the point of death – even death on a cross. Therefore, God also highly exalted him and gave him the name that is above every name.
>
> PHIL 2:5–9

Francis retraces in his life the self-annihilation during incarnation. He stresses the fact he was born naked and was going back into the earth naked. Nature in its immense womb takes us up in her protecting breast, thus enabling the absolute integration of men and women with God, who is father and mother full of goodness (Coniglione 2022, p. 71).

Coniglione points out that Francis defines himself *simplex et idiota* and follows (Mt 11:25) in casting doubt on the advantages of culture and studying (Coniglione 2022, p. 79). This is, Coniglione concludes, the paradigm of *sancta simplicitas* that brings about (Tommaso da Celano 2004, 2Cel, 775) the "true sister of wisdom" (Coniglione 2022, p. 73). As we read in (Francis of Assisi 1977, CompAs, 1627), Francis declares "blessed" those who will be "sterile for the love of God our Lord," whereby sterility means in ignorance, as relinquishing the habit of a science that has already been acquired. The emptiness of the mind and of all its preexisting cognitions, suggests Coniglione on the same page, is the condition of possibility of illumination (Coniglione 2022, p. 87).

2 Francis and the World of Nature

Francis subverts ancient and medieval views of nature as menacing and dangerous, e.g., earthquakes and wild animals. The legend of the wolf in Gubbio is a testimony of that. On the other side, it needs to be kept in mind that what we call "nature" in Francis' understanding means more precisely "creation." In sum, everything Francis says about the world of nature, of all creatures, must be seen in triangulation with the Creator, which is the basis of the fraternity of all beings, animated and unanimated, for all are creatures and sons and daughters (Coniglione 2022, pp. 136–138).

Here Francis talks to the twenty-first century. We need to start a conversation – Pope Francis has urged – that "includes everyone, since the environmental challenge we are undergoing, and its human roots, concern and affect us all" (Pope Francis 2015, no. 14). Biodiversity and cultural diversity are fundamental resources for nature and culture-based solutions aimed at addressing societal issues, solutions that allow a holistic approach to tackling challenges such as climate and water regulation, food quality and safety, and sustainable urbanization – at the same time, providing business and employment opportunities and promoting the protection, restoration, and sustainable management of ecosystems. By recognizing biodiversity and cultural diversity as engines for a sustainable economy and sustainable development, we philosophers can contribute to transform environmental and social challenges into opportunities for innovation impacts on environmental, social, and economic policies. We must continue investigating the relationship between biodiversity and cultural diversity and exploring methods to evaluate, develop, and implement large-scale solutions for continuing education. We must promote sensitivity towards biological and cultural diversity, interdependencies between species, and the continuum between nature and human beings.

The teachings of Francis of Assisi, Pope Francis (2015, no. 10–12) has argued, play a role in assessing our experiences of ecosystems with attention to biocultural diversity with the ultimate goal of enhancing the effectiveness of life in land and below water understood as part of the urban fabric and landscape. Both Francis and Pope Francis ask to end poverty and hunger, in all their forms and dimensions, and to ensure that all human beings can realize their potential with dignity and equality in a healthy environment. They require mankind to protect the planet from degradation through conscious consumption and production, managing its natural resources sustainably, and taking urgent action regarding climate change so that it can meet the needs of present and future generations. They renew hope that all human beings can enjoy prosperous and satisfying lives and that economic, social, and technological progress take place in harmony with nature. They promote peaceful, just, and inclusive societies that are free from fear and violence. There can be no sustainable development without peace, nor peace without sustainable development.

In conclusion, Francis of Assisi's thoughts on nature encompass indeed "the diversity of life in all its manifestations: biological, cultural, and linguistic – which are interrelated (and possibly coevolved) within a complex socio-ecological adaptive system" (Maffi 2007, p. 269; Pozzo 2021, pp. 173–175). Thinking of future generations, the preservation of biodiversity and cultural diversity, as natural heritage and public good, ought to be considered a moral duty. The means necessary to implement sustainability foster a spirit of strengthened global solidarity, focused particularly on the needs of the poorest and most vulnerable and with the participation of all countries, all stakeholders and all people.

3 Francis and the World of Spirit

Many authors speak of the mysticism of Francis under the category of "naturalistic mysticism," which has brought about a decisive turn in the history of Western mysticism (Coniglione, 2022, p. 188). On the other side, Coniglione stresses Francis' "mystic of the heart" and the request for a "new apostolic life" (Coniglione 2022, p. 218).

One of the most pronounced effects of the crisis and fall of grand ideological narratives has been a strong revival in world religions and their identitarian forms. This phenomenon is tightly connected – philosophically and historically – to the profound crisis of what Jürgen Habermas (1987, p. 2) has called the "secularization of values and norms" (*Säkularisierung von Werten und Normen*). Some fundamental changes are taking place. The first is the differentiation and dissolution of metaphysical philosophies within particular domains. The second is the disenchantment regarding the expectation that the progress of science shall lead to the discovery of truth, which shall make all enchanted forms of understanding the world disappear, and thus also religion (Melloni 2014).

What Francis teaches us on spirit is possibly the most beautiful course content a professor can dream of offering. Coniglione shows that Francis thinks about the question of God in the most radical way: because otherwise, philosophy gives up its vocation as soon as it abandons this question (Halfwassen 2010). If one looks into the connection of politics and religion in the context of the analysis of secularization, the nucleus is not simply the conviction that one could transform, based on the application of democratic principles, the signing up to a religious denomination from an element of conflict into a function of enhancement and stabilization of political democracy, but also and most importantly the highlighting of what conceptually and operatively follows from today's transfiguration of religion: the recognition of the pluralism of metaphysical, ideological, cultural, and religious stances.

Francis's teaching has a bearing on migration. There is a lot to say about Francis and historic figures. Coniglione mentions his dealings with Islam (Coniglione 2022, p. 322), his role in constructing the legend of Anthony of Padua the healer (Coniglione 2022, p. 327), and the establishment by Pope Nicolas IV on 18. August 1289 of the third Franciscan order, "the order of brothers and sisters of penance" (Coniglione 2022, p. 369). Francis has a say in improving measures and management through cultural crystallization of the migrant planet. Cultural migration mobilizes a plurality of factors, among other historical traditions, religious beliefs, imaginary or social models but also normative and organizational structures, media structures that inspire migration. The culture of migration is based on the imaginary of migration, which means the scene of the individual and collective representation of the migration experience. Two factors come into play: (a) persuasive communication, linked to new media (social networks), highlighting the elements of migratory success and (b) dissuasive communication practiced by traditional media (television and radio), which focus on the elements of dangerous and unsuccessful migration. This is why governments should define the issues in migration research around the digital economy and local development. Religious communities often develop programs that support and accompany migrants in their host countries, such as educational programs and food banks. Other elements that explain the link between migrants and religions come into play, including resilience when other modes of integration are possible.

The Beatitudes can be seen as a confirmation of the *United Nations 2030 Agenda* because they look to the future and give the certainty that conflict situations will find a reconciliation (they will be consoled; they will be satisfied; they will find mercy, etc.). It is a line of investigation that is based on the notion of fraternal dialogue, which was introduced for the first time in the social doctrine of the Church in *Gaudium et spes* to indicate the solidarity that exists between those who, despite their obvious differences and mentalities, are brothers because they are children of the same God, administrators of the same world, and they treat each other accordingly (Pope Paul VI 1965, no. 25). The 17 Sustainable Development Goals and their 169 targets indicate that the global commons pose challenges that require urgent and meaningful global cooperation. While waiting for justice to be fulfilled, for poverty to be abolished, for injustice to be remedied, how can all this be sustained? How can one be happy in adversity? From here, an apparent paradox: in these conditions, beatitude is a possibility of light, of redemption, not just a day – and it could be an important day, because a nuclear conflict would bring us close to the end of the world – but already now.

Never as in this beginning of this second quarter of century, with the tensions and dark clouds that clog the future of mankind, the teaching of Francis of Assisi has been so actual. Coniglione has made it clear that although he has come from another world, Francis speaks to the heart and the mind of men and women of this world, who are coping with their everyday activities and are too often unaware of what lies beyond the material goods by means of which, it seems, their lives ought to be filled. Francis' pointing out the way toward interior perfection without losing contact from effective caring for fellow humans, for nature, for an integral and rigorous exercise of the Gospel, is not a mission impossible for today's men and women. The exciting thing is that the world will step up to consider innovation, reflection, and inclusion in the direction of what Chinese scholars know for thousands of years under "spiritual humanism" 精神人文主义 (*jīngshén rénwénzhûyì*), which is now an "emerging global discourse" (Tu Weiming 2010, p. 7307). The next years will show how the teaching of Francis will trigger a mindset-change regarding reflection on locating culture (anthropology of space and place) for inclusion in education, life-long learning, healthcare, urban development, and regeneration during the decade that has just started.

References

Coniglione, F. (2022). *L'uomo venuto da un altro mondo: Francesco d'Assisi.* Acireale-Rome: Bonanno.

Frale, B. (2016), *La guerra di Francesco: Gioventù di un santo ribelle.* Turin: UTET.

Francis of Assisi (1977), *Fonti francescane.* Assisi: Movimento Francescano.

Gadamer, H.-G. (1975): *Truth and Method*, edited by Garrett Barden and John Cumming. New York: Continuum.

Habermas, J. (1987): *The Philosophical Discourse of Modernity: Twelve Lectures*, edited by Frederick G. Lawrence. Cambridge, MA: MIT Press.

Halfwassen, J. (2010): "Die Unverwüstlichkeit der Metaphysik." In: *Philosophische Rundschau* 57, 97–124.

Maffi, L. (2007), "Biocultural Diversity for Sustainability." In: J. Pretty, A. S. Ball, T. Benton, J.S. Guivant, D.R. Lee, D. Orr, M.J. Pfeffer and H. Ward (Eds.), *The Sage Handbook for Environment and Society*, pp. 267–277. Los Angeles: Sage.

Manselli, R. (2013), *San Francesco d'Assisi: Editio Maior.* Cinisello Balsamo: San Paolo.

Mansfield, M.C. (1995), *The Humiliation of Sinners: Public Penance in Thirteenth-century France*. Ithaca, NY: Cornell University Press.
Melloni, A. (2014), *Rapporto sull'analfabetismo religioso in Italia*. Bologna: Il Mulino.
Pope Francis (2015), *Encyclical Letter Laudato Si': On Care for our Common Home*. Rome: Libreria Editrice Vaticana.
Pope Paul VI (1965), *Pastoral Constitution on the Church in the Modern Word: Gaudium et spes*. Rome: Libreria Editrice Vaticana.
Pozzo, R. (2021), *History of Philosophy and the Reflective Society*. Berlin-Boston: De Gruyter.
Tommaso da Celano (2004), *Vita del beato Francesco: Vita prima e vita seconda*, edited by Claudio Leonardi. Milan: Fondazione Lorenzo Valla.
Tu Weiming (2010), "An 'Anthropocosmic' Perspective on Creativity," *Science Direct: Procedia Social and Behavioral Sciences* 2, 7305–7311.

CHAPTER 11

Some Considerations on Silence, Music, and the Essence of Things: an Opportunity to Explore Francesco Coniglione's "Southern Passage"

Emanuele Coco

Abstract

The scholarly work of Francesco Coniglione unfolds as a series of profound investigations into themes that, though seemingly disparate, are united by a common thread: the fundamental relationship between the individual and reality. This nexus is examined through multiple lenses – scientific and epistemological, mythical and mystical, as well as socio-political. To trace this connective strand, we might invoke the metaphor of the fabled *Northwest Passage*, here reimagined as the *Southern Passage* for reasons that will become clear. A particularly illuminating example of this intellectual trajectory is Coniglione's work on St. Francis of Assisi, where the concept of *silence* emerges as a gateway to inner revelation. Far from mere absence, silence becomes a dynamic space of contemplation, a medium through which images coalesce into understanding, and sound (or its negation) guides the seeker toward a deeper communion with reality. At its heart lies a dual aspiration: the discovery of harmony and the liberation from illusion. Here, the pursuit of external knowledge transforms into an introspective practice, a means of plumbing the depths of one's own interiority.

Keywords

Contemptus Mundi – cosmos – music – Schopenhauer – silence

1 Introduction

Aiming to pay tribute to the scholarly contributions of Professor Francesco Coniglione, in this reflection I revisit some of his studies. I am convinced that, despite their apparent diversity – which, nonetheless, reflects the breadth and richness of his intellectual pursuits – they ultimately converge on a central question: What paths are available to us for understanding reality, and what

 | HTTPS://DOI.ORG/10.1163/9789004756496-013

significance does this relationship with reality hold within our personal and collective experience?

This question, at once profound and pervasive, animates much of Coniglione's thought. It serves as a silent undercurrent, weaving together his explicitly epistemological writings, his later explorations of reason's "obscure" dimensions (that gnoseological realm often relegated to the domain of *anti-logos*), and his incisive reflections on societal and political concerns. Whether examining the plight of the modern university, or advocating for a vision of knowledge emancipated from utilitarian demands – one oriented toward spiritual elevation rather than instrumental gain – Coniglione's work returns insistently to this foundational tension between the desire for objective knowledge and the ineffable aspects of human experience. What emerges is not a mere catalog of ideas, but a coherent philosophical project: a quest to map the boundaries and possibilities of human understanding.

This tribute of mine aspires to map what in Coniglione's thought emerges as a conceptual route that enables fluid transitions across seemingly disparate domains of inquiry. These domains bring together his rigorous epistemological studies with his enduring fascination for Asian philosophy – a passion evident both in his specialized university courses and in recurring references throughout his writings, although never fully explored in dedicated monographs. The same intellectual current flows between his investigations into alternative truth paradigms (Coniglione 2017; 2018a, b, c) and his more recent, profound engagements with the thought and legacy of Francis of Assisi (Coniglione 2020; 2021; 2022a, b; 2023).

This trajectory reveals not merely eclectic interests, but rather a coherent philosophical sensibility attuned to the multiple dimensions of human understanding – from the analytical to the mystical, from Western rational traditions to Eastern contemplative practices. This explains the implicit reference to the Northwest Passage embedded in the expression *Southern Passage*: it is a metaphorical image used to understand the unexpected connections between apparently unrelated themes in Francesco Coniglione's scholarship. Much like that fabled maritime route – which offered explorers a transformative shortcut between distant oceans – Coniglione's work reveals conceptual channels linking intellectual territories that might otherwise require laborious circumnavigation. When viewed through this interpretive lens, his oeuvre discloses a distinctive *Southern Passage*: a theoretical express route bridging epistemological inquiries with humanistic concerns. The geographical resonance is deliberate – while Coniglione was profoundly influenced by his Polish experience, it was in the southern reaches of Sicily that he carried out most of his principal intellectual labor and civic engagement. This Mediterranean reorientation of the navigation metaphor that I have proposed (which we might also refer to as a *Southeast Passage*) serves as both homage to his academic milieu and as a recognition of how his thought transcends conventional boundaries: what appears as thematic diversity, in fact, constitutes an interconnected archipelago of ideas.

In tracing this thematic continuity across Coniglione's oeuvre, I found myself drawn to explore its intersections with my own scholarly preoccupations – particularly the philosophical dimensions of music, silence, and Schopenhauer's deployment of these concepts as mediators between consciousness and reality. This intellectual convergence mirrors Coniglione's own methodological approach, wherein the "Southern Passage" serves as both metaphor and methodological conduit, bridging spiritual inquiry with epistemological rigor, individual contemplation with collective flourishing.

Returning now to Coniglione's body of work, I seek to both revisit and substantially expand upon several key propositions advanced in my earlier study (Coco 2020). This re-examination aims to articulate with greater precision how these conceptual throughlines – between sound and silence, knowledge and transcendence – manifest throughout Coniglione's diverse intellectual engagements.

2 De Contemptu Mundi

First of all, I wish to draw upon two fundamental aspects of Francesco Coniglione's studies on Francis of Assisi: the *radical embrace of poverty* and the *paradoxical attainment of perfect happiness*.

The Franciscan path of poverty – epitomized by the saint's dramatic renunciation before the bishop – transcends mere material deprivation. As Coniglione (2020, p. 66) elucidates, this represents a profound spiritual *kenosis*: "This making emptiness within oneself, this stripping away of all worldly and human inclinations to create space solely for Christ in the naked purity of His unadorned Gospel words [...] constitutes not an end in itself, but rather the foundation for an entirely renewed relationship with reality and nature." This radical self-emptying stands as the cornerstone of Franciscan spirituality, giving rise to what Hughes (2013, p. 285) terms a "fraternal solidarity with all creation." The depth of this ontological communion finds its ultimate expression in Francis's final testament: at burial, he insisted on being placed naked upon the bare earth. As Coniglione (2022b, p. 23) poignantly observes, this act "reaffirms an essential solidarity with creation and ceremonially dissolves the artificial separation between humanity and the natural world, achieving a complete fusion of cosmic energies without mediation." Through this double nakedness – first spiritual, then physical – Francis articulates a theology of radical kinship that continues to resonate through Coniglione's interpretation.

As Leonardi compellingly articulates:

> In Francis, there is no disdain for nature or the body, as was common among the heretical movements of his time; no trace of a Gnostic shadow can be found in him. For him, experience originates in the senses and rises toward the spirit. There is no opposition, but rather continuity: through the body, the human being comes to understand himself. Both the faculties of nature and those of grace operate under the regime of grace, yet neither undergoes a change in its essential nature or structure.
>
> LEONARDI 2004, p. xxvii

Francis's distinctive relationship with external reality – including the *body*, *nature*, and the *cosmic* order they embody – moved Lynn White Jr. (1967, p. 1207) to identify in his thought a revolutionary quality, especially when viewed against the backdrop of the world-denying attitudes prevalent in the 11th and 12th centuries.

In Christian thought, the tension between the material world and the spiritual realm has often been colored by a tone of disdain for the former, as was evident among many of Francis' contemporaries. As Coniglione (2024) convincingly argues, this sentiment – known as *contemptus mundi* – reflects a yearning to retreat from worldly affairs, a stance traditionally embodied in the notion of *fuga saeculi*. While its origins cannot be directly attributed to the Holy Scriptures, it stems from misreadings of divine injunctions concerning the renunciation of material possessions. More likely, this attitude was reinforced by a pervasive sense of shame – not necessarily a modesty that diminished the value of physicality, thereby impeding access to higher spiritual awareness, but rather the weight of the burden inherited from Original Sin. As recounted in the Genesis narrative: "Before, he was naked [...] but he did not blush because his innocence was perfect. But when he had sinned, he realized he was naked and, out of shame, he ran to make a fig-leaf girdle" (Da Parma 1977, p. 1642).

Undoubtedly, the sermons of St. Paul and St. John contributed to deepening the perceived divide between the spirit and the external world. This need to flee from the vices of the flesh – encompassing the spectrum from temptation to eternal perdition, from avarice to concupiscence – is vividly echoed in the words of Abbess Herrada of Landsberg (12th century) in her *Hortus Deliciarum*, where she enumerates the enticements of the body: *lasciviousness, exuberance, temptation, flattery, love, pleasure, lewdness, profanity, filth, voluptuousness, fornication, amusement, ostentation, attractiveness, and discord* (von Landsberg 1979, p. 367).

However, as Coniglione aptly observes, contempt for the world "had already taken root in the classical tradition – beginning with the Pythagoreans, passing through Plato and the Stoics, and reaching an almost medieval-Christian tone in the writings of Marcus Aurelius" (2022b, p. 170). The devaluation of the body would ultimately come to dominate Western thought, shaped by influential figures such as Hilary, Jerome, Ambrose, and above all Augustine. This shift was made possible by what Coniglione – following the insights of Le Goff and Truong (2012) – calls a *grand retournement*: the transformation of Original Sin from a matter of pride, as presented in the Old Testament, into one of sexual concupiscence.

The medieval impulse of world-renunciation achieved its most authoritative expression in *De contemptu mundi* by Lotario dei Segni (later Pope Innocent III), a seminal text that profoundly shaped subsequent ascetic thought. Lotario's treatise advances a totalizing rejection – not merely of the flesh, but of worldly existence in its totality, condemning "human life in all its aspects" (Sanson 2009–10, p. 49). This radical repudiation of temporal reality resonated deeply with his contemporaries. The same profound existential disquiet finds equally powerful voice in Hugh of Saint Victor's *De vanitate mundi* (12th century) (Coniglione 2024, pp. 86–88), whose opening lament poignantly articulates:

> O munde immunde, quare sic dileximus te? Hiccine est fructus tuus? Haec promissio tua? Haec de te spes nostra? Quare in te speravimus? Quare tibi credimus? Quare de te cogitare voluimus? Ecce quomodo decepti sumus. Nihil reliqui habemus, inanes remansi sumus. O munde immunde, quare sic dileximus te?
>
> O impure world, why did we love you so? Is this the reward you offer? Is this what you promised? Is this the hope we were meant to place in you? Why did we trust you? Why did we believe in you? Why did we set our thoughts upon you? Thus are we deceived. We have nothing, we have remained empty. O impure world, why did we love you so?
>
> HUGH OF SAINT VICTOR 1880, p. 740B (translation mine)

It was precisely against this despair – and against the very causes of such despair – that Francis would direct his "revolution."

3 Making a Void within to Embrace the Whole

The body – and not the body alone – reappears in the second episode concerning the life and thought of Francis of Assisi that I mentioned earlier: the tale of *perfect joy*. In the *Fioretti*, we are told that Francis, traveling in winter with Brother Leo from Perugia to Santa Maria degli Angeli, suffering from the biting cold, called out to Brother Leo, who was walking ahead, and told him that perfect joy was not to be found even if a Friar Minor could speak every language, could prophesy and reveal not only future events but also the hidden secrets of hearts and souls. Nor would there be perfect joy, he continued, if the friar could speak the language of angels, know the course of the stars, the virtues of plants and birds, and the location of the earth's hidden treasures; not even if he succeeded in converting every non-believer to the Christian faith. When Brother Leo, now somewhat impatient, asked Francis what *perfect joy* truly was, Francis replied: if, upon reaching the convent of Santa Maria degli Angeli, they knocked and were turned away – rejected and left in the cold – and if they could accept this rejection without resentment or agitation, then they would have attained *perfect joy*. That is, if they could govern themselves so fully as to remain untouched by external events, they would have entered into a state of complete self-possession and mastery: here, the body's vulnerability becomes the crucible for spiritual triumph. As Coniglione writes, "perfect joy is rather an inner event, a step forward along the path of perfection; indeed, it is the moment in which this path reaches fulfilment, when man truly strips himself of all things in order to welcome only God" (Coniglione 2023, p. 171). This radical *kenosis*, Coniglione notes, finds startling resonance with Buddhist teachings on detachment.

Francis's call transcends mere resignation or self-negation. It is, rather, a summons to transfigure the obvious, to rise above the inertia of one's own nature and the world's weary patterns, which so easily degenerate into *contemptus mundi*. This is an invitation to *kenosis*: to empty oneself of the clamor of triviality and grievance, creating space for silent receptivity to both the inner self and the outer world. Only thus does one discover *perfect joy*. In this awakened state, reality is unveiled anew – no longer adversary but revelator. It demands an ear attuned to silence rather than noise, and an eye capable of true seeing. Here unfolds an ecumenical dialectic: for creation's splendor remains latent without a perceiver to apprehend it. As Goethe later observed, "The eye owes its existence to the light." The sun summons the eye into being, just as the eye, in turn, confers meaning upon the sun's radiance. Without this reciprocity, brilliance persists, but splendor goes unseen.

4 The Silent Cosmos

Our preceding reflections on Francis of Assisi lead us to a deeper consideration of a pivotal theme: the deliberate act of *distancing oneself from the world*. Such distancing takes shape in two fundamentally distinct forms. The first is *a path of rejection*: a stance of contempt and renunciation, epitomized by medieval *contemptus mundi* traditions. The second is a *path of purposeful withdrawal*, meaning a contemplative rite, a silence deliberately cultivated to foster deep listening and a detachment paradoxically aimed at *re-membering* one's place within the cosmic whole. This latter mode follows a transformative arc. It begins not in renunciation, but in *immersion*: a full embrace of worldly systems and values. Francis's early life exemplifies this trajectory. As a young man, he thrived within his father's mercantile world, internalizing its material logic, even aspiring to knighthood, that quintessential embodiment of feudal earthly ambitions (Coniglione 2022a, pp. 21–22; Cardini 2021, pp. 40–65). Only through such saturation could his subsequent withdrawal acquire its radical, counter-cultural power.

This *sonorous nakedness* – this *mystical silence* – is far from an effortless path. It eludes both the frenetic hedonist, who devours the days without pause or discernment, and the world-denouncing ascetic, whether medieval or modern, who flees creation's symphony – its clamor, its fervor, its jubilant colors – only to cloak silence in the hairshirt of penitence. Herein lies the paradox: *True silence is neither mute resignation nor aural oblivion.* The fearful turn it into a prison; the wise make it a sanctuary. It demands not the deafness of denial, but the acute attunement of a soul listening through stillness – a receptivity that transforms emptiness into plenitude, solitude into communion.

Silence – like nakedness – embarrasses and unnerves. In its presence, there is no shield against the unsettling possibility of discovering that we are not inherently exceptional. This revelation may arise when, with a pure and hopeful spirit, we strive to hear the voice of God, only to be met with apparent absence. The realization that the vast universe surrounding us may hold no special meaning for us in particular can be devastating. If a divine presence indeed hovers within these realms, it may remain elusive, offering no graspable sign. Such a silent cosmos haunted Pascal: "The eternal silence of these infinite spaces terrifies me" (Pascal 1995, p. 73). Earlier, however, the Pythagorean-inspired thinkers had been far more optimistic. For them, the cosmos was not silent. With a trained and attentive ear, one could hear – if not the voice of God itself – at least its manifestation in the form of geometrically unfolding laws. The cosmos moved in harmony, and the initiate – whether philosopher, mystic, or scientist – could, in grasping that harmony, feel himself part of a greater order. That order offered not only redemption from the chaos of earthly life, but also from the dread of a meaningless existence. According to this tradition, the cosmos was structured in elegant proportions. Its very movement produced a most delicate music, like the resonant tones of a well-tuned instrument, whose strings, when struck together, yield a sound that is pleasing to the soul.

This divine resonance – born of the cosmos's astonishing mathematical order – was perceptible only to those who had cultivated inner silence. Not the mute austerity of an indifferent deity, but a celestial hymn animating the universe itself: a vast, revolving love that sweeps all creation into its motion.

Does this render Pascal's terror before the "silence of infinite spaces" misguided? Not entirely. Even Aristotle, centuries earlier, had dismantled the Pythagorean vision of harmonic cosmology to advance his own physics of unordered celestial spheres. Where the Pythagoreans heard cosmic music, Aristotle perceived only the mechanical grind of ether, a universe stripped of transcendent melody. Pascal's existential shudder thus inherits this Aristotelian rupture – which he fully recognized in the universe proposed by Descartes –, while amplifying its spiritual consequences for the modern age:

> Some thinkers suppose that the motion of bodies of that size must produce a noise, since on our earth the motion of bodies far inferior in size and in speed of movement has that effect. Also, when the sun and the moon, they say, and all the stars, so great in number and in size, are moving with so rapid a motion, how should they not produce a sound immensely great?
>
> ARISTOTLE 1995, 290b, 15–20

The Philosopher of Stagira thus concluded: "That the stars are spherical and are not self-moved, has now been explained" (Aristotle 1995, 291a, 25).

The stars themselves stood motionless – eternal, fixed. It was the celestial spheres in which they were embedded that wheeled in silent procession, bearing their luminous charges through the void. For those who had cultivated inner stillness, listening intently for the promised music of the spheres, Aristotle's revelation was a theft of meaning. No attuned ear, however sensitive, could now redeem the cosmos from the silence imposed upon it.

Unable to alter the circumstances of his earthly existence as a mathematician in Graz – where he longed instead for the theological circles of Tübingen, for the company of his beloved mentor Michael Mästlin (Baumgart 1951, pp. 60–1) – Kepler turned his gaze upward, to the stars that captivated him not merely for their brilliance, but for the hidden order governing their motions. He could not embrace a mute universe, and so he refused this bleak cosmology. The spheres might be silent, the stars themselves unmoving – and yet, to Kepler, their very geometry *sang*.

The truth was that it was the celestial spheres, in which the stars were embedded, that orchestrated their motion. Yet for those who, ensconced within their private realms of stillness, eagerly awaited harmonious tidings from the uncharted expanses of the Universe, this revelation offered little comfort. Their attentive ear could not rescue the cosmos from the profound silence that Aristotle had unwittingly cast upon it.

This pursuit of cosmic harmony, as documented in Coco (2016), became his life's work. Through tireless study – no trivial endeavor – he filled lecture halls, derived his three enduring laws, and ultimately transcended the purely geometric model of celestial order, embracing instead a *musical* cosmology. In his 1596 *Mysterium Cosmographicum*, Kepler had sought harmony in nested Platonic solids dictating planetary orbits. But by 1619, with *Harmonices Mundi*, he uncovered a profounder truth: *it was not the shapes of orbits but the velocities of planets that sang in concord*. Their proportional speeds mirrored the very intervals that stir a musician's ear – major thirds, minor sixths – transforming the solar system into a grand polyphony.

For Kepler, the cosmos was no longer a silent mechanism but a celestial choir, a harmony he resurrected from the void. This was divine music, ordained by God, so exquisite in its consonance that it restored not just joy, but *wonder*. A profound solace for those content to hear it not with their ears, but through the mind's eye, tracing the elegant geometries of planetary motion across parchment. The music was, admittedly, more imagined than audible. Yet in that imagining, it accomplished something extraordinary: it shattered the oppressive silence that had haunted humanity's contemplation of the heavens. No longer were the stars mere indifferent wanderers in a meaningless expanse; their very orbits now testified to God's handiwork, a grand design where mathematics and melody converged. However, faint to mortal ears, this cosmic hymn offered what Aristotle's silence could not: *comfort*. The universe, Kepler insisted, was not orphaned. Its harmony was proof of a Creator who had tuned the world like a lute, and in that tuning, left humanity a lullaby written in the stars.

I advance a parallel here, in the guise of suggestion, without wishing to argue a historical thesis or anything that can be supported on the basis of documents. We might claim that Kepler transposed to a scientific plane the very harmony and worldview that suffused the spiritual vision of Francis of Assisi. Where Francis heard the *Canticle of Creation* in the laughter of sunbeams and the rustling of leaves, Kepler detected its mathematical counterpart in the elliptical orbits of planets, their velocities singing in perfect Pythagorean intervals. Both men, though separated by centuries, shared a fundamental conviction: that the cosmos was *legible*, its order revealing divine artistry. Francis embraced poverty to hear creation's song; Kepler wielded logarithms to transcribe it. The friar found God in the humility of a sparrow; the astronomer, in the precision of Mars' orbit. This was no mere analogy, but a profound continuity: Kepler's *Harmonices Mundi* (1619) systematized what Francis had intuited – that nature's laws were not cold mechanisms, but love letters from the Creator, inscribed in geometry and audible to those who listened with disciplined wonder. Where medieval *contemptus mundi* saw a fallen world, both Franciscan and Keplerian thought discovered a *sacramental* one, where every equation and earthworm alike bore witness to grace. Both were mystics of measurement: Francis counting the notes in a bird's call, Kepler the days of Mars' revolution.

5 Schopenhauer and "The Sound of Silence"

Thus far, my discourse has engaged deeply with Francesco Coniglione's nuanced meditations on the spiritual journey of Francis of Assisi – particularly those subtle allusions to the interplay between the world and its manifold appearances. Now, however, I propose to depart from these theological frameworks, leaving behind the Christian tradition's long-standing preoccupation with *peeling away the veil of appearances* in pursuit of spiritual authenticity. Instead, I turn to a secular voice – one no less eloquent, though unmoored from doctrinal imperatives – to illuminate this perennial dialectic. The tension between material reality and transcendent truth persists as a defining dyad, shaping not only how we conceptualize the physical world but also how we derive meaning from it. Where medieval Christianity framed this dichotomy as a call to renunciation, modernity has often recast it as a creative paradox: the world's surfaces are neither mere illusion nor obstacle, but the very medium through which deeper truths might shimmer into view.

The secular voice to which I now allude is that of the German philosopher Arthur Schopenhauer – whose life, in some respects, echoes the very anguish experienced by Kepler, and whose leitmotif was marked by profound isolation and, as I will argue, a peculiar *destiny toward silence*. "I trust only in myself, in this matter as in any other; this is my fate" (Brianese 2013, p. xxi). In his magnum opus, *The World as Will and Representation*, Schopenhauer acknowledged that even metaphysics has its limits. The world-as-will could only be apprehended up to "the borders of the expressible," beyond which – as Wittgenstein would later echo in the *Tractatus* – "the rest is silence." Yet this silence, as Giorgio Brianese observed, was "anything but devoid of philosophical value" (Brianese 2013, p. xxi). It was, rather, a charged void – a space where language fails, but meaning persists.

If the *Southern Passage* charts a conceptual route from epistemology to spiritual contemplation, then its deepest station may well lie in the silent space where thought ceases and listening begins. In each of the three authors mentioned, for different reasons and with distinct attitudes – Schopenhauer, Francis and Kepler – we find not the clamor of certainty, but the resonance of wonder. Each, in their own register, gestured toward a reality that speaks through silence: the geometry of the heavens, the humility of a sparrow, the metaphysical void beyond representation. This silence is not the absence of meaning, but its radical concentration. It is the echo of a universe made legible not by speech, but by attention; not by mastery, but by reverence. Whether through the harmony of celestial spheres, the beauty of Creation, or the charged stillness at the limits of metaphysics, Coniglione's thought reminds us that to understand is not only to explain – it is also to listen. In this listening, the *Southern Passage* finds its destination: a space where philosophy and contemplation meet, and the sound of silence resounds with meaning.

Yet in Schopenhauer, silence does not merely resonate as contemplative stillness. When examining silence in Schopenhauer's philosophy, we confront a fundamental paradox: the very ground from which he attempts to grasp the essence of reality is itself a roaring torrent of ceaseless striving. The "silence" at the heart of his system is not an absence, but rather the ineffable space where the deafening clamor of the Will becomes manifest. Two crucial passages from *The World as Will and Representation* reveal this tension with particular clarity:

> But now, what kind of knowledge is it that considers what continues to exist outside and independently of all relations, but which alone is really essential to the world, the true content of its phenomena, that which is subject to no change, and is therefore known with equal truth for all time, in a word, the Ideas that are the immediate and adequate objectivity of the thing-in-itself, of the will? It is art, the work of genius. It repeats the eternal Ideas apprehended through pure contemplation, the essential and abiding element in all the phenomena of the world. According to the material in which it repeats, it is sculpture, painting, poetry, or music. Its only source is knowledge of the Ideas; its sole aim is communication of this knowledge. Whilst science, following the restless and unstable stream of the fourfold forms of reasons or grounds and consequents, is

> with every end it attains again and again directed farther, and can never find an ultimate goal or complete satisfaction, any more than by running we can reach the point where the clouds touch the horizon; art, on the contrary, is everywhere at its goal. For it plucks the object of its contemplation from the stream of the world's course and holds it isolated before it. This particular thing, which in that stream was an infinitesimal part, becomes for art a representative of the whole, an equivalent of the infinitely many in space and time. It therefore pauses at this particular thing; it stops the wheel of time; for it the relations vanish; its object is only the essential, the Idea.
>
> SCHOPENHAUER [1818–19] 1969, pp. 184–185

It is art, then, that permits us to grasp the true content of phenomena. And among all the arts, music is the most significant. For, contrary to what Leibniz believed, music is not the product of a mind that calculates without knowing it, but rather the expression of a mind that – through music – philosophizes almost unknowingly. Music, in this sense, becomes the vehicle through which the essence of things is disclosed, not through rational deduction, but through an intuitive, almost metaphysical insight:

> we find that there is yet another fine art that remains excluded, and was bound to be excluded, from our consideration, for in the systematic connection of our discussion there was no fitting place for it; this art is music. It stands quite apart from all the others. In it we do not recognize the copy, the repetition, of any Idea of the inner nature of the world. Yet it is such a great and exceedingly fine art, its effect on man's innermost nature is so powerful, and it is so completely and profoundly understood by him in his innermost being as an entirely universal language, whose distinctness surpasses even that of the world of perception itself, that in it we certainly have to look for more than that *exercitium arithmetic occultum nescientis se numerare animi* which Leibniz took it to be. Yet he was quite right, in so far as he considered only its immediate and outward significance, its exterior. But if it were nothing more, the satisfaction afforded by it would inevitably be similar to that which we feel when a sum in arithmetic comes out right and could not be that profound pleasure with which we see the deepest recesses of our nature find expression. Therefore, from our standpoint, where the aesthetic effect is the thing we have in mind, we must attribute to music a far more serious and profound significance that refers to the innermost being of the world and of our

> own self. In this regard the numerical ratios into which it can be resolved are related not as the thing signified, but only as the sign.
>
> SCHOPENHAUER [1818–19] 1969, p. 256

And again:

> Thus, music is as immediate an objectification and copy of the whole will as the world itself is, indeed as the Ideas are, the multiplied phenomenon of which constitutes the world of individual things. Therefore, music is by no means like the other arts, namely a copy of the Ideas, but a copy of the will itself, the objectivity of which are the Ideas. For this reason, the effect of music is so much more powerful and penetrating than is that of the other arts, for these others speak only of the shadow, but music of the essence.
>
> SCHOPENHAUER [1818–19] 1969, p. 257

It would seem the right path: if one seeks meaning in existence and in the universe, if one longs for an intuition of the self and a sense of fullness in life, one must shed the world – the external, the redundant, the noise. In a manner akin to what Francis of Assisi pursued, and as is common to every mystical tradition, it becomes necessary to create a vacuum, to embrace silence itself. Only then may one encounter the essence of things, an essence that, for Schopenhauer, can be grasped solely through music. Yet here one must be wary: not to fetishize the instruments, the symbols, or the experiences that may help pierce the "veil of Maya" (Schopenhauer [1818/19], 1969, p. 352). It would be an error to believe that reciting a specific mantra, contemplating a singular image (the cross, for Francis; or, in Asian mysticism, any sacred figure), or listening to a particular sequence of sounds (which is, indeed, the nature of music for Schopenhauer), is in itself the path to truth. These are but itineraries, anchoring points meant to liberate the mind from all overlays and accretions that obscure its transparency and prevent it from gazing upon the world with unblemished eyes. As Giovanni Pascoli expressed poetically in *Il fanciullino*, and Blaise Pascal grasped with spiritual clarity: "Wisdom leads us back to childhood. *Nisi efficiamini sicut parvuli*" (Pascal [1670], 1995, p. 29), thus offering a mystical reading of the Gospel injunction, "Unless you become like children, you will never enter the kingdom of heaven" (Matthew 18:3). For Francis, this essence was identified with the intuition of God, reached through an arduous spiritual ascent nourished by compassion for others' suffering and by love for all creation. For the secular Schopenhauer, however, the purification of the mind was no less essential in apprehending the essence of the world. Influenced by Indian philosophies and the *Upanishads*, he believed such a state could be approached through compassion and love for all beings who suffer under the dominion of time and causality. "For the veil of Maya," he wrote, "has become transparent for the person who performs works of love, and the deception of the *principium individuationis* has left him" (Schopenhauer [1818/19], 1969, p. 373).

The silence we invoke here is not merely an absence of speech, but an *interior kenosis*, a purgation of mental clutter that renders the mind a polished mirror, capable of reflecting, for Francis of Assisi, the whispered voice of God in creation's fabric and, for Schopenhauer, the primal murmur of the Will beneath phenomena. This cultivated void constitutes philosophy's supreme irony: those who have spoken most profoundly about silence – the mystics and metaphysical poets – have invariably been history's most prolific articulators of the ineffable. Their voluminous writings stand as testament to silence's fertile paradox: only the mind that has been scoured of trivialities can become a conduit for cosmic meaning:

> Mystics simply have not been silent. Many have spoken without restraint, and others have written voluminously. The genre of mystical literature is not only quantitatively vast, but linguistically luxuriant. In mystical discourse, language runs riot: it leaps, it vaults, it sings. It speaks in prose and poetry; it gives objective descriptions of experience and flies on the wings of ecstasy; it guides neophytes with gentle care and cuts through illusion with razor-sharp arguments. Mystical language can be kataphatic in the extreme, chanting the ninety-nine beautiful names of Allah, or evoking the images of the "three-million" gods and goddesses of the Hindu pantheon. No genre of writing can rival it for complexity, subtlety, and linguistic virtuosity.
>
> COUSINS 1992, pp. 238–239

And yet, unlike in the work of Francis of Assisi, Schopenhauer's music does not resonate as an ecumenical choral closeness to others. It does not translate into a practice of solidarity and service; indeed, many episodes attest to the hardness of the philosopher's heart. For Schopenhauer, music can only be truly grasped by those who, in meditative solitude, have silenced the ways of "ordinary existence" – that existence defined by words such as 'office,' 'home,' 'social life,' 'wife and offspring,' and all the other noisy trappings of the world (Schopenhauer [1921] 2003, p. 29). Francis of Assisi, by contrast, resolved his own existential uncertainty – whether to devote himself to hermitic meditation or to go out into the world to preach the Gospel and serve others – with a courageous break from the eremitic model of his time. He chose to walk the streets without home or security, living day by day, yet remaining among others, sharing their suffering and alleviating it through word and example (Coniglione 2023, pp. 11–15, 35 and passim). In Schopenhauer, however, the pendulum swings toward solitary meditation – a personal practice of asceticism that distances itself from the masses and ordinary people, aspiring instead to reach an intuition of the essence of things. It is a theoretical turn marked by a considerable gap between what he professed and the life he actually lived.

The philosopher believes he has experienced this loneliness in the course of his own life:

> As soon as I began to think, I felt cut off from the world. [...] I am left alone. And yet, in all sincerity, I can say it was not of my choosing: I have never rejected or shunned anyone who, in mind and heart, was truly human. But I have found only miserable gnomes – limited in intellect, malevolent in spirit, base in sentiment – with the exception of Goethe [...] and a few others
>
> SCHOPENHAUER [1921] 2003, pp. 65–66

Once again, we encounter a distancing from the world, a departure from the ordinary, a chaste abstention from life's time-consuming entanglements, marriage first among them. "He that hath wife and children hath given hostages to fortune," Bacon wrote ([1597] 2008, Essay viii). In following this path – the path of one who sets himself apart to accomplish "great works" – Schopenhauer appeals to Giordano Bruno: "He who seeks and finds the truth, from an ordinary man becomes a wild one, like a stag or a hermit, and all those in the world who would have aspired to a higher life would say in unison: '*Ecce elongavi fugiens et mansi in solitudine*'" ["Lo, I fled far off and dwelt in solitude"; cf. Ps. 55:8]. "Indeed, their preoccupation with divine matters has rendered them as if dead to the common people" (Schopenhauer [1821] 2003, p. 55). How can we blame him? And yet, a touch of irony is permitted. Perhaps Schopenhauer might have chosen a different philosopher in defense of his monastic inclinations (on his complicated relation to women, and his platonic acquaintance with Elisabeth Ney, see Safranski 1990, pp. 347–8). After all, *withdrawing from the world* had not prevented Bruno from indulging in certain worldly pleasures, at least if one is to believe the accusations of Giovanni Francesco Mocenigo. And it is, tellingly, to female figures that Bruno entrusts the most significant lines of his *Eroici furori* (Coco 2023, pp. 85–101).

And yet, perhaps solitude is not always the product of *hybris* or philosophical affectation. Perhaps, as one might be inclined to believe, Schopenhauer's isolation was not a chosen posture, but a necessity. Few philosophers of the modern era seem to have possessed such strength – such implacable resolve – in refusing to entangle themselves with the trivialities of everyday life.

It becomes thinkable, then, that a *fuga saeculi* might be practiced without falling into *contemptus mundi*. But this, too, carries its own risks, as Lord Byron wryly noted in a remark that Schopenhauer himself was fond of quoting: "The more I see of men, the less I like them. If only I could say the same of women, all would be well" (Moore 1830, p. 499).

•••

In *The World as Will and Representation*, Schopenhauer – alone, in his silence – discerned a form of music capable of dispelling illusion: an art that pierces the veil of appearances and leads to what he called "reality," the innermost essence. As he wrote in the passage cited earlier, "The other arts speak to us only of shadows; music, instead, of the essence of things."

Thus, paradoxically, the essence of the world can only be reached by stepping away from it – yet through something that still occurs within it: music. And here, if we again want to look for a reference to Coniglione's research, we can go back again to what happens in Francis of Assisi, albeit in very different ways and in profoundly different eras and cultures. For him, distancing oneself from the world did not mean scorning it, but cherishing its creatures, all of which bore witness to the glory of God.

The implication is clear: should asceticism become a withdrawal so complete as to sever all ties with the world and its movements, the philosopher would fail to understand, would cease to feel – and therefore would lose any real access to the essence of things.

The experience *simpliciter et sine glossa* of Francis of Assisi, as illuminated in the studies by Coniglione mentioned earlier, proves far more revealing than the naïveté of the *simplex et idiota* religious figure might initially suggest. Francis – who made of nudity not only a metaphor but a physical act, who transformed his relationship with the world and nature into a gesture of deprivation rooted in love – leads us to a place far removed from the banality of ordinary life, yet unwilling to escape into an abstract ideality that forgets the former.

After all, distance from the physicality of existence would be of little use. Not only as evidenced by the lives of Francis and Giordano Bruno – each with his own distinct methods and customs – but also because the overcoming of the tension between the reality of the world and the shadows of illusion must necessarily pass through the human: it is a human experience, one that concerns both the external and internal realms. It is in the vibration of this duality that music is born – and music, as Schopenhauer writes, possesses "a much more serious and profound meaning, one that touches on the most intimate essence of the world and of ourselves."

Without that essence of the self, there would be no reason to seek truth, no access to what lies beyond illusion. And perhaps only through self-knowledge can one rise above the drama of the impossible reconciliation between ghosts and reality. It is, then, a matter of reaching a much more arduous place, one suspended between the world and the idea. A place where music becomes clearer. A place that requires boundless courage: what Bruno once called *heroic fury*.

References

Aristotle (1995). *On the Heavens (De caelo)*. In: *The Complete Works of Aristotle*, ed. by J. Barnes, Princeton: Princeton University Press.

Bacon, F. ([1597] 2008). *Complete Essays*. Mineola, NY: Dover Publications.

Baumgardt, C. (1951). *Johannes Kepler: Life and Letters*. New York: Philosophical Library.

Brianese, G. (2013). La parola dell'enigma e il demone dell'uomo: la cognizione del dolore secondo Schopenhauer. In: A. Schopenhauer, *The World as Will and Representation*, ed. by G. Brianese. Turin: Einaudi.

Cardini, F. (2021). *L'avventura di un povero cavaliere di Cristo. Frate Francesco, Dante, Madonna povertà*. Rome-Bari: Laterza Publishers.

Coco, E. (2016). L'eredità pitagorica nella costruzione della Natura. Alcune considerazioni critiche attorno a un'idea filosofica mai tramontata. *Bulletin of the Italian Philosophical Society*, 217, 59–76.

Coco, E. (2020). Schopenhauer and "the essence of things". Silence and music in relationship with reality. *The Protagoras*, year XLVII, sixth series, 33–34, 99–110.

Coco, E. (2023). The end of the spirits. Nature as a philosophical investigation of the Self. Milan-Udine: Mimesis.

Coniglione, F. (2017). "Rejected Knowledge" and the Tradition of Reason. A few Hints on their Necessary and Inextricable Link. In: R.L. Cardullo and F. Coniglione (Eds.) *Reason and no-reason from ancient philosophy to neurosciences*, pp. 211–231. Sankt Augustin: Academia Verlag.

Coniglione, F. (2018a). Hybris illuminista e accesso alternativo al vero nel mondo greco. In: M. Albana and C. Soraci (Eds.), *Amicitia res plurima continet*, pp. 139-162. Acireale-Rome: Bonanno Editore.

Coniglione, F. (2018b). L'intreccio antropologico tra mito e logo: una indispensabile (ma non pacifica) coesistenza. In: E. Coco and S. Vasta (Eds.), *Il mito tra filosofia e scienza. Temi e prospettive dall'antichità a oggi*, pp. 109–140. Catania: Malcor D'.

Coniglione, F. (2018c). Some Remarks on the Meaning of Esotericism and Plato's Unwritten Doctrines. *Mondi. Movimenti simbolici e sociali dell'uomo,* I, 1, 5–51.

Coniglione, F. (2020). Misticismo e nuovo sentimento della natura in San Francesco. In: R.L. Cardullo and F. Coniglione (edited by), *Vivere in Assisi. Natura e cultura dopo San Francesco*, pp. 61–100. Acireale-Rome: Bonanno.

Coniglione, F. (2021). Povertà, "minorità" e apertura al mondo in Francesco d'Assisi. *Annali della facoltà di Scienze della formazione*, 19, 231–247.

Coniglione, F. (2022a). Da un mondo all'altro. Ciò che Francesco d'Assisi può ancora dire all'uomo d'oggi. *Mondi*, 1, 18–43.

Coniglione, F. (2022b). *L'uomo venuto da un altro mondo. Francesco d'Assisi*, Acireale-Rome: Bonanno.

Coniglione, F. (2023). *La perfetta letizia. L'itinerario spirituale di Francesco d'Assisi*, Acireale-Rome: Tipheret.

Coniglione, F. (2024). *Contemptus mundi* e disprezzo del corpo nel basso medioevo. La svolta di san Francesco. In: *Il vento di Eros. Atti del XIII Colloquio Internazionale Medioevo romanzo e orientale* (Dipartimento di Scienze Umanistiche, Catania, 21–22 settembre 2023). Soveria Mannelli: Rubbettino, pp. 81–106.

Cousins, E. (1992). Bonaventure's Mysticism of Language. In: S.T. Katz (ed.), *Mysticism and Language*, New York-Oxford: Oxford University Press.

Da Parma, G. ([1227], 1977). Sacrum Commercium Sancti Francisci cum Domina Paupertate. In: *Fonti Francescane.* Assisi-Padova: Movimento Francescano – Edizioni Messaggero.

Goethe, J.W. (1810). *Zur Farbenlehre*. Tubingen: JG Cotta.

Hughes, K. L. (2013). Francis, Clare, and Bonaventure. In J.A. Lamm (ed.), *The Wiley-Blackwell Companion to Christian Mysticism*, pp. 282–296. Chichester: Wiley-Blackwell.

Le Goff, J., Truong, N. (2012). *Une histoire du corps au Moyen Âge*. Paris: Liana Levi.

Leonardi, C. (ed.) (2004). *La letteratura francescana*, vol. I, *Francesco e Chiara d'Assisi.* Milan: Fondazione Lorenzo Valla-Mondadori.

Moore, T. (1830). *Letters and Journals of Lord Byron, with Notices of His Life*, vol. I. New York: J. & J. Harper.

Pascal, B. ([1670], 1995). *Pensees and Other Writings*. Oxford: Oxford University Press.

Sanson, M. (2009–10). *Il corpo nell'opera di Francesco d'Assisi e di Iacopone da Todi*. Ph. Thesis, Trento: Università degli Studi.

Hugh of Saint Victor (1880). *De vanitate mundi*. In: *Patrologia Latina*, vol. CLXXVI.

Safranski, R. (1990). *Schopenhauer and the Wild Years of Philosophy*. Cambridge, MA: Harvard University Press.

Schopenhauer, A. ([1854] 1959), *Ueber das Sehen und die Farben.* Trad.it., *La vista e i colori. E carteggio con Goethe*. Turin: Bollati Boringhieri.

Schopenhauer, A. ([1818/19], 1969). *The World as Will and Representation*, Translated from the German by E.F.J. Payne. New York: Dover Publications.

Schopenhauer, A. ([1821] 2003). *L'arte di conoscere se stessi, ovvero Eis Heautón*. Milan: Adelphi.

Von Landsberg (von Hohenbourg), H. ([1175] 1979). *The Hortus Deliciarum of Herrad of Hohenbourg (Landsberg)*. A Reconstruction by Rosalie Green, Michael Evans, Christine Bischoff, and Michael Curschmann. With contributions by T. Julian Brown and Kenneth Levy, ed. by R. Green. London: The Warburg Institute.

White Jr. L. (1967). The Historical Roots of Our Ecological Crisis. *Science. New Series*, 155 (3767), 1203–1207.

Woollard, W. (2017). *Il buddhista riluttante. Viaggio di un occidentale alla scoperta del buddhismo*, Milan: Esperia.

CHAPTER 12

How Myths Help to Deal with Complexity

Santo Di Nuovo

Abstract

The aim of the article is to examine how myths can contribute to understanding complex phenomena, difficult to explain with logical reasoning, reductionism, and linear causality typical of classical science. The myth reconstructs complex reality using two important mental functions, imagination, and metaphorical thinking. Through the imagination, different domains, and levels of the interpretation of reality are connected with great hermeneutic power. Metaphorical thought creates elaborations not linked to the consolidated organization of the language, connecting different domains of knowledge. Examples of understanding myth through intersubjectively shared imagination and metaphors are Jung's archetypes, as exemplified in his "Red Book". The mythology founded on imagination and metaphors also could find reference in culturally shared "memes". The imaginative and metaphorical forms that flow into myths can lead to the breaking of consolidated schemes, and be a source of planning change, aimed at promoting social innovation.

Keywords

complexity – Jung – knowledge – myth – science.

In an essay on the responses to the challenge of complexity in contemporary science, published with Coniglione (Di Nuovo and Coniglione 2002), we wrote that ever since science has begun to face "chaos" and complexity, many disciplines had to reconsider the traditional cornerstones of empirical science: reductionism, the analysis of systems in terms of their component parts, determinism, and linear causality. The aim was to try to reach a more holistic approach to reality, for a better understanding and transformation of it. Subsequently, although each of us followed distinct scientific paths, we gradually realized the profound significance of myths and the disciplines often intertwined with them (such as religion and mysticism), not only in scientific

 | HTTPS://DOI.ORG/10.1163/9789004756496-014

thought (an aspect that Coniglione particularly addressed in his work on the history of epistemology), but also within the sphere of psychology and neuroscience (the area of my competence) and in human life more broadly. Thus, our different trajectories ultimately converged on this very theme, unfolding in harmony yet independently, yielding reflections that resonate in numerous shared aspects.

Bearing in mind Coniglione's work on this topic (2017; 2018; 2019a, b; 2021; 2022a) I wish to offer insights from a psychological perspective, also with some critical suggestions in relation to his stance. Specifically, this article aims to scrutinize how myths attempt to reconstruct complex and otherwise unspeakable or unexplainable realities, using cognitive and linguistic structures such as imagination and metaphor.

1 Myth as a Means for (Re)constructing Reality

For Ancient Greek μῦθος means "narrative", fiction aimed at explaining the development of social customs, with their cultural rules and taboos. According to Eliade (1998), also in archaic societies myths were exemplary stories about reality, but with characteristics of sacrality. But in some indigenous cultures myths have a negative meaning, opposite to empirically stated truth. Also, the Judaic and Christian tradition considered the myth as an "illusion" derived from fantasy, opposed to λόγος (i.e., word, discourse) and to history.

Can this sharp opposition be overcome in some way?

Ethnologists and historians of religions have underlined the sense of sacred tradition, considering myths as exemplary models of aspects of reality difficult to understand (Coniglione 2019a). Lévi-Strauss (1978) wrote that myths attempt to face and understand what every society, in a certain period of its history, finds contradictory and inexplicable. Moreover, a function of the myths is to provide a basis for legitimizing cultural norms and social institutions (Malinowski 1926).

The myth attempts to bring order to the chaotic nature of a reality that cannot be traced back to unique relationships of cause and effect, and therefore explainable with the rationality of history and science. For example, mythological explanations have been devised to explain the cosmogonies, the flood and submersion of the earths, the original state of the world as an original "paradise," the physical death, and the afterlife.

On a psychological level, the myth explains essential components of normal and pathological human psychic development. Freud's interpretations of classical myths (e.g., the myth of Oedipus who unknowingly kills his father) attempt to explain psychological realities valid in the contemporary world. Subsequent psychoanalysis continued along this path.

But what does the myth use to discover the complexity of life? Two essential components are the cognitive functions of "mental imagery" and metaphorical thinking, which will be focused on this paper, to arrive at the mythology expressed in the Jungian Red Book, which is an exemplary expression of these functions.

2 Mental Imagery, between Analogic and Verbal

The mental image is an association of verbal and figurative cognition. The ancient debate on "images in the mind" contrasted *propositional* theory (based on concepts and verbalizations) and *analogical* theory (pictorial, analogous to perceptual elaboration). Coniglione (2008) effectively demonstrated the role of imagery even within scientific thought, underscoring its significance in the post-Popperian turn of epistemology marked by Kuhn, and subsequently continued in post-positivist epistemology. He particularly emphasized the value of "tacit knowledge" in this context, knowledge that eludes explicit verbalization and propositional formulation. Pioneering contributions in this realm have also come from Polish epistemologist Ludwik Fleck, whose epistemological writings were edited for their Italian edition by Coniglione (Fleck 2019; Coniglione 2020).

However, both the approaches to the reality – the propositional and the analogic-visual are valid, in different conditions and contexts (Paivio 1986). By "image" we mean a representation – not necessarily visual – constructed and embodied in the mind, which in the imagination represents (simulates) itself in action (Kosslyn 1980; Marks 1985).

Imagination plays an important role in daily life as much as ideas, bringing the mind back to the past and projecting it into the future. It allows to see and mentally re-elaborate what refers to reality, but, in many cases, re-constructs it by inserting internal contents significant on a cognitive and emotional level. Therefore, the imagery process comes out of the reductive stimulus-response constraint and constitutes the foundation of what in our mind does not exactly correspond to reality, but builds new stimuli and new responses.

The creative project also involves the recall from memory of the "imagined" perceptive and semantic elements, their manipulation and recombination in a new and original way (Piaget and Inhelder 1966). The importance of imagination in designing and creating original realities is well known: the artistic productions of painters, sculptors and architects, the staging by directors, actors, choreographers, derive from it. Many inventions and scientific discoveries are translations of images generated in the mind. The integration of the iconic and symbolic functions takes place above all in the creative project. Flexibility and openness to novelties are dimensions of creativity that are strongly correlated with imagination. E.g., the artist realizes and communicates through his product a particular vision of reality, or the novelist constructs new scenes, effectively using the projection of his own internal images.

In these cases, and in the others of creative production, the "algorithmic" proceeding – i.e., based on the application of pre-established rules – which is typical of formalized thought and language, must be integrated and overcome (even if not cancelled) by holistic procedures, in part based on intuition, in which imagination necessarily plays a fundamental role. Indeed, the holistic nature of mythical thinking has been emphasized by Coniglione in contrast to the approach of rational thinking: "In myth, there is a holistic view of reality, a synthetic approach where it is comprehended as a whole and in the close interrelation among its parts (it is 'dialectical'), whereas on the other hand, logos tends to distinguish, separate, in order to better control and explain more accurately (it is 'analytical')" (Coniglione 2022a, p. 12). This characteristic of "overview" allows myth to stimulate creative production, as also highlighted by Ernst Cassirer ([1923] 1955), for whom myth "is one of the two fundamental expressions of thought, the one of the creative dimension, which is parallel and irreducible to that of discursive logic, having its most exemplary manifestation in science" (Coniglione 2019, p. 115).

Thus, the imagination constitutes a foundation of mythological production, where different domains and levels of the interpretation of reality are connected without an apparent logical thought, but in a semantic of great hermeneutic power.

Similarly, in metaphorical thinking something is created which is not directly connected to the perceptions and consolidated elaborations of the stimuli; it passes from one domain of knowledge to another and connects the different domains.

3 The Metaphor, from Rhetoric to Linguistics to Psychology

Among the rhetorical figures, the metaphor is perhaps the most common. Coniglione has not specifically focused on this aspect of mythical thinking, although the use of metaphor is mentioned by him several times in various contexts (Coniglione 2018, p. 148; 2021, pp. 178, 184).

Metaphor represents the substitution of a term or concept with another whose literal sense bears some resemblance to the sense of the one it replaces. In the contraction of a comparison between two terms or concepts, an entity comes to identify, confuse, with the one with which it is compared.

Latin rhetoric considered the metaphor a *similitudo brevior*, a shortened comparison. While in the simile the term of comparison is made explicit, instead in the metaphor it is not directly predicated ("That man is strong, like a lion" becomes: "That man is a lion"). Language is basically metaphorical, and linguistic rules serve to regulate or reduce (in many cases impoverish) the metaphorical richness that characterizes man – according to the definition shared by Cassirer (1944, pp. 410–411) – as an "animal symbolicum", concept later revived and developed on semiotic grounds by Eco (1984). Averill (1994) has focused on how metaphors express the emotional world: exemplary ways to understand, and to say in other words, emotional realities that are not clear and cannot be transmitted with logical expressions. From these studies it is possible to deduce the central role that the metaphor has in human knowledge, also opening highly innovative paths to understanding and action. According to Lakoff and Johnson (2003) the metaphor serves to structure reality and, through this (re)structuring, to develop thought.

The metaphor is intermediary between semantic and analogical thought; it constitutes a category of the mind even before that of language. As it allows to understand a domain of reality using ideas and conceptual terms of another domain, it serves to "go beyond" the logical-sequential thinking that is dominant (literally: belonging to a single domain), and to unhinge the consequent crystallized mental scheme in its reassuring simplicity and precision (Di Nuovo 2015).

In the Piaget's (1950) theory of continuous connection between the scheme to be adapted to the data and the data to be assimilated to the scheme, the cognitive scheme facilitates the understanding of reality data, highlights rationality, lights up with scientific approach: but often loses sight of the complexity of life that cannot be schematized, nor reduced to conceptual enclosures. The scheme reassures from the dispersive complexity, but at the same time forces and obliges the vital reality in a prejudicial, almost virtual cage. To truly grasp reality, we need to put conceptual schemes in brackets, avoiding that they become "pre-concepts", and go further, towards the essence of things as proposed by the phenomenological approach (Ray, 1994). To avoid being misled by the scheme, we have to deceive the schematized and simplified reality, and find a "structure that connects" (according to the Bateson's expression) this reality with a new, different, "out-of-the-box" reality. As Coniglione (2022b) argued in describing the Poznań School's modelling conception of science and in relation to the importance of complexity, it would be a fatal and paralyzing error for scientific thinking to "live in the model", i.e. to exchange the ideal model – that helped us to schematize reality and make us orientate ourselves within it – with reality itself, thus inhibiting other ways of thinking outside the model, as was also emphasized by Watzlawick (2000, pp. 112–121). In this sense, as Hobbes (1651) had already glimpsed, the metaphor with its conceptual "blending" is a deception, because it tends to construct a different reality from what the "data" tell us: this is not to falsify, but to create a new reality, the "truth to be built."

The metaphor has an inherent dynamic element of change, intrinsic to the term itself: The Greek μεταφορά derives from the verb μεταφέρω, to transfer, to carry over, to change. Metaphorical thought (and language) can create a new reality and help a mind to be not only repetitive and reflective of facts, but to become productive and creative (McReynolds, 1994). This goes beyond the reduced and simplified data of reality and produces another complex and difficult reality to "say" except through art or, if it loses all contact with reality, through psychotic delirium, metaphorical thought for excellence. Thinking in metaphors – like the imaginative capacity – is typical of artists, poets, creative scientists. Aristotle recognized in it the "sign of genius," other authors the figure of madness. But genius and madness are not that far apart.

Metaphor and imagination constitute aspects of the same process of simulation and re-construction of reality (Di Nuovo 2011). The psychodynamic approaches, from Freud onwards, had already brought these two realities closer together; but according to more general models, the metaphor is conceived as a "figurative language" that merges propositional knowledge and mental images.

In the process of detachment from the mental scheme crystallized on the mirroring of the fact of reality, and of opening towards a different reality, the metaphor makes use of the imaginative function. When we speak of metaphors based on unforeseen and non-repetitive images, we are therefore always dealing with individual, idiographic experiences, and as such necessarily "different" and potentially innovative. Through the imaginative tool, the emotion that is connected to it, and the novelty that the combination of these elements can produce, the metaphor becomes a powerful means of transforming experiences from the past to the future.

If the same linguistic structure can "represent" another, the univocal vision of reality is replaced by one that is flexible and open to diversity and change of perspective. Therefore, I feel in tune with the perspective advocated by Coniglione of moving "towards a non-imperialistic view of knowledge", to which – in his reconstruction – Feyerabend's reappraisal of myth also leads, namely, "the reevaluation of a dimension of human beings tied to their imaginative faculty, to storytelling, to fantasy, to a way of engaging with reality that employs various symbolic tools" (Coniglione 2022a, p. 29). Among these tools, the symbols and images promoted by the metaphor – which integrates the logic of words with the analogue of images – favor a fluid emotional climate that facilitates change, which is based precisely on flexibility.

This is what happens in successful psychotherapies, modifying those myths that the human mind creates in a maladaptive sense, and that psychoanalysis has identified as collective constructions.

4 The Collective Construction of Myths: The Archetypes in Jung's "Red Book"

An important step in the vision of the metaphorical imagination as a cognitive and propositional innovation is the possibility that it can have interpersonal dimensions. This step is supported by the *Red Book*, an exceptional "unfinished" in the history of literature, elaborated over sixteen years, in which the author states that it was a "terribly inadequate" enterprise (Jung 1930). Around 1913 when Jung began writing the *Red Book*, he was exploring the "mythopoietic" function of the psyche, analyzing imaginations and fantasies, which he called "my most difficult experiment" because before he considered daydreaming as an "impure" form of thought (Jung 1925).

Following James, in his initial writings Jung (1912) distinguished the "purposive thinking" typical of science from the "imaginative, associative and passive daydreaming" typical of mythology (although not all myths have these characteristics). Images were undervalued and considered "illegitimate, demonic, pagan" if spontaneous and not derived from "doctrine" (Hillman, 1983).

In his scientific autobiography, Jung (1963) wonders if the fantasies that emerged at the time had to do with science or not with art, and if his unconscious had given shape to a different personality from his own individuality and expressed with "his own visions" (p. 202). These visions, primordial images (*Urbilder*) are myths that describe the complexity of humanity, made concrete in the *Red Book*, an encyclopedia of images and metaphors. A blend of rational and irrational, of dream and reality, of sanity and madness, of past and future: all merged into the symbols that contain all these aspects.

Starting from the *Red Book*, the story of the discovery of the imaginal, "the royal road" to access the depths of the individual and collective psyche is considered the active imagination: in this process an image enters the mind, freed from the "noise" of everyday life, and it is amplified through writing, painting, dance (von Franz 1978).

But are imaginative and metaphorical symbols destined to remain forever in the domain of subjectivity, or can they be read and interpreted by others, or even shared? Does simulating a "subjective" reality different from the "objective" reality (in the metaphor based on original, creative images tending towards change) allow it to be translated into "intersubjectivity"?

These aspects are not taken into consideration by Coniglione, despite his interest in the collective bases of esotericism. However, it is worth delving into.

Jung (1934) recalled how the image contained in metaphors is the main organizer of the human mind; myth and archetype are universal metaphorical images immersed in a collective unconscious. In the symbol that founds the myth, the image is not born in the individual mind: it is the result of a shared social representation in the context from which the subject draws the inferences and the symbolic and emotional referents. The myth that is based on shared – or sharable – images have a collective, social dimension, learned during development. For example, representations of symbolic opposites (father/mother, male/female, life/death, love/hate, etc.) and the "complexes" studied by psychoanalysis express universal dualisms of the human mind (on this aspect, also Coniglione 2019b), which are verified but not produced by individual experience. For example, the Freudian "Oedipus complex" interprets infantile relationships with metaphorical intentions through images that do not concern childhood, but an ancient myth handed down by Greek tragedy, and attempts to translate a universal symbology into material and biological experience. For Jung, the "complex" can be represented in the form of myth, but the images from which mythology draws do not refer to life experiences, not to a theory (Kranefeldt 1942).

The myth, as connected to existential experiences, allows, and promotes a collective use of metaphorical images that can extend in a conservative sense of values and traditions, of cognitive biases and prejudices, but also – if properly guided – in the direction of innovation and of social change, as for the metaphor (Lakoff 2002).

In the Jungian *Red Book*, we find a continuous unfolding of images and metaphors, in an interdisciplinary encounter between philosophy, alchemy, anthropology, literature, and psychology, capable of overcoming the realities crystallized in the patterns rooted in the social mentality. As previously stated, this re-proposition of myths in the experience of everyday life allows innovation to be achieved with respect to thought and action limited to a single domain. People and social groups can be activated towards a multidimensional openness that interprets complex realities and requires equally complex tools of knowledge and action. This "openness" is precisely the condition for viewing knowledge not in an imperialistic way but as the possibility of constructing multiple models that frame and schematize reality in different ways, from different points of view (Coniglione, 2022b, p. 55*ff.*). It avoids indulging in the "theory of everything" (Hawking, 2005; Nicolaides, 2020) whereby knowledge progresses in an ever-higher pyramid increasingly encompassing empirical territory, eventually culminating in an equation that purportedly explains everything. Indeed, myths enable the creation of an original collective and multidimensional framework of knowledge.

5 New Myths among Neurosciences and Media

Mythology based on imagination and metaphors, and archetypal theory, find further evidence in contemporary psychological and social sciences, e.g., in the theories of "memes" (Dawkins 1976), an aspect we address here as an addition to Coniglione's perspectives. A meme is an idea or image that spreads among people within a culture; it is a self-diffusing unit of cultural evolution, analogous to what the gene is to the evolution of the species. It is therefore an element of civilization transmitted by non-genetic means, especially by imitation. This schematic information unit stabilizes in the brain, can influence the environment by translating it into action by the people who carry it, and can be propagated by founding the cultural transmission of the "extended phenotype" (Dawkins 1999).

When a meme is expressed in the same way and without modifications by several people there is "memetic inertia", while when it changes in the transfer from one person to another a "memetic drift" takes place: i.e., a process through which a meme changes while it is transferred, a frequent circumstance when a complex reality has to be represented.

The objections and criticisms of Dawkins' theory (including those of Dennett 1995) are aimed at the analogy with genes, not at the existence of memes, which is another name to talk about socially shared linguistic and cultural constructions, as we have seen to be the myths and archetypes.

Since ancient times images and metaphors, now defined as memes, have been contained in epic poems, but also in literary and cinematographic works, proverbs and aphorisms, fairy tales, nursery rhymes, lullabies, religious beliefs, and political ideologies.

Memes generate, transmit, perpetuate, or change the myths of the modern era: about health, ecology, and the global economy. They report and disseminate the cultural phenomena proposed in the media regarding nutrition and body care, the uses of technologies, and their modifications over the ages and cultures (e.g., cell phones, telematic communications).

The Internet and social media are the main vehicles of memes, often reduced to satirical cartoons, made up of images plus a mostly metaphorical text, with the risk of trivializing the great creative potential of dynamic and alternative cultural productions.

The theory of a co-evolution of genes and culture, affirming that the fundamental units of culture biologically correspond to networks of neurons that function as nodes of semantic memory, was supported by the biologists Lumsden and Wilson (1981). The information structure that reflects the mental content based on a meme network is defined as a *memeplex*, replicating together such as religions and cultural values. According to Blackmore (2000), memes explain complex problems such as the origin of language, the development of the human brain, sexual phenomena, the Internet, and the notion of self, which can be defined as the collection of memetic stories of the person (*selfplex*).

The flow of new information chains with the built-in traces of memory; organizes itself to create memes that attract brain activation patterns representing memories and concepts. In neural networks, such states are called *attractors*. Each excitation of memory strengthens its activation pattern, causing stronger associations of different, even the most distant information. This mnemonic trace creates a symbolic "basin of attraction" that connects with other similar basins.

Connectomics developed methods to describe structural and functional connectivity, with which neurosciences explain dynamic models of cognitive and affective processes. Brain connectivity is formed by genetics and social development processes; thanks to neuroplasticity it is shaped by life experiences, learning processes, social interactions, and culture (Sporns and Betzel 2016). In this sense, from a neuroscientific point of view, beliefs are also defined as the neural product of the perception of objects and events in the external world. They are the product of brain processes that attribute affective meaning to concrete objects and events and of an internal affective state, reflecting personal symbolic meaning (Seitz, Paloutzian, Angel, 2018).

In this regard, Coniglione (2017) proposes an explanation for the mythical – and also – mystical function of the brain (including creativity, holistic character, ability to elaborate metaphors, etc.) and the analytical function (discursive and sequential, "logical" thinking) as the result of an evolutionary strategy that led the human brain to specialize, allocating distinct functions to the two hemispheres. He supports this thesis with recent studies on bilaterality (e.g. Rogers and Vallortigara 2017; Rogers, Vallortigara, and Andrew 2013; Goldberg 2009, among others). These studies no longer solely focus on humans, but also extend to the brain functions of animals, including less advanced species like mice or chickens. Coniglione's thesis finds resonance in various traditions of thought, as well as in the basic conceptions of certain cultures (e.g. the Yin and Yang distinction in Chinese culture). It has also garnered support from contemporary philosophers and intellectuals (e.g. Watzlawick 2000), even though they might not directly refer to hemispheric specialization. He underscores that this perspective doesn't treat the two hemispheres as separate entities or fetishize their functions: "we must avoid the mistaken thinking that the different brain functions are exclusively localized in only one side (right or left) of its parts, inasmuch as it has a plasticity that allows its functional reorganization, as happens in the case of diseases". However, this does not negate the possibility of recognizing "the fact and the general recognition [...] that there are functionally different ways in which the brain works, wherever its various functions are localized" (Coniglione 2017, p. 219).

This is not the place to delve into this perspective and discuss – in light of the studies on Neural Networks and simulation of the whole Neural System – whether cerebral laterality indeed possesses the explanatory power attributed to it by Coniglione. What matters here, from my point of view, is the shared belief that the individual brain is capable of reworking cultural contents in a personal way and offering them to semantic, symbolic, and existential interpretations to be shared in turn with the other people of the social group, creating a collective construction of culturally active contents, in which mythic thought plays an essential role.

6 Conclusions

Myths are not just a remainder of the past, or a manifold explanation of a present complex reality. The imaginative and metaphorical forms that flow into the myths, and into their contemporary media versions, have the propulsive force for overcoming consolidated schemes, for planning, and change. They can be used to foster social innovation when they are not the fruit or refuge of an individual mind, but rather the foundation and stimulus to the action of a collective mind, because it is in the social sphere that the image and word represented from the myth can be realized and find the paths to develop experience and action. As Coniglione argued:

> Reason exercised exclusively on an analytical-rational basis would lead to madness if not compensated for by uninhibited holistic moments. Conversely, the mere holistic disinhibition would lead to the loss of

> reason and thus to the diminuendo of humanity if there were no rational procedures, able somehow to concretize and make holistic insights operational. Analytical reason is driven by holistic creativity to new horizons, which are, however, accessible only to the extent that they become rationally manageable. And vice versa, explosive creativity fertilizes analytical reason, which would otherwise wither by eternally repeating the identical. In the same way as wakefulness needs sleep, sleep must be translated into an effective and operational wakefulness
>
> CONIGLIONE, 2017, pp. 224–225

Neurosciences can define the biological foundations of these processes, while social media can contribute by offering the tools to implement them in shared forms of knowledge. In any case, myth (and other related types of experience) proves to be not a "dross" to be eliminated with the progress of knowledge and science, but a fundamental element not only in the mind of each individual, but also for humanity itself. Without it we would have halved men and social communities.

However, the regressive risks inherent in any change of perspective must be avoided. If collective mental constructions (whether these connected contents are defined as myths, archetypes, or memes) uncritically receive, accept, and disseminate *fake news* and *big lies* – e.g., "conspiracy" theories on socially disturbing phenomena (Duch 2021) – they end up transmitting distortions of social memories (Schacter, Fischbach, and Coyle, 1997) that lead to maladaptive and uncomfortable forms of social and civil attitudes and behaviors.

To this end, the formation of people and social groups in the appropriate use of both science (and its dissemination) and the media is essential. The aim is to direct the "memetic" change of collective mythologies towards progressive goals for the well-being of humanity.

References

Averill, J.R. (1994). Six metaphors of emotion and their theoretical extensions. In: D. Leary (ed.), *Metaphors in the history of psychology*, pp. 104–132. Cambridge: Cambridge University Press.

Blackmore, S. (2000). *The meme machine*. Oxford: Oxford University Press.

Cassirer, E. ([1923] 1955). *The Philosophy of Symbolic Forms*, vol. two: *Mythical Thought*. New Haven: Yale University Press.

Cassirer, E. (1944). *An Essay on Man. An Introduction to a Philosophy of Human Culture*. Garden City, NY: Doubleday Anchor Books.

Coniglione, F. (2008). *Popper addio. Dalla crisi dell'epistemologia alla fine del logos occidentale.* Acireale-Rome: Bonanno.

Coniglione, F. (2017). 'Rejected Knowledge' and the Tradition of Reason. A few Hints on their Necessary and Inextricable Link. In: L.R. Cardullo, F. Coniglione (Eds.), *Reason and no-reason from ancient philosophy to neurosciences*, pp. 211–231. Sankt Augustin: Academia Verlag.

Coniglione, F. (2018). Some Remarks on the Meaning of Esotericism and Plato's Unwritten Doctrines., *Mondi. Movimenti simbolici e sociali dell'uomo*, vol. I, n. 1, 5–51.

Coniglione, F. (2019a). L'intreccio antropologico tra mito e logo: una indispensabile (ma non pacifica) coesistenza. In: E. Coco and S. Vasta (Eds.), *Il mito tra filosofia e scienza. Temi e prospettive dall'antichità a oggi*, pp. 109–140. Catania: Malcor D'.

Coniglione, F. (2019b). I due volti dell'umano. In: E. Frasca (Ed.), *Il valore e la virtù. Studi in onore di Silvana Raffaele*, pp. 69–82. Acireale-Rome: Bonanno Editore.

Coniglione (2020). Conoscenza tacita: Ludwik Fleck nostro contemporaneo. In: A. Czajka, G. Cunico, E. Colagrossi (Eds.), *Cento anni di filosofia e di cultura polacca,* pp. 225–251. Sesto San Giovanni (MI): Mimesis Edizioni.

Coniglione, F. (2021). La penultima verità: il mito tra *pistis* e *logos* al tramonto del mondo antico. In: R.L. Cardullo, F. Coniglione (Eds.), *Mythos e Logos. Tra archetipi antichi e sguardi sul futuro*, pp. 178–205. Pisa: Edizioni ETS.

Coniglione, F. (2022a). Dal mito alla scienza e ritorno. Per una visione non imperialistica della conoscenza. In: E. Coco (Ed.), *L'invenzione della realtà. Scienza, mito e immaginario nel dialogo tra psiche e mondo oggettivo. Una prospettiva filosofica,* pp. 11–30. Pisa: Edizioni ETS.

Coniglione, F. (2022b). Complessità e pluralismo cognitivo. *Complessità*, n. 2, 32–72.

Dawkins, R. (1976). *The Selfish Gene.* Oxford: University Press (3rd ed. 2006).

Dawkins, R. (1999). *The Extended Phenotype: The Long Reach of the Gene.* Oxford: Oxford University Press.

Dennett, D.C. (1995). *Darwin's Dangerous Idea: Evolution and the meanings of life.* New York: Simon and Schuster.

Di Nuovo, S. (2011). In cammino verso nuove realtà: metafora e immaginazione. In: M. Maldonado (ed.) *Fenomenologia della scoperta*, pp. 17–26. Milan: Mondadori.

Di Nuovo, S. (2015). La metafora come strumento mentale per costruire l'innovazione. In: R.L. Cardullo, D. Iozzia (Eds.) *Bellezza e virtù*, pp. 819–831. Acireale-Rome: Bonanno.

Di Nuovo, S., Coniglione, F. (2002). Il caos e la regola: una sfida alla complessità. Riflessioni sulla metodologia della ricerca clinica e psico-sociale. In: M. Bellotto, A. Zatti (Eds.) *Psicologia a più dimensioni*, pp. 102–130. Milan: Franco Angeli.

Duch, W. (2021). Memetics and neural models of conspiracy theories. *Perspective*, 2, (11), 1–13.

Eco, U. (1984). *Metafora e semiosi.* In: *Semiotica e filosofia del linguaggio*, pp. 141–198. Turin: Einaudi.

Eliade, M. (1998). *Myth and Reality*. Long Grove (Ill): Waveland Press.

Fleck, L. (2019). *Stili di pensiero. La conoscenza scientifica come creazione sociale*, edited by F. Coniglione. Milan-Udine: Mimesis.

Goldberg, E. (2009). *The new executive brain. Frontal lobes in a complex world.* Oxford: Oxford University Press.

Hawking, S. (2005). *The Theory of Everything. The Origin and Fate of the Universe*, Beverly Hills, CA: Phoenix Books.

Hillman, J. (1983). *Puer Papers*. Washington: Spring Publications.

Hobbes, T. [1651 (2012)] *Leviatan*. Oxford: Oxford University Press.

Jung, C.G. [1912 (1956)] *Wandlungen und Symbole der Libido*. Engl. Ed.: *Symbols of Transformation*. In: *The Collected Works of C. G. Jung*, Vol. 5. Princeton: Princeton University Press and London: Routledge & Kegan Paul.

Jung, C.G. [1925 (1989)] *Analytical Psychology: Notes of the Seminar Given in 1925*. Princeton: Princeton University Press.

Jung, C.G. [1930 (2009)] *The Red Book – Liber Novus*. New York: Norton.

Jung, C.G. [1934 (1969)] *The Archetypes and the Collective Unconscious* In: *The Collected Works of C. G. Jung*, Vol. 9 – part. I. Princeton: Princeton University Press and London: Routledge & Kegan Paul.

Jung, C.G. (1963). *Memories, Dreams, Reflections*. New York: Pantheon Books/Random House.

Kosslyn, S.M. (1980). *Image and Mind*. Cambridge: Harvard University Press.

Kranenfeldt, W.M. (1942). Komplex und Mythos. In: C.G. Jung, *Seelenprobleme der Gegenwart* (Kap. XII). Olten: Walter Verlag.

Lakoff, G. (2002). *Moral politics: How liberals and conservatives think*. Chicago: Chicago University Press.

Lakoff, G., Johnson, M. (2003). *Metaphors We Live By*. Chicago: Chicago University of Chicago Press.

Lévi-Strauss, C. (1978). *Myth and Meaning*. London: Routledge & Kegan.

Lumsden, C.J., Wilson, E.O. (1981). *Genes, Mind, and Culture: The Coevolutionary Process*. Cambridge, MA: Harvard University Press.

Malinowski, B. (1926). *Myth in Primitive Psychology*, New York: Norton.

Marks, D.F. (Ed.) (1985). *Theories of image formation*. New York: Brandon House.

McReynolds, P. (1994). Motives metaphors: a study in scientific creativity. In: D. Leary, *Metaphors in the history of psychology*, pp. 153–171. Cambridge: Cambridge University Press.

Nicolaides, D. (2020). *In Search of a Theory of Everything. The Philosophy behind Physics*. New York: Oxford University Press.

Paivio, A. (1986). *Mental representations: a dual code approach.* New York: Oxford University Press.

Piaget, J. (1950). *Introduction à l'épistémologie génétique. Tome III: La pensée biologique, la pensée psychologique et la pensée sociale.* Paris: P.U.F.

Piaget, J., Inhelder, B. (1966). *L'image mentale chez l'enfant.* Paris: P.U.F.

Ray, M.A. (1994). The richness of phenomenology: philosophic, theoric, and methodological concerns, in J.M. Morse (Ed.) *Critical issues in qualitative research methods* (pp. 117–133). Thousand Oaks: Sage.

Rogers, L.J., Vallortigara, G., Andrew, R.J. (2013). *Divided Brains. The Biology and Behaviour of Brain Asymmetries.* Cambridge: Cambridge University Press.

Rogers, L.J., Vallortigara, G. (2017) (Eds.). *Lateralized Brain Functions. Methods in Human and Non-Human Species.* New York: Humana Press.

Schacter, D.L., Fischbach, G.D., Coyle, J.T. (1997). *Memory Distortion: How Minds, Brains, and Societies Reconstruct the Past.* Cambridge, MA: Harvard University Press.

Seitz, R.J., Paloutzian, R.F., Angel, H.-F. (2018). From believing to belief: a general theoretical model. *Journal of Cognitive Neuroscience,* 30, 1254–1264.

Sporns, O., Betzel, R.F. (2016). Modular brain networks. *Annual Review of Psychology,* 67, 613–640.

Von Franz, M.-L. (1978). Bemerkungen zur Aktiven Imagination. *Analytische Psychologie,* 9, 161–171.

Watzlawick, P. (2000). *Il linguaggio del cambiamento. Elementi di comunicazione terapeutica.* Milan: Feltrinelli.

CHAPTER 13

Objectivity, Knowledge, and Rights

Fabrizio Sciacca

Abstract

This paper recalls Francesco Coniglione's thematic strand of work devoted to the philosophy of knowledge and its practical implications in the fields of politics and law. I try to show here how the idea of narrative relates to that of knowledge, and whether this connection gives rise to any form of objectivity. Furthermore, I also ask whether the fruit of this connection is of interest to rights and politics.

Keywords

knowledge – objectivity – politics – rights – truth

1 Objectivity and Narratives

> There are thus more than enough reasons to seriously criticize the belief in the objectivity of values. In part, for the benefit of philosophy itself; in part, because in this way it might perhaps once again be possible for philosophy to exert a fruitful influence on specialized science.
>
> ALF ROSS

Francesco Coniglione's studies devoted to epistemology and the philosophy of knowledge amply demonstrate how useful and topical is a reflection on the limits and possibilities of knowledge, as well as on the relationship between the idea of objectivity and the problem of normativity (Coniglione 2015a; 2015b; 2015c; 2016). In this paper I intend to build on these themes to propose an analysis of some practical implications between knowledge and the social sciences, with particular reference to political science and law.

I would like to point out an assumption by Francesco Coniglione about the difficult semantic nature of knowledge. Coniglione writes that the irruption of time in modern science leads to the declaration of the impossibility of the existence of reproducible experiments and puts an end to the concept of objectivity of measurement (Coniglione 2015a).

 | HTTPS://DOI.ORG/10.1163/9789004756496-015

Let's start with the problem of the objectivity of narratives. "Narrative" is also a word in common use in the language of philosophers, but it belongs only to a very small extent to the language of philosophy. "Narration", in a broad sense means "discourse", "exposition". An exposition that is generally personal, and which, however correct and bona fide it may be, is not necessarily presumed to be objective. Only in ancient rhetoric, as we know, *oratio* had an expository part – the *narratio*, in fact – that had to be objective.

If I were to propose some semantic form of this word, by "narration" I would mean a form of elaboration of facts about the present or recent past that are communicated without being explained (and pass the scrutiny of acceptance without an actual explanation), more or less as happens with the tradition of myths (which are also certainly the subject of narration). Unlike myths, however, which have an infra-historical suitability (they are not limited to a specific present time but belong to every present, because they are not chronologically identifiable in time), modern narratives belong to a specific present and do not possess the autonomous (and transcendent) force of myth. They are not myths, nor *a fortiori* social myths. Myth does not recognize man's dominion over the world, it sees the world as a set of mysteries and enigmas that cannot be answered. Myth gives transcendent reality an immanent objectivity to this world. And certainly, this does not belong to narrative, although there is a narrative component in myth (myth has a narrative form, but not every narrative tells a myth).

In this sense, as Coniglione pointed out, "post-modernism is first and foremost an analytical reflection and a critical stance regarding the notion of 'modern', characterized by its most typically ideological manifestations, the most fundamental aspect of which relies on the 'great narratives' (the metanarratives) to give some meaning to history and indicate the place that humanity occupies in it" (Coniglione 2015b, p. 8). A reaction to Enlightenment modernity, albeit from a rational standpoint, occurs in Alasdair MacIntyre's. "It is surprising that in order to prove that there are no true narratives" we write narratives (MacIntyre ([1981] 2007), p. 214). In this sense we could say that our beginning has not begun. Our narrative beginning never begins. We are immersed in what flows with the beginning of being the author of our actions. In the narration of the self, it does not matter so much the narrative form, which we might say is imaginary or not. Moreover, when our subjective figure immerses itself in the process of objectification of action in a shared space, it immediately participates in what is already flowing. Our beginning is thus a paradoxical conceptual moment, and we could say that everything flows in this narrative whether it is true or not. It does not matter. We wanted the situation in which we find ourselves. As is obvious when it comes to a pandemic, yet we suffer its effects or constructions or constraints. The fact of unpredictability only announces to us that we cannot somehow determine what will happen to us precisely because being the author of a narrative does not equate to the ability to determine destiny. We can therefore contribute to the transformation of something. But there is always a kind of unpredictability in which we are embedded. Unpredictability and teleology coexist as parts of our lives, giving rise to the need to have an end, even an end that cannot be conditioned or conditionable, an end that is therefore not subject to existential verification, but can be pursued without knowing anything about it or its manifestation in the future. I do not believe that today more than ever we can speak in a strong sense of objectivity, and of this I think we can all be quite sure. By objectivity I mean in a weak sense (the only one in which I could speak of it) impartiality in the formulation of judgements (Habermas 1968),[1] not as universalization but as acceptability: appropriateness [*Angemessenheit*] to the context (Günther 1988).

2 Knowledge Knots

As Coniglione notes, "that of knowledge is a problem that philosophy has always posed throughout its history, even if not always at the dependence of science" (Coniglione 2016, p. 13). Certainly, the problem of knowledge belongs to philosophy. Turning to the relationship between knowledge and narratives, knowledge is that series of information, notions, etc. that one has acquired through study or experience, or otherwise possesses: "ostentation of one's knowledge," "culture," "doctrine," "erudition." Is knowledge that manifests itself through narratives rather than arguments (i.e., that tells without explaining) objective knowledge or rather ostensive knowledge? In its broadest sense – whether we like it or not – politics is delegated the task of guiding and managing a country or community in situations where there is an impending external threat. A war, a natural disaster, an economic crisis, an epidemic. The rational, but also emotional, expectation of citizens is that the people entrusted with leading the community will do their best to minimize the adverse consequences of such events. In principle, everyone would probably agree. However, a first problem arises when one realizes that any measure taken by the public

1 Habermas (1968) replaces objectivity with disclosure of interests that guide the knowledge process.

authority has asymmetric and unequal effects on the members of the community itself. Recent examples include the imposition of the wearing of masks, the closure of shops, the restriction of freedom of movement and other restrictions on constitutional rights.

To overcome this difficulty, again in principle, it could be agreed that the "right" measures are those that minimize harm to the community "as a whole." On this point, too, agreement can probably be reached. The next intellectual and political hurdle is to ensure a proper consensus on which measures are the best option for the community as a whole. And here the irruption of "data" and "facts" seems inevitable, with all the problems that this entails, which in difficult times such as these appear in all their evidence. One of the effects of the enormous mass of information created and disseminated by the official and independent, alternative, or even "underground" media is – rightly so – that of inducing skepticism. If some say A and others not-A, it undoubtedly means that at least one option is wrong. And that perhaps both are. So, the problem of some objectivity probably arises. And of the use of "data and facts" in political action. And I think this problem must be tackled at its root: does the search for objectivity make sense? Can politics reduce its action to the application of purely fact-driven measures? By using very simplified option models, a famous experiment showed how difficult it is to assess the objectivity of an apparently "factual" description: alternative options that are absolutely identical from a logical, numerical and statistical point of view are interpreted and perceived differently by a sample of people with even high cognitive levels.

Let us imagine that during an emergency in which human deaths become apparent, two different teams are entrusted with the elaboration of solution strategies. The question is how the strategies are perceived by non-experts, by ordinary people we would say. The strategy of saving lives is preferred over the number of people who might die, even though the number of saved and dead is the same. However, it is easy to see that the two cases are effectively identical.

> The only difference between them is that the outcomes are described in problem 1 by the number of lives saved and in problem 2 by the number of lives lost. The change is accompanied by a pronounced shift from risk aversion to risk taking. We have observed this reversal in several groups of respondents, including university faculty and physicians. Inconsistent responses to problems 1 and 2 arise from the conjunction of a framing effect with contradictory attitudes towards risks involving gains and losses.
>
> TVERSKY AND KAHNEMAN 1981, p. 453

Does it make sense to speak of objectivity, in a practical sense, when the phenomena are extremely complex and not 2–3 variable models? Perhaps not. But perhaps neither does the opposite instinctive reaction, that facts do not exist or that no assertion can be disproved or falsified. It seems that lately everyone has "their own facts". And those in power seem to want to give the impression that they are acting on the basis of facts. But is this so?

Technique is dominated by facts; politics is dominated by values. And if facts are in themselves "neutral" and "objective" (with all the problems and limitations of interpreting them), values are fully subjective categories.

Is it a mistake, then, to think we can reduce politics to technique (doing the right thing by being guided by facts)? Part of this misunderstanding is perhaps also lexical. Everything that pertains to the public sphere is placed within the category of "politics": legislative power, executive power, administration, local government bodies. Is it all always just "politics"? Are those "task forces" of "experts" made up of private citizens appointed ad hoc by the executive power also "politics" insofar as they can steer decisions in the public sphere (without its members being democratically elected)?

Nichols (2017) posed this problem and said that expertise and knowledge imply competence, and competence implies credentials, such as scientific training and objectively competitive entry criteria, such as careful entry selections, but this is only the first step, not the end point. Specialization is necessarily exclusive, and this flies in the face of democracy.

One of the reasons why experts and laymen are at odds with each other is that they are both human beings. No one has, to the end, a crystal ball. And no one has, in the end, an objective notion of objectivity. What is more, sometimes experts are not objectively experts: a fraudster or a boaster certainly cannot be one.

But let us ask ourselves: how is a mayor called upon to lead a city a politician rather than a technician? How does he differ from a CEO of a company called upon to run it in the best possible way? And a mayor is only political in appearance, perhaps, and could do better if he had a technical approach to his actions.

> This is the worst of all worlds, in which both democracy and expertise are corrupted because neither democratic leaders nor their expert advisers want to tangle with an ignorant electorate. At that point, expertise no longer serves the public interest, but the interest of whatever political clique is taking the temperature of the public at any given moment.
>
> NICHOLS 2017, p. 223

3 Truth and Rights

Another topic I would like to bring up is that of freedom, rights, and risk. Do I have the right not to wear a mask? Let's say yes, in the first instance. But what if my behavior can cause harm to others? Let's say no, in the second instance. Of course, we are talking about 'potential' risk: there is no certainty that my not wearing a mask will cause infection to others, especially if I keep my distance. If, however, one continues to lean towards no (you have no right not to put it on), one would have to explain why one has the right to fly a small private plane over other people's homes, for example. The concept of risk and its quantitative assessment, perhaps, is the discriminating factor. Sending a helicopter to fly over a dwelling in order to "break up" a barbecue on a terrace is also a dangerous and risky activity. Why is it considered lawful (or more lawful than the grilling itself)?

This is a revisiting of the dilemma between *positive freedom* (delimited by rules of law) and *negative freedom* (without interference with public power). A problem well present in Coniglione, which he also and correctly reads in light of another dilemma, that between science and prudence. Indeed, Coniglione argues that the question of the fulcrum between individual liberty and public liberty "no longer has anything to do with the conformity of the decisions made with respect to constitutional dictate and the principle of legality, but instead concerns the values put at stake – threatened or guaranteed – by political decisions" (Coniglione 2021, p. 10).

Let's imagine that I listen to an engineer who is an expert in technological innovations, blockchain while being interviewed on the radio, and he responds to the interviewer that first of all he needs to understand the importance of the Peace of Westphalia and the system that resulted from it. The interviewer then admits his ignorance and has a few things explained to him. Surely it might give me satisfaction that at the end of the day we both came, in different ways, to reason about the nation-state system and its importance, trying to get to the root of the concept. Globalization, as I see it, is an economic pressure to expand massively into physically distant areas of the world that technology has made virtually close. To complement my idea of globalization, I can add the character of a legal negotiation between those who seek to ride these pressures and those who fear the consequences. The resultant of these forces has so far gone in the direction of pandering to expansion (so to speak), but it has also undoubtedly produced tensions of another kind that prompt reactions (nationalism, protectionism on the one hand, attempts to harmonize the 'rules of the game' on the other).

Globalization may not have a global ethos, but it is the outcome of a process of thickening relationships and information that, compared to the past, has the novelty of very powerful technological support.

Here is the Westphalian antithesis: production, consumption, capital, do not follow borders. Work is probably (at least the traditional one) more territorially delimited. When we were children, we wondered why the world was divided so precisely into states. Everything so precise to be spontaneous. Each with a government, currency, capital, religion, language, flag, national anthem. What is the interesting and topical aspect? The fact that globalization is highlighting how artificial it is and also how misleading it is to think in terms of: China, Germany, Korea, Italy, and so on.

Let's say that a trade between companies in the ski industry develops across the border between Austria and Italy. Boots, bindings, clothing, equipment. Trade between manufacturing companies but also consumer goods. And let's say it is mainly a local phenomenon, limited to the Alpine valleys of South Tyrol and Carinthia. What would be the point of talking about import-export between Italy and Austria? The impact on the Sardinian economy, just to mention an area not affected, would be zero. But then it is undeniable that all this data is aggregated and presented with the division of the world into nation-states in mind. And the astonishing fact is not that this is completely wrong, but that by doing so, one completely loses sight of what the socio-economic phenomenon in question is: an activity developed in an Alpine area, involving operators administratively falling within Italy and Austria but which are much more closely related and interconnected with each other than with the rest of their respective countries. In other words, the only acceptable meaning of globalization would be (i) recognize its economic nature divorced from the political fact and purged of ethical pretensions and (ii) to accept its pure fact, endowed with human but not ethically reducible elements.

Whether everything is worthy of value in one's life as an individual system of ultimate ends in a well-ordered society (Sciacca 2012, p. 80), questions about the truth-value (in the ethical sense) of rights are beyond the scope of the assessments that political justice theory is called upon to make. Rights are essentially factors in the equilibrium of well-ordered societies and are essential to the survival of the respectable idea of political justice as constitutional justice: a concept that is inclusive and morally grounded, though hopelessly lacking in ethical truth. On the other hand, liberalism requires (implies) a discourse on truth and thus a meta-ethical use of truth. It is useful to ask in what sense.

Our intuitions about the objectivity of truth suggest that truth is to be sought, in the sense that it is worthy of being sought. In this sense, truth has something normative about it: "a proposition's being true makes that proposition prima facie good to believe" (Lynch 2004, p. 57).

It probably makes sense to distinguish truth from falsehood because one needs to separate appropriate beliefs from inappropriate beliefs. Thus, in his latest book on the value of truth, Bernard Williams argues that a lie is "an assertion, the content of which the speaker believes to be false, which is made with the intention to deceive the bearer with regard to that content" (Williams 2002, p. 96). Truthfulness is for Williams inherently good, only if we observe the belief that it is. Of course, it is true what Williams says, that it is a sufficient condition for something (e.g., trustworthiness) "to have an intrinsic value that, first, it is necessary (or nearly necessary) for basic human purposes and needs that human beings should treat it as an intrinsic good" (Williams 2002, p. 92).

Michael Lynch has argued that Williams is right when he states that truthfulness worth cultivating helps one to be more truthful. However, he then objects to Williams that he fails to understand what this has to do with truth being intrinsically good. "It only shows that it is instrumentally good to believe that it is." (Lynch 2004, p. 154).

Based on this, I pose the following question: I do not see what is good or socially useful about it being instrumentally good to believe that truth is intrinsically good. Not only do I disagree with it, but I consider it a far cry from being good (Allah's fanatics actually believe that they are acting at Allah's behest, thereby assuming for truth that this is their supreme and unbreakable duty; but I do not see how such a belief and such an attitude of sincere consistency could be intrinsically good or even socially useful). How, politically, are truth and belief (assuming religion is a form of belief) reconciled? And if actions performed in the name of beliefs (e.g., religious beliefs) become a political tool, what relationship is possible between truth and politics?

For the understanding of this point, it is very useful what Coniglione argues about beliefs. It is the task of philosophers to understand what rules must be followed to arrive at knowledge, understood as *justified true belief*. Usually, such justification is understood as a search for those secure *foundations* that can guarantee the entire edifice of knowledge or, alternatively, as a search for conditions of *consistency* (Coniglione 2016, p. 15).

An insincere attitude does not mean disregarding the power and importance of the truth: the trait is captured by Lynch with the example of the "zealous missionary" who, in order to convert the natives to his religion, makes use of false connecting elements between his own religion and the tribal rites of the natives. And this, even, to make the "greater truth" of his religion triumph over that of the natives, with a false argumentative device (Lynch 2004, p. 156).

This is not to deny that in the liberal perspective truth from a political perspective does not matter. From the fact that the interest in truth is a constitutive part of liberal democracy, Lynch raises two questions (Lynch 2004, p. 160): (i) because it is almost always wrong for an elected government to lie to its citizens; (ii) to what extent should a liberal democratic government value truth (what is the maximum or minimum, depending on perspective, value that can be assigned to truth).

Sub (i), I can say that the problem has to do with an evaluative function of truth. The question arises with reference to the problem of consensus. In essence, it is a question of political justice that translates into a test of fairness. If we want to know what a political society is like, we necessarily have to evaluate it: asking ourselves what kind of political society we live in implies making a reflexive use of our political judgement; therefore, we have to find out, first of all, whether our political society satisfies our demands for fairness (but also our preferences for utility). Subjective political judgement implies evaluative reflection. It is certainly almost always wrong for an elected government to lie to the voters who have given their consent. This is, as I have called it elsewhere, the lie detector of the political system. Feelings of approval or discomfort weighed against voter expectations tell us exactly this: there is something wrong here, and that is exactly the same as saying how right, or unfair, a system is. If expectations are betrayed, the polygraph of the political system will register a violation of the liberal democratic conception of the state by the so- called liberal rulers.

Sub (ii), I can return to the problem evoked by Lynch of the normative function of truth, which has a clear social dimension (i.e., truth would have a social value). The question arises with reference to the problem of dissent. Dissent involves (admits) at least two positions (e.g., A and B) that the bearers believe to be true. A and B have a choice. In arguing for p, A asserts that p is correct. If A believes p and B does not-p, this may obviously create dissent, but the two positions, in relation to the specific context, may both be justified.

The fact that liberal institutions should always tell citizens the truth is a meta-ethical problem of political justice, not a problem of political ethics. It is a test of how well the political system works. The people are the source of the government's authority and in a sense, one could say, with Williams, that government is a trust: there is "a special relationship between government and people, and it is a violation of this conception for secrecy or falsehood to come between trustee and people" (Williams 2002, p. 210).

The idea that one cannot abstractly attribute a truth-value to evaluative-prescriptive propositions therefore seems to me to be agreeable. If evaluative prescriptions are not truth/falsity predicates, they cannot be configured as conclusions of arguments deducible from alethic premises. In this sense, the possibility of a logic of normative propositions in the strict sense, i.e., a logic applied to norms, is questionable, since to do so would require normative propositions to be predicated of truth or falsity.

Understood as social sciences, however, law or politics are systems governed by rules and normative imperatives to follow rules. Argumentative rhetoric contains, communicates, and intends meaning: the question is therefore whether the truth of meaning exists objectively or has a claim to objectivity. In this sense, the problem of meaning and the problem of truth are connected but are not the same thing.

There are different interpretations of truth: for example, (i) a realist position would say that truth is a correspondence of a certain type to subsisting states of affairs; (ii) a verificationist position would instead argue that truth is what would be verified under ideal conditions of investigation. From this it follows for (i) that a truth could be non-objective, but also that for (ii) objectivity could also be untrue. Of course, there are also alternative positions to the dualism between realism and verificationism, e.g. (iii) the Deweyan idea that the very notion of truth as metaphysics must be rejected, whereby one can assert something but not say that it is true. It is well known that the philosophical circle of the Wiener Kreis fully shared this cautious attitude (variously and improperly defined as skeptical, relativistic, etc.) towards the acceptance of a semantics of truth (in this sense, the positions of Ludwig Wittgenstein and Hans Kelsen are analogous).

In an alternative position to the correspondentist and verificationist positions, and albeit far removed from the Wiener Kreis view, also stands the Polish scientist Ludwik Fleck – very well studied by Coniglione – for whom "the true and false do not arise from a relation between judgment and the state of affairs, but from a ternary relation in which the current state of scientific knowledge and culture in general, that is, the style of thinking within which the individual researcher places himself" (Coniglione 2019, p. 34). So, "[...] epistemology that is separate from historical and comparative research, that does not consider psychological and sociological methods, that is not concerned with economics, technology, art and even politics [...] is for Fleck an empty play on words: it has little to do with an autonomous and 'internist' logic of science" (Coniglione 2019, p. 22; Fleck [1936] 2007, p. 235).

I would like to draw attention to this third philosophical attitude (which I would only describe as 'cautious') towards truth in order to argue that, in the social sciences and especially in the domain of political theory, we are faced with a very specific point: the fact of being governed inevitably translates into being subject to rules. This point is a real conundrum, because it poses the problem of whether freedom and moral equality are compatible with being subject to rules, hence to normative propositions. Most answers to this problem rely on the argument of the lesser evil: authority is a necessary idea, since it is hard to see how in a society without rule of law categories of individuals who are potential victims of crime could be protected.

Political authority is the lesser evil, but still an evil, since there is always the risk of the state abusing its power. A necessary evil, however, because its absence would deprive individuals of their freedom. What are we to make, however, of political skepticism towards the truth? If this question does not make much sense for totalitarian theories, it does for political liberalism: in particular, the question is whether liberalism implies an objective notion of truth. If the answer is yes, the question should probably be rephrased as follows: to what extent must a liberal democratic government value truth, that is, what is the highest value that can be assigned to truth. For, in other words, if truth is assignable some value, this does not mean that truth is objective, but that it has a normative function, i.e., it is understood in a more acceptable sense – veridical, i.e., credible.

I do not believe then that liberalism implies that an objective notion of truth must necessarily be upheld. To demonstrate this, I will use the classic example of the viability of the conception of the good. If liberalism implied that an objective notion of truth had to be upheld, we would counterfactually have to admit that liberal neutralism in order to defend its liberalism would have to defend any freedom of choice of a conception of the good, hence also the "freedoms" of fundamentalist groups or in general those conceptions of the good that entail the flourishing of one part of the human race to the detriment of others. If liberalism therefore implied an objective notion of truth, it would logically have to accept the danger of its own self-destruction, and thus the crisis of liberalism itself. This, then, is clearly not the case for liberalism.

The conception of liberalism that is considered here to be compatible with the only possible notion of truth is that of a truth that is credible, that is, acceptable to liberalism itself: the liberalism of rights understood as constitutional. *Therefore, we say that what is credible is what is justifiable as veridical* (hence as sensible, compatible) with the system of fundamental rights on which all liberal philosophical theories rest.

In this sense, fundamental rights are such because they do not bend to the interests of politics and thus to anything that the majority of a government of a liberal state, for example, believes to be true.

If the majority believed the idea of abolishing a freedom guaranteed to all, i.e., restricting the right to dispose of it to a few, or restricting its use, would we say that what the majority believes is true? Evidently not. Liberalism cannot logically accept that everything assumed to be true is objectively so. In this sense, the only form of objectivity accepted by liberalism is legal objectivity at the highest level, i.e., constitutional. As constitutional rights, fundamental rights are a part of liberal theory itself.

It seems to me that this does not mean that the same notion of rights presupposes that of objective truth. I do, however, believe that the idea that there are constitutional rights that cannot be circumvented by political beliefs of governments is a form of objectivity of law that we can assume to be veridical in a legal sense – which is quite different from claiming that there is an objective form of truth.

References

Coniglione, F. (2015a). La complessità al crocevia tra teoria e realtà. In: G. Sapienza (ed.) *Processo alla complessità*, pp. 15–47. Caltagirone: letteredaQalat.

Coniglione, F. (2015b). Science and the knowledge society in Europe. *Nauka*, 2, 7–23.

Coniglione, F. (2015c). Quale conoscenza per la "Società della conoscenza?. *Bollettino della Società Filosofica Italiana*, 216, 4, 3–24.

Coniglione, F. (2016). La sfida della scienza alla filosofia: dal descrittivismo al normativismo. In: E. Coco, F. Coniglione, S. Vasta (eds.), *I limiti naturali della conoscenza. Storia e prospettive di un dialogo tra filosofia e biologia*, pp. 13–48. Catania: Malcor D'.

Coniglione, F. (2019). Lontano da Vienna, lontano da Leopoli. Ludwik Fleck: l'uomo giusto nel posto e nel tempo sbagliati. In: L. Fleck, *Stili di pensiero. La conoscenza scientifica come creazione sociale*, pp. 7–93. Milan-Udine: Mimesis.

Coniglione (2021). Scienza e prudenza. Ovvero epistemologia e politica del coronavirus. In: *Dialoghi Mediterranei. Periodico bimestrale dell'Istituto Euroarabo di Mazara del Vallo*, 50, 80–91.

Fleck, L. ([1936] 2007). Zagadnienie teorji poznawania [The Issue of Theory of Cognition]. In: C. Werner, F. Zittel, F. Schmaltz (eds.), *Style myślowe i fakty. Artykuły i świadectwa*, pp. 215–251. Warszawa: Wyd. IFiS PAN.

Günther, K. (1988). *Der Sinn für Angemessenheit. Anwendungsdiskurse in Moral und Recht*. Frankfurt am Main: Suhrkamp.

Habermas, J. (1968). *Erkenntnis und Interesse*. Frankfurt am Main: Suhrkamp.

Lynch, M.P. (2004). *True to Life. Why Truth Matters*. Cambridge (MS) and London: The MIT Press.

MacIntyre, A. ([1981] 2007). *After Virtue. A Study in Moral Theory*. Notre Dame, IN: University of Notre Dame Press.

Nichols, T. (2017). *The Death of Expertise. The Campaign against established Knowledge and why it matters*. Oxford: Oxford University Press.

Sciacca, F. (2012). Human Rights, Justice and Pluralism. In: T. Cushman (ed.), *Handbook of Human Rights*, pp. 77–84. London-New York: Routledge.

Tversky A. and Kahneman D. (1981). The Framing of Decisions and the Psychology of Choice. *Science* (*New Series*), 211, 4481, 453–458.

Williams, B. (2002). *Truth and Truthfulness. An Essay in Genealogy*. Princeton: Princeton University Press.

Index of Names

Abbagnano, Nicola 60
Abbes Herrada of Landsberg 234
Adorno, W. Theodor 3–4
Agazzi, Evandro 51, 102
Ajdukiewicz, Kazimierz 1, 176
Alfieri, Francesco VIII
Alighieri, Dante 23
Allen, Arthur 174
Ambrose of Milan 235
Anaximander 199
Aristotle 8, 65, 85, 89, 91–92, 97, 206, 238, 255
Augustine 235

Bachelard, Gaston 66
Bacon, Francis 245
Banfi, Antonio VIII, 49–56, 58, 60–62, 65, 67–68
Barbour, Julian 199
Barié, Giovanni Emanuele 60
Bastide, Françoise 181–183, 186–187
Bateson, Gregory 254
Bergoglio, Jorge Mario (Pope Francis) 133, 226
Bergson, Henri 78–79, 82–84, 86
Berkeley, George 8
Berlinguer, Luigi 129
Bernard, Claude VIII, 74–80
Bersellini, Guido 51
Bertin, Giovanni Maria 67
Betti, Arianna 2
Biuso, Alberto Giovanni IX
Blackmore, Susan 259
Bloor, David 17, 195
Bohm, David 195
Bohr, Niels 192
Bontadini, Gustavo 51
Borbone, Giacomo VII–VIII, 8, 35, 37, 129, 143
Burbelka, Jolanta 7
Boyle, Robert 100
Brach, Jean-Pierre 26
Brentano, Franz 11–12
Brianes, Giorgio 241
Brouwer, Luitzen 32
Brzechczyn, Krzysztof VII–VIII, 7
Brzeziński, Jerzy 1, 7
Buczkowski, Piotr 7
Bunge, Mario 12
Bruno, Giordano 26, 33, 245–247

Cambi, Franco 53
Cantoni, Remo VIII, 52–53, 55, 58, 67–68
Cardullo, R. Loredana IX, 27
Carnap, Rudolf 8, 12, 22, 68, 98, 102–103
Cartwright, Nancy 102
Casamassima, Olimpia 1
Cassirer, Ernst VIII, 8, 28, 89–97, 99–103, 253–254
Cattaneo, Carlo 49–50
Ceccato, Silvio 51
Charles, Sébastien 35
Chomsky, Noam 126
Coco, Emanuele X, 35
Colantoni Stevani, Angela 67
Collins, Randall 122–124
Coniglione, Emanuele 1
Coniglione, Marialuisa 1
Cristaldi, Mariano 3–5
Croce, Benedetto 10, 55–56, 66, 82

Dal Pra, Mario 50, 52, 54, 57, 59, 63–64, 67–68
Darwin, Charles 125, 127
Dawkins, Richard 258
de Broglie, Louis 192, 195
Della Volpe, Galvano 6–7, 10
De Mauro, Tullio 176
Dennett, Daniel 258
De Saussure, Ferdinand 175, 177, 180
Descartes, René 78, 237
Dewey, John 182
Di Nuovo, Santo X
Di Tommasi, Rosaria 35
Dick, Stephanie 148
Dilworth, Craig 102
Dirac, Paul 192
Dodds, Eric R. 25–26
Duhem, Pierre 96, 99
Durkheim, Émile 79, 107

 | HTTPS://DOI.ORG/10.1163/9789004756496-016

Eco, Umberto 181, 254
Einstein, Albert 102–103, 192–193
Eliade, Mircea 28, 251
Empedocles 23
Enriques, Federigo 66
Escher Di Stefano, Anna 1
Evola, Julius 2–5

Fabbri, Paolo 182
Fadda, Emanuele IX
Faivre, Antoine 26
Faraday, Michael 193
Feyerabend, Paul 6, 15–16, 20–21, 33, 99, 174, 192, 255
Feynman, Richard 191
Fleck, Ludwik VII, IX, 15–16, 20–21, 90, 174–177, 180–184, 186–187, 252, 274
Formaggio, Dino VIII, 53–54, 56, 58, 67–68
Fracastoro, Girolamo 66
Francese, Anna Benedetta 35
Francis of Assisi X, 31–33, 80, 223–229, 232–234, 236–237, 239–241, 243–244, 246
Frazer, James George 28
Freud, Sigmund 255
Friedman, Milton 144

Gadamer, Hans-Georg 224
Galilei, Galileo 8, 10, 59, 96, 100, 125
Galli, Giorgio 26–27
Garin, Eugenio 55
Gaul, Marek 7
Gay-Lussac, Joseph Louis 100
Gelmini, Mariastella 139
Gentile, Giovanni 3, 10, 55–56, 66
Geymonat, Ludovico 48, 50–52, 59, 62–64, 67
Giere, Ronald 102
Gödel, Kurt 32
Goethe, Johann Wolfgang 58, 236
Goldberg, Elkhonon 28
Goldfrank, Walter 116
Goldstone, Jack VIII, 109, 116–124
Gonseth, Ferdinand 66
Goodricke-Clarke, Nicholas 26
Goodwin, Charles 177, 184–186
Gorgias 20
Gramsci Antonio 10
Greimas, Algirdas Julien 182

Habermas, Jürgen 227, 266
Hanegraaff, Wouter J. 26
Hassabis, Dennis 151, 153, 161
Hegel, Georg Wilhelm Friedrich VII, 8–10, 23, 67–69, 82, 92, 155
Heidegger, Martin VII, 32
Heis, Jeremy 102
Heisenberg, Werner 192, 195
Hempel, Carl G. 8
Herder, Johann Gottfried 101
Hesiod 214
Hilary of Poitiers 235
Hilbert, David 102
Hinton, Geoffrey 150, 159
Hobbes, Thomas 255
Homer 214
Howitt, Peter 196
Hoyningen-Huene, Paul 35
Hume, David 8, 66, 107
Husserl, Edmund 8, 50, 56, 69, 92–94

Iamblichus 208, 219

James, William 78, 182
Jaruzelski, Woiciech 6
Jasińska, Anna 7
Jastrow, Joseph 179
Je, Ke 151
Jerome 235
Johnson, Mark 254
Jung, Carl X, 256–257

Kahneman, Daniel 28
Kant, Immanuel 13, 54, 69, 101
Kasparov, Garry 151
Kaufmann, Matthias 35
Kepler, Joannes 238–241
King, Richard 24
Klawiter, Andrzej 7, 35
Kmita, Jerzy 90
Kuhn, Thomas 6, 15–16, 20–21, 64, 99, 174, 177, 186, 192, 252
Kupracz, Andrzej 7

Labriola, Antonio 10
Lakatos, Imre 6, 15, 174
Lakoff, George 254
Łastowski, Krzysztof 7
Latour, Bruno 182

LeCun, Yann 150
Lee, Rebecca S.K. 116
Le Goff, Jacques 235
Leibniz, Gottfried Wilhelm 242
Lévy-Bruhl, Lucien 28
Lévy-Strauss, Claude 28, 251
Licata, Ignazio 35
Lippmann, Walter 85
Locke, John 6
Lord Byron (George Gordon Noel Byron) 246
Lotario dei Segni (Pope Innocent III) 235
Lumsden, Edward O. 259
Lycan, William 12
Lynch, Michael 272–273

Mach, Ernst 96–97
MacIntyre, Alasdair 266
Malaguzzi, Daria 51
Mandela, Nelson 153
Mandelbrot, Benoît 8
Manzoni, Piero 66
Marconi, Diego 65
Marcus Aurelius 235
Marcuse Herbert 3–4
Marinus of Neapolis 217, 220
Mariotte, Edme 100
Martinetti, Piero VIII, 50, 53, 60, 63
Marx, Karl 9–10, 82, 92, 107, 126
Mästlin, Michael 238
Maturana, Humberto 17
McCarthy, John 148
McLelland, James L. 149
Menicanti, Daria 51, 56
Migliorini, Ermanno 49–50
Mill, John Stuart 92, 125
Minazzi, Fabio VII–VIII
Mocenigo, Giovanni Francesco 245
Mondolfo, Rodolfo 52
Montesquieu, Charles-Louis de Secondat 107
Monti, Augusta (Lella) 51
Monti Rossi, Mario 64
Montuschi, Eleonora 35, 143
Morpurgo Tagliabue, Guido 51

Nagel, Ernst 8, 73
Neri, Guido 52
Neri, Peter 149
Neurath, Otto 8, 68
Newell, Allen 148
Newton, Isaac 32, 92, 96, 197
Ney, Elisaneth 245
Nickles, Thomas IX, 35
Nicols, Tom 269
Nietzsche, Friedrich 25, 52
Nowak, Leszek 2, 6–7, 15, 90, 99, 101, 125–126
Nowakowa, Izabella 6

Otto, Walter F. 25

Paci, Enzo VIII, 52, 54–56, 58, 61, 67–68
Papi, Fulvio 50
Parrini, Paolo 12
Pascal, Blaise X, 237–238, 243
Pascoli, Giovanni 243
Pauli, Wolfgang 192
Pearson, Karl 96
Peirce, Charles S. 175, 177–181, 186–187
Phytagoras 24
Piaget, Jean 66, 254
Plato VII, IX, 8, 54, 65, 82, 91–92, 100, 197, 204–210, 212–213, 220, 235
Plotinus 207
Polanyi, Michael VII, 16, 21, 174
Poli, Roberto 1
Pope Nicolas IV 228
Popper, Karl R. VII–VIII, 3, 7–8, 48–49, 60–61, 73–76, 79–86, 90, 192
Pozzi, Antonia 50, 55, 58
Pozzo, Riccardo X
Preti, Giulio VIII, 48–50, 53, 55–58, 61–62, 64–65, 67–69
Prieto, Luís J. 175
Prigogine, Ilya 8, 17
Proclus IX, 204, 209–211, 214, 216–220
Przybysz, Piotr 1, 7
Putnam, Hilary 18

Quine, Willard Van Orman, 99

Ramjoué, Celina 164
Ravaisson, Félix 80
Reichenbach, Hans 73, 102
Rensi, Giuseppe 52
Ricardo, David 127

Richards, Ivor Armstrong 177
Riemann, Bernhard 193
Rizza, Cinzia 35
Romer, Paul 144
Roni, Riccardo VII–VIII
Rorty, Richard 18, 24
Rumelhart, David E. 149
Russell, Bertrand 2, 8, 11, 27, 32, 98
Russell, Stuart 161

Sapienza, Giuseppe 35
Schelling, Friedrich Wilhelm Joseph 66
Schlick, Moritz 8, 32, 34, 68, 102–103
Schopenhauer, Arthur X, 240–241, 243–247
Schrödinger, Erwin 192, 194
Schumpeter, Joseph 144
Sciacca, Fabrizio X
Scirè, Giambattista 133–134
Sedol, Lee 151
Sereni, Vittorio 51
Simon, Herbert A. 148
Sini, Carlo 52, 54, 67
Skocpol, Theda VIII, 109–117, 120–125
Smith, Adam 84, 127
Smith, Dennis 108
Smolin, Lee IX, 192–193, 198
Snowden, Edward 146
Socrates, 82, 205
Spinoza, Baruch 32–33
Sri Aurobindo, 23
Syrianus 208

Tambiah, Stanley Jeyaraja 28
Tatarkiewicz, Władisław 11
Thom, René 8
Togliatti, Palmiro 50
Topolski, Jerzy 90
Toqueville, Alexis de 107
Toulmin, Stephen 186
Trimberger, Elen K. 116
Trubeckoij, Nikolaj Sergeevič 175
Trump, Donald 166
Truong, Nicolas 235
Tuchańska, Barbara 7
Turing, Alan 148
Twardowski, Kazimierz 11
Tylor, Edward 28

Vaccari Pietro 56–57
Vailati, Giovanni 52
Vallortigara, Giorgio 28
van der Broek, Roelof 26
Van der Waals, Johannes Diderik 100
Varela, Francisco 17
Vernant, Jean-Pierre
Versluis, Arthur 26
Vico, Giambattista 63
Villa, Guido 56
Viola, Enrico 35, 143
von Foerster, Heinz 17
von Glasersfeld, Ernst 17–18
von Helmholtz, Hermann 96
von Humboldt, Wilhelm 101
von Wasserman, August 183

Watzlawick, Paul 255
Weber, Max 107
Weizenbaum, Joseph 158
White Jr., Lynn 234
Widmar, Bruno 51
Williams, Bernard 272
Wilson, Charles J. 259
Wittgenstein, Ludwig 16, 27, 32, 34, 177, 179, 185
Wolniewicz, Bogusław 176–177
Wolters, Gereon 14, 35, 64

Yates, Frances A. 26

Zambelloni, Franco 57
Zecchino, Ortensio 129
Zielińska, Renata 7

Subject Index

Derived or related terms are also indexed when relevant (e.g., positive, positivism, positivistic, etc., but not when they appear in phrases or expressions such as "...the positive responses...").

Abstraction VIII, 7–10, 12, 20–21, 29–30, 57, 68, 73, 81, 83–84, 89, 91–95, 97–98, 102, 154–155, 198, 225, 246
AI (Artificial Intelligence) IX, 142–143, 146, 148–151, 154–155, 157–165
Approximation 29, 100, 197
Astronomy
  Platonic Astronomy 204–206, 210, 214
  Proclus Astronomy 209, 214
  Ptolemaic Astronomy 100, 214
  Chaldaean and Orphic Astronomy 215

Buddhism 18, 236

Cerebral hemispheres 28, 259–260
Complexity X, 16, 21, 86, 92, 101, 191, 193, 198, 200, 244, 250–257
Concretization 92, 99–100, 121, 123, 125
Constructivism 17
Critical rationalism VII, 2, 48–50, 54–55, 59–65, 67–69, 82, 90
Critical realism VII, 47–69
Critical ontology 62, 69

Democracy 114, 143, 146, 155, 164–166, 227, 269, 273
Determinism 77–78, 91, 98, 183, 193, 198, 250
Dialectics 4, 9, 60, 67, 204

Eastern thought 23–24
Empiricism 9–10, 61, 77, 89–90, 92–93, 97, 99, 178
Enlightenment 8, 22, 25, 49, 59–60, 62, 65, 266
Epistemology
  Cassirer's Epistemology 91, 96–97, 99
  Polish Epistemology 10
  Popperian Epistemology 6, 12, 14–16, 20, 48, 73–74, 252
Esotericism IX, 22–23, 26–29, 31–32, 35, 155, 175, 204, 210, 212–213, 220, 257

Experiment VIII, 28, 57, 74, 77–78, 99–100, 148, 163, 179, 181, 194, 196, 256, 265, 268

Falsificationism VIII, 7, 68, 73–77, 99
Frankfurt School 3–4, 63, 65
Function 28–29, 33, 68, 73–74, 76, 89–91, 94, 96–99, 101–103, 193, 195, 211, 252–253, 255–256, 259–260, 273, 275

Hermeticism 26
Holism VII, 2, 28–30, 80–81, 115, 153, 226, 250, 253, 259–261

Iceberg metaphor VII, 19–20
Idealism and neo-idealism 4, 9, 48, 55, 63–64
Idealization VIII, 6–10, 16–17, 21, 29, 89–94, 97, 99–107, 109, 120, 122–128
Incommensurability 64
Indian thought 18, 23–24, 244
Induction 9, 73, 93–94, 97, 126
International Insubric Centre 49–51
Intuition 13, 20, 30, 34, 75, 77, 79, 81–87, 153, 156, 242–243, 245, 253
Irrationalism 22, 31, 57, 82–83, 196
Italian Communist Party 6, 50, 58

Jainism 18
Justice 84, 134, 271, 273

Knowledge society 3, 35–36, 38, 142–143, 145–155, 157–158

Logos VII, 17, 20, 22–23, 25, 30, 34–35
Lvov-Warsaw School 7–8, 10, 12, 15, 66

Marxism 4, 6–7, 9–10, 93
Metaphysics VII, IX, 4, 7, 21–22, 35, 59–60, 67–68, 75, 101, 191–193, 199, 207, 209, 214, 217, 227, 240, 242, 244, 274
Milan School VIII, 47, 49–69

 | HTTPS://DOI.ORG/10.1163/9789004756496-017

Mysticism IX–X, 9, 16, 18–19, 23, 28–35, 79, 81–83, 85, 192, 197, 200, 213–214, 223, 227, 231–232, 237–238, 240, 243–244, 250–251, 259
Myth VII, IX–X, 3, 19, 21, 23–25, 27–30, 83, 199, 213, 220, 231, 250–253, 255–263, 266

Narration X, 24, 47, 63, 67, 213, 227, 251, 265–267
Neoplatonism IX, 204, 206–209, 212–213, 219–220
Neopositivism VIII, 2, 8, 11–12, 14–15, 57, 68, 90–91, 98, 102–103

Objectivity 57, 61, 67–68, 96, 241, 243, 265–271, 274, 276
Open society VIII, 73–74, 79–85

Pantheism 33
Phenomenology 8, 21, 47, 53, 56, 58, 61, 67–68, 92, 147, 191
Platonism 196–197, 203, 206–207
Pluralism 14, 16–17, 21, 227
Polish Philosophy 2, 7, 15, 90
Positivism, post-positivism 9–10, 12, 48, 59, 63, 65, 90–91, 98, 103, 174–175, 252
Post-Popperian epistemology 6, 12, 14–15, 20, 252, 254
Poznań School of Methodology VIII, 2, 6–10, 12, 15, 17, 22, 89–90, 99–100, 254
Problematicism 62–63

Quantum mechanics IX, 147, 191–202

Rationalism VIII, 24, 47–48, 60–61, 65, 69, 81–82
Realism IX, 47–49, 60, 62, 92, 99, 181, 191–193, 195–196, 274
Reason VII, 18, 22–24, 27, 30–31, 34–35, 57, 59–63, 65, 67–69, 75, 82–83, 86, 143, 154, 212–213, 224, 232, 241, 260–261
Received View 15–16, 20, 174
Relativism 17–18, 274
Rejected knowledge 3, 26–27, 41, 224
Revolution, scientific (epistemic, cognitive, methodological) 10, 15, 24, 26, 30, 41, 150
  astronomical 61, 65, 218
  Industrial 150–151, 155, 157–158, 165
ROARS (Return on Academic Research and School) 36

Scientific method VIII, 10, 13, 29, 94, 100, 108–109, 126, 197
Scientific philosophy 3, 7–8, 10–15, 29, 31, 48, 90
Scientific rationality 15
Solar theology IX–X, 207–211, 214, 216–217
spiritual X, 23, 31–32, 37, 77, 132, 209, 213, 217, 220, 223, 229, 232–234, 236, 239–241, 243
Substance 76, 91–92, 94–95, 210–211
Symbolism 4, 21, 33, 84, 86, 91, 94–96, 99, 101, 103, 148, 150, 154, 207, 212–213, 219, 224, 243, 253–254, 256–257, 259–261
Sympathy 79, 84, 158, 218–219

Tacit knowledge (dimension, learning) VII, 16, 20–21, 32, 252
Theory of revolution VIII, 107–111, 113–125, 127–128
Theurgy IX, 203–220
Thought style IX, 90, 100, 102, 175–177, 180, 184
Transcendental 47, 54, 59, 61–62, 64, 68–69

University system IX, 37, 129–140
Unwritten doctrines IX, 204, 213

Verification 61, 76–77, 99, 136, 274
Vienna Circle VIII, 11–12, 15, 61, 68

www.ingramcontent.com/pod-product-compliance
Ingram Content Group UK Ltd.
Pitfield, Milton Keynes, MK11 3LW, UK
UKHW022219051026
15624UKWH00001B/237

9 789004 727717